Environment, Technology, and Health

Human Ecology in Historical Perspective

Environment,

Technology, and Health
Human Ecology
in Historical Perspective

Merril Eisenbud

New York • New York University Press • 1978

TD
170
.E37

Contents

Contents

Preface

The surge of public interest in environmental matters that developed in the 1960s was a welcome development to those who, like this writer, had already labored professionally for decades in the environmental field. It was satisfying to know that government officials, the general public, and the electronic and printed media had at last recognized the need to control pollution, rehabilitate spoiled land, and preserve wild life.

Much has since been accomplished. The National Environmental Policy Act, the Occupational Health and Safety Act, and the Toxic Substances Control Act are only a few of the many important laws that have come into being during the past decade and which have provided the basic legal and administrative machinery by which environmental control can be achieved. Billions of dollars have been appropriated for environmental protection and rehabilitation, and public interest remains high, as evidenced by the continuing attention given the subject in the daily press and other media.

However, there is also reason for disappointment. The subject of "environment" has become highly politicized. Discussions take place in an atmosphere of advocacy in which complex issues tend to be described in terms of black and white, rather than in the required shades of gray.

This book is an appraisal of the contemporary environmental movement insofar as public health is concerned. The subject is so ramified that it is not possible to examine all of the important issues in a single volume. However, the subjects that have been selected for discussion have been chosen in order to illustrate the reasons for the inherent complexity of environmental issues.

As is true in other fields of knowledge, an understanding of the past assists one to understand the present and plan for the future. It is for this reason that so much of this book is concerned with the history of the human environment.

I am indebted to many colleagues and organizations for assistance during preparation of this book. The broad coverage of the subject matter required efficient library support, for which I thank Mrs. Christine Singleton, Assistant Curator of the New York University Medical Center Library, and the staff of the Tuxedo Public Library.

It is also a pleasure to acknowledge my indebtedness for many hours of stimulating discussions with colleagues and friends who reviewed portions of the book. Among these were David Axelrod, Joseph Delibert, Leonard Goldwater, Marvin Kuschner, Morton Lippmann, Jacqueline Messite, Joseph O'Connor, Lin Root, Arthur Stern, and Guenther Stotzky. I am also appreciative to Nat LaMar, who edited my manuscript with insight and helped to structure it into final form. I acknowledge the assistance of these and other colleagues and friends, but in no way do I wish to dilute the responsibility an author must assume for the material presented.

Roger Sparling prepared the illustrations, and my assistant, Eleanor Clemm, was invaluable in an editorial capacity and in coordinating the extensive library work required, as well as by typing every word of the four to six drafts from which the final manuscript of this book evolved.

Merril Eisenbud

New York University
September 1978

PART I

Setting the Stage

CHAPTER 1

What is the Environmental Movement?

The modern environmental movement was born in the mid-1960s, which was a period of great turbulence. Unrest in the black community, the rise of feminism, and opposition to the war in Vietnam were examples of divisive issues concerning which the public expressed itself with great intensity, and sometimes with violence. Together with the widespread popular concern that developed about pollution, population control, and resource conservation that together came to be known as "environmentalism", these movements have made a lasting impact on this period of our history.

Many environmental problems had attracted attention in previous times, and much had been accomplished towards their solution. Air and water pollution were being brought under control in many cities in the United States, and many occupational diseases had virtually been eliminated during the first half of this century. The danger of unlimited population growth had been considered as early as 1798 when Malthus published the first of his often-mentioned but little-read essays. Rules for land conservation began to be established as far back as biblical times, when fields were required to lie fallow at intervals to restore the diminished productivity from uninterrupted cultivation.

These accomplishments of the past are important, but the history of human relationships to the environment has, more often than not, been characterized by failures. It will be seen that early man turned vast fertile landscapes into deserts. His way of life was partly responsible for the great plagues that decimated the populations of Europe in the Middle Ages; rats and lice could not have spread epidemic disease if congested and unsanitary communities had not offered such hospitable habitats to vermin of all kinds. People would not have been

3

decimated by typhoid if they had not discharged human wastes into the same waterways that provided their drinking water.

For thousands of years, life was harsh and remained unchanged for most humans. Except for a few idyllic exceptions, where a combination of mild climate and abundant food supplies favored a relatively comfortable existence for the ordinary person, life required strenuous exertions, and survival was constantly threatened by the vicissitudes of nature.

Until comparatively recently, only a small fraction of the population has had the opportunity to become educated, or has possessed the means to acquire more material possessions than those needed for bare survival. It took all the energy and skill a person could muster to meet the basic requirements for food, clothing, and shelter during a lifetime that rarely reached beyond the modern definition of early middle age.

Beginning only a few centuries ago, men became increasingly aware that knowledge could be used to achieve a better way of life. Philosophers who, from classical times through the Renaissance, had sought knowledge for its own sake were superseded in increasing numbers by artisans and practitioners—the engineers, chemists, physicians, technologists, and managers—who used knowledge to temper nature's harshness. Human inventions and initiatives gradually improved the productivity of soil, facilitated transportation and communication, and used the natural products of the environment for better housing, clothing, tools, machines, and countless aids to cultural development. But humans were slow to realize that while their way of life was improved initially by the applications of human intelligence, the environment struck back frequently in subtle ways. This process of human action and environmental reaction has been going on for thousands of years. It is only in recent decades that the full extent and significance of these interactions have begun to be realized.

Even the most innocent of inventions has initiated chains of events that have had major influences on human ecology. Sometimes the influence has been for the good of humankind, but sometimes not, and the wisest of men might disagree in many instances as to whether it was the one or the other. Consider the subtle role of the horse collar. It may seem absurd to include it among inventions that have had important effects on the balance between humans and their environment, but when the horse collar was evolved in the twelfth or thirteenth century, it became possible to increase the productivity of farms greatly by improved plowing and to facilitate haulage of fresh produce to market.[1,2] By contributing to better nutrition, higher fertility, and lower death rates, the horse collar was among the earliest reasons for population increases in Western Europe.

The environmental impact of the steam engine was far less subtle when it was introduced in England about 200 years ago. Steam engines made possible the Industrial Revolution which brought about many of the changes that improved the human condition—and some that did not.

Environmental effects such as those resulting from the invention of the horse collar or steam engine are readily identifiable in retrospect, but how can their effects be anticipated? Is it reasonable to expect that James Watt should have been able to foresee the impact of the Industrial Revolution that would result from his invention of the steam engine? Certainly not! Could Thomas Alva Edison, when he constructed the first central electricity generating station at the end of the nineteenth century, have been expected to understand the long-range effects, mostly good but some bad, that would result from development of the electric utility industry? We now see it as essential that the environmental impact of all technical innovations be considered thoroughly. We also should recognize that some effects may be so subtle that they are not likely to be predicted and may escape detection for some time after they have occurred.

Should a technical innovation be blocked because there are identifiable deleterious side effects? This simplistic approach is taken by many people at present. Some would outlaw manipulation of genes, nuclear energy, and supersonic transport. This is not a new phenomenon: the sixteenth-century physician Georgius Agricola, in his classic book about metal mining and metallurgy, devoted several pages to a defense of metals in response to widespread concern that metals were being used to kill: "Several good men have been so perturbed by these tragedies that they conceive an intensely bitter hatred towards metals, and they wish absolutely that metals had never been created, or being created, that no one had ever dug them out."[3] The burning of coal was banned in England as early as the fourteenth century because of the nuisance created, and it is said that at least one man was executed for violating that law.[4]

Some problems associated with technical innovations can be easily foreseen and, in many cases, eliminated by taking proper precautions. Other side effects may not be foreseen because of their subtlety. We now see it as obligatory to investigate the risks and benefits of an innovation before it is initiated, and to hold a human agency responsible for damage to the environment. That is the purpose of environmental-impact analysis. It is an approach that has been required by law for the first time in this decade.

While the environmental-impact statement is an important and valuable requirement, its use must be exercised with wisdom, humility, and caution. It must be applied with wisdom because of the technical complexities of the assessment process, humility because one must recognize one's limited abilities to forecast the consequences of technical innovations; it must be applied with caution because, although we should not permit human activities in which the environmental damage seems to greatly outweigh the expected benefits, humankind should not be denied the benefits of new developments when the benefits greatly outweigh the risks.

WHAT IS THE ENVIRONMENT AND WHO ARE THE ENVIRONMENTALISTS?

The word "environment" has developed a new meaning since the mid-1960s. It is a word that is so broad in its meaning, and so inherently subjective, that it invariably reflects the interests, biases, perspectives, and motives of the user.

The environment is more than the air, water, land, and presence of living things. Environmental deterioration threatens our well-being when air, water, or food become contaminated, but also when government breaks down, or our cities are destroyed by war, or free passage in the streets is interrupted, whether by muggers, choking traffic, a burst water main, or a poorly maintained road surface. Literally defined, the environmentalist should be a universalist with interests that go beyond natural systems, pollution, and preservation of green space. This is of course unrealistic, and it becomes necessary to define the environment and the role of the environmentalist in accordance with modern perceptions of the subject.

Neither logic nor an analysis of the structure of the modern environmental movement permit definition of its bounds. Because of the importance of pollution, some environmentalists confine themselves to that subject alone. But preservation of endangered wildlife is equally important to other environmentalists (who may prefer to be identified as "conservationists"), and to the extent that a species is endangered because of pollution, the two groups have a common purpose—reduction of pollution. However, preservation of endangered species may go far beyond pollution. Porpoises are endangered because they are being caught in great numbers in the nets of the tuna fishing fleets. Whales are endangered because they are overhunted. Protection of marine mammals requires less attention to pollution than to marine-resource technology and international politics Power-plant emissions produce pollution, control of which is a laudable objective of environmentalists. However, when environmentalists attempt to solve the problem of power-plant pollution by advocating that new power plants should not be built because we do not need more energy, they become involved in complicated questions of national policy for which their training may not prepare them. Many other examples could be given to demonstrate that *some* environmentalists respect no limits to what they perceive as the logical scope of their activities.

In this discussion, it is important to differentiate between that with which one becomes involved in a professional way and that which is of avocational concern. Every citizen should be interested in the major issues of the day, and he has a right (perhaps a duty) to take a position on one side or the other. However, no one is entitled to speak *authoritatively* on all matters. Many scientists in recent years have not recognized this distinction. That is, they do not disqualify themselves as experts when they speak publicly on issues outside their field of specialization.

There are many inconsistencies, not only in what some environmentalists insist on including in their movement, but also in what they exclude. If a toxic household chemical is left below a sink within easy reach of a curious infant, who dies as a result of eating it, have we not identified an environmental problem? Many infants die each year in this way, yet this is one of many environmental problems that has not attracted the interests of the modern environmentalists. Why should not environmentalists advocate programs of parent training with the same fervor as they advocate the banning of a toxic pesticide? If parent education will prevent infants from being poisoned, is that not as important as reducing the sulfur dioxide concentration in the air they breathe?

The contemporary environmental movement has assumed a form that reflects the interest and perceptions of those who have provided leadership to the movement. The movement has been characterized by a relatively narrow perspective, despite the fact that it includes some advocates of zero energy growth, and other objectives that involve the total structure of our economy. This is because the movement is led by the economically elite, who put a high priority on clear skies, clean water, and green space. The "environment" of a mother struggling to support a family in the decayed core of a large city bears no relationship to the "environment" as perceived by an upper-middle-class family with an apartment in the better part of town and perhaps a cottage in the country. Blue skies, clean air, and clean water mean very little to disadvantaged men and women who live in rat- and roach-infested homes and are subjected to all the evil by-products of their underprivileged existence. The very environmentalists who take pride in their roles as global ecologists and who do not hesitate to take firm positions in matters of broad economic or political policy have ignored the environmental consequences of allowing the housing stock of our cities to deteriorate beyond the point of rehabilitation. Not that it is up to environmentalists to restore rundown buildings—that is the function of the appropriate agencies of local, state, and federal government. But contemporary environmentalists are activists who achieve their purposes by means of the literature they publish, their influence on the media, and their access to public officials. The environmental problems of the ghettos receive all too little attention on the agenda for environmental rehabilitation.

The narrow perspective of modern environmentalism is seen also in its limited concern with the problems of the streets and highways. The considerable amount of effort that has been expended in this area has been motivated mainly by pollution from automobile tailpipe emissions, a matter of far less importance to human well-being than other environmental effects of the automobiles (traffic accidents and congestion, for example), and the defects in planning that have created such a high degree of dependence on the automobile. The modern environmental movement includes many examples of misplaced emphasis.

THE ROLE OF MODERN TECHNOLOGY

Modern technology has made it possible for people to live longer in more comfort and with greater leisure. It is estimated that if the death rates due to infectious diseases in 1900 had persisted to the present time, one-fourth of the inhabitants of the developed world would not be alive today.[5] Yet the contemporary environmental movement often decries technology and longs for a return to the simpler life of former times. It overlooks the many positive contributions to our well-being and condemns all technologies for their negative effects. Technologies have brought intellectual and material privileges to ordinary people that kings could not have enjoyed only a few generations ago. Certainly the undesirable by-products of our technological development are everywhere in evidence and present us with problems that will become increasingly complex as time goes by. The basic question is whether we can restrict the deleterious effects to an acceptable level.

From an evolutionary point of view, human technology is an essential environmental adaptation, equivalent to those made by other species in the struggle to survive. Human environmental adaptation has, of necessity, been achieved by intellectual rather than physiological or anatomical means. Since every species must interpret the world around it from the point of view of its own needs for survival, technology is both good and necessary from the point of view of the human race. Man cannot compete with other species because he runs faster, has superior eyesight, or is endowed with great muscular strength. Humans are inferior in these respects and can survive only by the products of their intellect. It has been only in the relatively few recent millennia that we have progressed from wandering food gatherers, to farmers, toolmakers, philosophers, artists, poets, scientists, or engineers. And only in the last few hundred years has technology begun to release us from enslavement by an environment in which we are at a disadvantage in so many respects.

At least six major interrelated by-products of technology have negative environmental impact: the dangers of modern war, exponential growth of population, waste of raw materials and other natural resources, inadequacies of urban design and organization, poverty, and pollution. Of the six, pollution is probably the only one for which society has both the technical knowledge and social institutions with which the problem can be managed. Moreover, to a major degree, pollution results secondarily from the others.

The capability that humans have developed to wage thermonuclear war must certainly head the list as the greatest single environmental threat. An hour or two of all-out war using contemporary nuclear armaments could cause destruction of life and property on a scale that could preclude, perhaps forever, a return to the kind of life that mankind has evolved over the centuries. The environmental and social effects of such a war would be so complex as to defy comprehension. The threat is especially dangerous because of the suddenness with which war could come and its long-lasting effects. Other environmental stresses, such as overpopulation, attrition of raw materials, or global pollution, will develop over a period of decades or centuries. There will be time to debate, time for public attitudes to change, and time to take corrective action. But there will be no time for corrective steps if a thermonuclear war is launched. Nuclear war is a cardinal example of an environmental catastrophe that must be prevented, but this must be the goal of all the people, whether they are identified as "environmentalists" or not. Another way of stating it is that all people are environmentalists in one way or another. The important point is that solution of the most important environmental problem in history is one that the "environmentalists" do not and cannot preempt.

Population growth is another such example. The "good life," made possible by sanitation, more food, and medical advances, has resulted in a population explosion. The world's population, which grew slowly during biblical times and did not double in size until about 1650, suddenly doubled again in only 150 years, and was to double again, and again, in successively shorter periods. The time in which the world's present population will double is estimated to be only about 35 years. It is clear that this trend cannot continue much longer. There may be disagreement about the number of people that can be supported by the world's resources, but there can be no disagreement that there is a limit. In fact, population size has already outrun the availability of natural resources in many parts of the world.

With respect to raw materials and energy needs, it is probable that future generations will look back unkindly at the twentieth century for its failure to use its resources in a more responsible manner. We can debate whether a given mineral resource will last 30 years or 500 years, but it is clear that at some point, at a not-too-distant time from the perspective of human history, the abundant supplies of many seemingly essential raw materials will be depleted. We already face shortages of petroleum and natural gas. The reserves of other necessary resources are bound to shrink faster as the world's population grows and more people arrive at a standard of living comparable to that in the United States and Western Europe. Humans may someday be able to obtain raw materials in new ways, and methods will eventually be developed for producing energy that will no longer require fossil fuels. Nevertheless, as raw materials and energy become more costly, it will be necessary to evolve a society that uses less.

The manner in which cities are designed and function presents environmental problems that have thus far defied solution. Cities are essential to the well-being of developed nations but, once built, they are subject to obsolescence both in function and form. The newly built cities that rose from the ashes of bombed-out Germany and Japan, and the cities of the southern and western United States that have undergone their greatest development in the past two decades have not yet encountered the major problems of obsolescence. Yet even these new cities are already congested by automobile traffic, troubled by air-pollution problems, and are beginning to sprawl in the manner that has created such difficulties in the cities of the eastern United States.

Obsolete forms of local government, inability to maintain the quality of housing stock, and transportation systems that cannot adapt to changing needs present enormous problems to urban planners and make the urban environment increasingly unpleasant for all but the wealthy few. The environmental problems of the cities deserve a high priority and they must be dealt with by legislators, economists, engineers, and concerned citizens. The environmentalists must become involved, but not in the fragmented ways that have characterized their efforts to date. There has been much concern about air pollution in cities, but little or no involvement with the more fundamental environmental problems that literally threaten the future of our cities and the health and well-being of millions of their inhabitants.

Of all the socioeconomic problems related to the environment, the most neglected are those associated with poverty. It is ironic that some of the most attractive natural environments in the world provide the habitats for impoverished people whose way of life stands in contrast to the beautiful settings in which they live. In Appalachia, on the hills of Rio de Janciro and Hong Kong, and on Indian reservations located in the spectacular scenery of southwestern United States the inhabitants have been trapped by generations of poverty from which they are unable to escape for one reason or another, usually beyond their control.

Although poverty is a major problem in many rural areas, it is more visible and widespread in the large cities. The difficulties of modern city living and the lack of amenities have driven many middle-class city dwellers to the suburbs, leaving voids into which the poor from rural areas soon moved. Their presence in such great numbers detracts further from the attractiveness of city living and provides additional incentives for the more fortunate to leave. The poor have been unable to maintain the tax base, and for lack of skills they cannot easily become productive members of the community. This process creates basic social and economic instabilities, and it is not at all clear by what process this trend will be slowed or reversed.

The contemporary environmental movement has focused on pollution, but in many cases the pollution is secondary to other defects of our technological society such as overpopulation, wastefulness, or poor urban design. The streets of a city may be dirty because they are cluttered by too many automobiles to permit cleaning by mechanical equipment or because the compound effect of greater population density and burgeoning per capita production of solid waste has overwhelmed the municipal capability to keep the streets clean. The subject of pollution covers a broad range of topics, from the highly visible but superficial consequences of soda pop bottles strewn along a roadside to the invisible but potentially disastrous effects of the improper use of the toxic chemicals on world ecology.

It will be seen that society is being required to deal with pollution problems of increasing complexity and subtlety. We often do not understand the public health and ecological consequences of new pollutants being introduced into the environment. This is a problem that has been greatly exacerbated by the huge number of new synthetic chemicals and their use in ways that can pollute the world from pole to pole. The essence of the new environmental danger is that if a mistake is made, the effects may not be restricted to local areas as in the past but may involve the entire world in some unpredicted irreversible disaster. The probability of this happening is no doubt very small, but the consequences would be very great.

NOTES

1. White, Lynn, Jr. "Medieval Technology and Social Change," Oxford Univ. Press, New York (1962).
2. Jope, E. M. "Vehicles and Harness," in: "A History of Technology" (Charles Singer et al., eds.), Oxford Univ. Press, New York (1956).
3. Agricola, Georgius. "De Re Metallica" (1556) (trans. by Herbert and Lou Hoover). Reprinted by Dover Publications, Inc., New York (1950).
4. Marsh, Arnold. "Smoke: The Problem of Coal and the Atmosphere," Faber and Faber, Ltd., London (1947).
5. President's Science Advisory Committee. "Chemicals & Health," Report of the Panel on Chemicals and Health, Science and Technology Policy Office, National Science Foundation. U.S. Government Printing Office, Washington, D.C. (September, 1973).

CHAPTER 2

The Preindustrial Centuries

More than 3 billion years have elapsed since the first forms of life appeared on earth. During almost all of this period (more than 99.99 percent), human beings did not exist. When they did evolve, it was into a world of intricate ecological relationships in which the new species would be greatly influenced by an environment which it, in turn, would affect.

The influence of humans on the environment was at first no greater than that of the wild beasts, although it is thought by some that, while they were still in the food-gathering stage of existence, they developed methods of hunting that created great grasslands and perhaps speeded the extinction of giant animals. This remains a matter of speculation. What we do know is that, when humans shifted from a life of hunting and gathering to one of farming and herding about 10,000 years ago, they began to make an impact on the environment that has continued with increasing complexity to the present time.

EVOLUTION OF ECOSYSTEMS AND
THE EMERGENCE OF THE HUMAN SPECIES

We are told by geologists and biologists that the earth came into being about 5 billion years ago, and that the fossil remains of early life forms have been found in rocks more than 3 billion years old. It is believed that life began when the physical and chemical conditions that had evolved in the atmosphere and oceans were conducive to the synthesis of the basic chemical structures from which all living matter is formed. In those primeval times, conditions on earth were fundamentally different than now. The atmosphere, which is today rich in nitrogen and oxygen, is believed to have then consisted mainly of methane, ammonia, water vapor and hydrogen.[1]

13

The basic organic compounds that are essential for the first steps in the synthesis of living systems could have been formed originally in the atmosphere by the action of ultraviolet light or by bolts of lightning. It is assumed that the compounds formed in this way were washed into the oceans, where they accumulated as a nutritive broth in which the organisms yet to be formed would initially subsist.

It was a long way, by chemical pathways as yet unexplained, from the thin mixture of simple organic compounds to the self-replicating process we call life, but life did eventually evolve, in the form of a simple bacterialike organism. Higher forms of life have since evolved by the process of genetic mutation. Each reproducing cell is endowed with genetic determinants such as chromosomes that, if replicated faithfully, assure that the progeny will resemble the parents in both form and function, within the relatively narrow limits of normal biological variability. From time to time, the structure of a chromosome is altered by the action of heat, natural radioactivity, or perhaps certain chemicals present in nature. This may result, ultimately in the development of mutant forms. In most cases the mutant is not viable, but in those cases where the organism does survive initially, it is immediately tested to determine whether the characteristics of the mutated form are advantageous or disadvantageous in the competition to survive in that particular environment. If the genetic change is desirable with respect to environmental competition, the mutated form will succeed in reproducing and, because of its survival advantages, may eventually dominate other species within the ecological niche it occupies.[2]

Thus, the first single-celled organisms mutated to more complex single-celled plants and animals, and then to multicelled plants and animals that in time became progressively more elaborate in structure and function. This evolutionary process gave rise to humanlike creatures a mere 2 million years ago.

Thousands of geophysical and geochemical factors have influenced the course of evolution. The physical and chemical composition of soil, the distance of the earth from the sun, the characteristics of the earth's perisolar rotation, the composition of the sun's radiations, the presence of certain chemical elements, the thermodynamic characteristics of the atmosphere, the remarkable properties of carbon and water, and an endless list of other fundamental properties of our world on earth constitute the physical-chemical environment within which life as we know it has evolved. These chemical and physical factors vary only within relatively narrow limits which must not be exceeded if life is to continue. Within these limits diverse environments exist for which biological evolution by the processes of trial, error, and natural selection eventually found inhabitants.

Over eons of time, such disparate environments as the muds at the bottom of lakes, mountain crags, deserts, and ocean islands have provided hospitality for the diverse life forms that evolved. The species in each habitat were at once competitive and interdependent. In subtle ways that are not yet clearly understood, the extraordinarily complex mixture of living forms known as the biosphere became interwoven into an ecological system in which each species plays a unique role and each organism affects every other, either directly or indirectly.

From the first seemingly unimportant specks of living matter produced in the ancient period of geological history known as the pre-Cambrian, the earth's biological forms have evolved to an enormous number. More than 1 million species have been identified by now, and it is estimated that the total number may reach 2 million when the inventory of the plants and animals on earth is completed.

Living things not only adapt to their environment; they also modify it and, particularly in the case of man, control it to some extent. Life processes have produced highly significant geophysical changes. The most important of these has been the production of atmospheric oxygen by the photosynthetic activity of algae and higher plants. Prior to the gradual release of oxygen in this way, it was chemically bound in water and the earth's crustal materials. Soils, too, can be changed significantly by living things: The addition of the organic debris produced by their dead remains makes it possible for the soil to hold more moisture and adds to its capacity to store nutrient materials in a form available to plants and soil organisms. The great beds of limestone were produced by the gradual accumulation of calcareous remains of sea organisms in ocean sediments. The coral reefs and atolls are developed from the skeletons of marine organisms.

The fossil record is far from complete, but one humanlike species that walked fully erect, and which used fire and primitive stone tools, is known to have existed more than 2 million years ago and to have survived for several hundred thousand years, evolving in the meantime to other pre-human species that populated many parts of the world. Until this point, evolutionary change depended mainly on the slow process of mutagenesis, but the emergence of human intellect made it possible for the evolution of human beings to proceed more rapidly.

It is likely that the level of primitive intelligence was distributed in a mathematically "normal" way, as are other biological characteristics. Individuals with greater-than-average intelligence were able to cope more effectively with environmental stresses of all kinds and thus had a higher probability of surviving and reproducing. Since the quality of intelligence is at least in part a genetic characteristic, the more

intelligent human beings endowed their progeny not only with greater
cultural achievement, but also with a higher-than-average intelligence
relative to the norm of that period.

One can easily see that cultural evolution could proceed more
rapidly than biological evolution. Fires of natural origin are certainly
common enough, and when early man first learned to transfer fire to
his shelter it became possible for him to move to cooler climates. In
this he was also aided by primitive garments, another cultural development. He soon learned to preserve food by cooking, drying, or
smoking. Simple stone tools made it possible for him to hunt and to
withstand predation. Primitive speech in the form of simple grunts
facilitated coordination among individuals during hunting and fighting.
With these relatively few achievements, human evolution broke away
from the slow rate of change dictated by the gradual alterations in
genetic characteristics. Human evolution was now accelerated by
cultural inventions that aided humans in the competition for survival
and which, like genetic characteristics, were passed from one generation to the next. Primitive cultural achievements gave early *Homo
sapiens* a sufficient survival advantage to permit substantial population growth, allowing him to spread from his place of origin in East
Africa to other parts of the world.

While cultural influences now accelerated the rate of change of
human physical and behavioral development, it still took hundreds of
thousands of years for *Homo sapiens* to become the dominant human
species, approximately 100,000 years ago. Until then, other manlike
creatures speciated and spread to various parts of the world, but
they became extinct. Whether they became extinct because they were
not sufficiently intelligent to enjoy lasting environmental adaptability
or whether they were killed off by disease or in the earliest of the
human wars is a matter for speculation.

Homo sapiens had many characteristics that assisted his ascen-
dancy and domination. Our species was capable of inductive and
deductive reasoning. We possessed an articulate vocal apparatus,
which became increasingly effective for purposes of communication
and self-expression. Human ability to adapt to varied environments
was assisted by many other advantageous anatomical and physiological characteristics, among these a remarkable hand at the end of a
rotatable arm and characterized by an opposing thumb. This appen-
dage, which had functioned as an aid to locomotion among antecedent
species, could now be used as a powerful and versatile tool which, in
conjunction with the more highly developed intellect, would give the
human species primacy over other living beings. Another advantage
was a highly versatile digestive system. No other animal (except

perhaps the rat, with which man has more in common than we care to admit) can utilize so wide a range of plant and animal foods. This would be a characteristic of obvious advantage in adapting to the diverse environments mankind was destined to inhabit.[3]

The food-gathering phase of human development accounts for more than 99 percent of the time since the first manlike creatures appeared. Not until about 10,000 years ago did man begin to cultivate edible plants and to breed cattle and other stock animals. This was the period of the Neolithic revolution.

Man's shift from food gatherer to food grower did not take place simultaneously all over the globe. In fact, even today there are a few tribes that have been so isolated that they have not developed beyond the food-gathering stage.

THE PREINDUSTRIAL COMMUNITY

It takes no feat of the imagination to understand the basic changes in the human environment that must have occurred when humans ceased their nomadic wanderings and settled down to an agrarian existence. For the most part, the sedentary way of life provided a more compatible environment. But new problems did develop. Nomads need not be concerned with pollution to the same degree as persons living in permanent shelters assembled into congested communities. An overriding source of pollution must have been the combustion products from indoor fires. From the time he first built fires in caves until the time he began to use ventilated space heaters and cooking stoves, in relatively recent times, man has been exposed to a noxious mixture of carbon monoxide and smoke. An account of the winter habitations in England during Roman times describes them as "deep caves dug into the earth, where [the inhabitants] resided, surrounded by their provisions for the winter, almost wholly concealed from casual view, and suffocated by smoke."[4]

The lungs of an ancient Egyptian mummy have been found to be heavily burdened with carbon—undoubtedly the result of exposure to the smoke of fires within the home.[5] The kinds and amounts of pollution to which primitive peoples have been exposed due to fires maintained under primitive conditions remain a neglected field of anthropological research. The air within Eskimo igloos, the tents and tepees of North American Indians and Arab bedouins, Navajo hogans, and the shacks, sheds, and caves in which primitive peoples live (or once lived) in many parts of the world can still be studied to document the polluted atmospheres to which humans have been exposed within their places of habitation. Many of the primitive home environments exist today in forms that have changed little over the centuries. They should provide fascinating and useful research opportunities for anthropologist, ecologist, and epidemiologist alike.

The water-pollution problems of ancient communities are something we know more about. The Romans learned how to bring pure water to their communities more than 2,000 years ago, but their hydrological arts were neglected following the decline of the empire, and water supplies became a major source of infectious disease during medieval times. In 1713 Bernardino Ramazzini called attention to the fact that the Romans had used public and private baths, but that they had fallen into disuse.[6] He was one of the first physicians to be concerned with environmental influences on health, but his references to the Roman baths gives one the impression that he regarded them as another example of Roman-style luxury—without obvious hygienic benefits. As late as 1873 the residents of Munich were said to be content to use only one liter (approximately one quart) of washing water per day, and home bathing facilities were unusual.[7] In contrast, the use of water in a modern American city is in excess of 400 liters per person per day.

The filth with which most people lived has been perhaps the most unpleasant aspect of the human environment until comparatively recent times. Much of the filth was the result of the congested way of life associated with the poverty that has existed throughout history. The word "poverty" in technologically advanced societies is applied to that fraction of the population whose income, nutrition, and educational level are less than the minimum modern standards. But there was a time in Europe and the United States when all men except for a tiny aristocracy, were impoverished. All of the energy of poor people was devoted to obtaining food, providing clothing and shelter, and raising a family. The food supplies were inadequate more often than not. There was little in the way of surplus wealth. The life-style in the preindustrial era changed hardly at all from medieval times until the nineteenth century in many parts of the world. The medieval home was, for most people, a single unfloored, unheated room that lacked furniture and which the family shared with poultry and pigs.[8] Descriptions of the lives of the early eighteenth-century Scottish peasantry seem little different.[9] And it is reported that as recently as 1837, in an Irish town of 9,000 inhabitants, there were only ten beds, 93 chairs, and 243 stools![10] The streets of the principal cities of Europe and the United States were generally unpaved until the nineteenth century, despite the fact that the Romans recognized the importance of pavements 2,000 years ago.

Modern domestic sanitation requires the availability of ample supplies of hot and cold running water, time for chores associated with personal hygiene and home care, waste-water plumbing connected to sewers, and systems of garbage collection and disposal. Home cleanliness is also encouraged by paved and well-drained streets, and materials of indoor construction that facilitate cleaning. All these prerequisites for household hygiene were lacking until the industrial nineteenth century. Moreover, during preindustrial times people worked long hours, either in the field or at handicrafts, and had no time for such time-consuming activities as carrying excrement and kitchen wastes to prescribed disposal areas. The absence of hot water made bathing impractical during much of the year. Many people owned only a single set of garments which would not be changed except when they were beyond repair. The houses and inhabitants alike were ridden with vermin for which modern chemical remedies were not available.

The problem of vermin was not confined to the impoverished. Samuel Pepys, the aristocratic seventeenth-century English diarist, made frequent references to his discomfort due to the lice that infested his body, clothing, and wigs.

Even the absence of artificial illumination was a handicap to household cleanliness. Chores that can now be performed after nightfall were not possible in homes illuminated by primitive oil lamps. Moreover, oil was an unavailable luxury to all but a few.

EARLY EXAMPLES OF LAND ABUSE

During the hundreds of thousands of years during pre-Neolithic times, the human race was too small to have had a significant effect on the physical and biological environment with which it was in ecological equilibrium. During part of this period, man could control his environment to some degree through the use of primitive tools, clothing, and fire; and in some places the forests were burned to facilitate hunting, or to promote vegetative growth. The use of fire was the first means by which man could alter large land areas.[11,12] But this influence was probably insignificant compared to the effects of fires of natural origin, such as lightning.

The hunting practices that early man used may have exterminated large game animals over limited areas, but this is a matter of speculation.[13,14] Any environmental changes man may have caused were not brought about to a significant extent until the Neolithic period, about 10,000 years ago. After this, man's influence was no longer that of another animal species competing in the biological world. Men cleared forests, introduced domestic animals, and caused soil to erode. Unfortunately, the relationship between man and his environment during most of the time since he became a food grower is very much a matter of conjecture, since recorded history began only about 3,000 years ago. For the first 7,000 years of the new sedentary existence, agricultural civilizations were developing and the populations were increasing in size in many parts of the world without any record of the extent to which human activity impacted on the natural environment. However, we do know that man's early activities of farming and herding began to make major changes on the face of the earth.

Enormous areas of arable land, one of mankind's most valuable resources, were devastated for lack of knowledge of basic principles of agricultural ecology. Deforestation, soil exhaustion, and erosion formed a sequence that occurred in North China, Mesopotamia, Iran, North Africa, and Palestine.[15,16] It occurred also, though to a much lesser extent, in Europe and the United States. Many of the deserts that now exist in parts of the world that were once known to be agriculturally productive were not caused by climatic changes but by the increased rate of evaporation and lowered levels of ground water associated with erosion of soil. In a cynical reference to the misuse of land, it has been said that the early home of Chinese culture "now resembles a huge battlefield scarred by forces far more destructive than any modern engines of war. The sculpturing of that fantastic landscape is the greatest work of Chinese civilization."[17]

The foraging of the goat was well advanced by the time of the classical Greek civilization,[18] and the writers of that period provide us with a number of references to the advanced state of soil erosion.[19] Among these is one by Plato, who wrote:

By comparison with the original territory, what is left now is, so to say, the skeleton of a body wasted by disease; the rich, soft soil has been carried off and only the bare framework of the district left. At the time we are speaking of these ravages had not begun. . . . What we now call the plains of Phelleus were covered with rich soil, and there was abundant timber on the mountains, of which traces may still be seen. . . . Not so very long ago trees fit for the roofs of vast buildings were felled there and the rafters are still in existence. There were also many other lofty cultivated trees which provided unlimited fodder for beasts. Besides, the soil got the benefit of the yearly "water from Zeus," which was not lost, as it is today, by running off barren ground to the sea; a plentiful supply of it was received into the soil and stored up in the layers of non-porous potter's clay. Thus the moisture absorbed in the higher regions percolated to the hollows and so all quarters were lavishly provided with springs and rivers.[20]

Many of the present deserts and semideserts in countries along the semiarid Mediterranean coasts once supported dense populations. Now, they are no longer capable of doing so. The predominant opinion of climatologists is that there has been no marked change in the climate of the Mediterranean basin during the past 3,000 years. The extensive desertification of these areas is probably due to the fact that the once-fertile soils were eroded because of unrestricted practices of clearing forests, poor tillage of soil, and the introduction of grazing animals.

Deforestation opened vast areas for either farming or goat grazing, and the latter not only prevented reforestation, but generally denuded the land of vegetation. Both uses had the effect of exposing the soil to the eroding effect of the occasional torrential rains that are characteristic of many parts of the Mediterranean. The loss of vegetative cover increased both evaporation and runoff, and the soil would be washed away during the rains and blown away during the droughts.

The situation in Palestine, the biblical "land of milk and honey," had deteriorated to such an extent by 1937 that the British Royal Palestine Commission found only 76 square miles of forest and farms out of 6,250 square miles of land fit for such purposes.[17,21] Resettled Jews began to reforest the land successfully by applying modern agricultural methods to soil management. The population has grown from about 250,000 in 1948 to about 3.5 million. Whereas the soil had been destroyed to such an extent that only 1.2 percent of the land was fit for forestation or cultivation in 1937, about 20 percent of Israel is now under cultivation. We thus see that at least in some semiarid regions, the land can be restored in a relatively short period of time by using modern agricultural methods.[24]

Examples of extensive land abuse by primitive people can be found even in modern times. A classic case is to be found in the equatorial highlands of Kenya and Tanzania where the Masai live a seminomadic life on an area that is currently about 36,000 square miles in extent. Cattle are the basis of the Masai economy and cattle blood is used for food. The sale and slaughter of cattle are discouraged, however, for the number of head is a measure of the wealth, prestige, and security of the owner.[22]

Until recently, the numbers of the Masai cattle were controlled by several factors. The first of these was the prevalence of livestock diseases, but in recent years these have been reduced by government veterinary services. Second was predation, but the number of predators such as lions and leopards has been greatly reduced by European hunters and the encouragement given to poachers by the high price of animal skins. Finally, the size of herds was controlled by the amount of water available in the dry season, but the national governments have added to water supplies by digging wells, building dams, and constructing pipelines. Whereas the herds were kept under control naturally by disease, predation, and limitations on the water supply prior to the European influences, the herds have now increased to the limits of the new water supply, thereby increasing the devastation due to overgrazing. The Masai herdsmen did not understand that control of livestock disease and more dependable water supplies could not be of lasting benefit unless they limited the size of their herds.

The experiences of two of the earliest American colonies present an interesting contrast between foresight and neglect in their land-management practices. When the English first settled in Massachusetts in 1620, the home islands had long been deforested, and because the Royal Navy was dangerously short of timber for ship-building, it was required to buy its wood from Scandinavian countries. To protect the New England forests, an order was issued in 1691 which prohibited cutting of trees more than 24 inches in diameter without permission of the British government. This was the so-called "Broad Arrow policy," which derived its name from the manner of marking the trees with three axe blazes in the form of a broad arrow, the symbol of the British Navy. By 1711, the policy had been extended to all of New England, New Jersey, and New York.[23-25]

This policy contrasts with the lack of foresight in the use of land for growing tobacco in Maryland and Virginia, from 1606 until the mid-nineteenth century. The early settlers soon found that there was such an enormous demand for tobacco that the return per acre of tobacco was six times that for any other crop. The pressure was so great to grow tobacco that no land was set aside for stock animals and therefore no manure was available. Since this was at a time when artificial substitutes for manure were not yet in use, the soil could not be fertilized. Accordingly, the settlers began to open new fields, working them for a few years to the point of exhaustion, and then moving West. By the end of the eighteenth century, the lands in Maryland and Virginia were totally spent, and most of the settlers had migrated to Kentucky and Georgia. As a result, the region became a depressed area, and by 1850, one-third of the Virginia- and Maryland-born families had left what were potentially among the most productive agricultural tracts in the country for new land in the West and Southwest.[26]

SOME ECOLOGICAL ASPECTS OF COMMUNICABLE DISEASE

With the establishment of the first permanent human dwellings, there began a process of community evolution that is continuing at the present time. It has had environmental implications throughout the world.

The pre-Neolithic human wanderers probably moved in groups of 50 to 60 people who rarely came in contact with other humans, since it is estimated that each of these tribes required about 500 to 1,500 square kilometers for survival.[27] During the long period of wandering, it took the total effort of the tribe to make a few simple tools and to gather enough food for a marginal existence.

A more productive way of life came with the evolution of towns. Food growing and animal husbandry were carried out more efficiently than previously. Agricultural functions then could be delegated to a portion of the community, leaving others to make tools, exploit local mineral resources, and make clothing and other handicrafts. Some settlements might be situated near an excellent supply of flint. Others might be located near clay of good quality, or close to sources, perhaps, of furs, or stones, or metals, that could be used for ornament. Thus began the need for trading among communities. As a result, people in the Neolithic period more frequently came into contact with those from other communities, and this was further encouraged after the villages had been connected by networks of trails that later became roads. Trading was, of course, greatly facilitated among communities located on navigable water.

The epidemic diseases were probably the first of the major ecological crises to arise out of the evolving style of urban life. It has been suggested that prior to the establishment of cities, the human race was relatively free of most pathogenic (disease-causing) organisms specifically adapted to the human species.[28,29] In contrast to most life forms that exist external to other organisms and either derive their food by photosynthesis or by grazing or hunting, there are many bacteria, fungi, viruses, protozoans, and worms that live within the bodies of other organisms. Their primary ecological niche is the environment within the cells or organs of other, usually higher, species. All living things are subject to invasion by such parasitic forms, which in some cases perform beneficial functions required by the host. The bacteria that inhabit the intestine of the cow, for example, are essential to its digestive processes. However, parasites can also be injurious to the host, and can cause sickness and death.

The relatively simple organisms that cause disease undergo frequent genetic mutation, and survival of the mutant depends on whether it can find a compatible environment. Alternatively, it is possible for the environment to change, and the microorganisms will die unless they can adapt or mutate to an adaptable form. The simpler organisms undergo spontaneous genetic mutations at a sufficiently high rate that it is not unusual for new diseases to develop from time to time.[30] This may explain why some diseases have suddenly appeared for the first time in relatively recent years. For example, there is no record of epidemic poliomyelitis before about 1840. There were no known cases of encephalitis before the eighteenth century, when at least three types of this viral disease of the central nervous system appeared in a space of 20 years.[31]

Infectious diseases require the existence of clearly defined ecological relationships among the external environment, the host, the pathogen, and in some cases an intermediate organism (vector) such as an insect or the rat. It will help to illustrate this concept as it applies to the disease malaria. Malaria can be caused in various mammals by single-celled microorganisms known as plasmodia (types of protozoa) that can be transmitted to humans only by the bite of an infected anopheline mosquito.

When the mosquito has ingested infected blood, the plasmodia penetrate the mosquito's stomach and form cysts on its outer wall. The cysts rupture, releasing plasmodia which make their way through the mosquito's body cavity to the salivary glands, from which the plasmodia can be passed to an uninfected human at the time of the mosquito's next blood meal. The mosquito's range is no more than one mile. Thus, the disease cannot be transmitted to humans located at greater distances unless there are means by which the mosquito or the person can be transported. This was unlikely when wandering tribes were located at great distances from each other.

Malaria can only exist where the environment is acceptable to the anopheline mosquito. Stagnant pools of water within a specific temperature range must be available in which the anopheline female can lay her eggs. Human activities frequently modify the environment in ways that cause such small pools to accumulate, thereby facilitating anopheline reproduction. Until recently, malaria was one of the most widely distributed diseases, and a major cause of death. As late as 1943, it was estimated that not less than 3 million malarial deaths and 300 million cases of malarial fever occurred each year throughout the world.[32] The World Health Organization of the United Nations in 1955 adopted the objective of eradicating malaria, and it is one of the great triumphs of public health that substantial progress towards this goal has been achieved in the twentieth century. When the WHO program started in 1955, about 2 billion people lived in malarious areas of the world. By the end of 1975, 20 years after the program was announced, more than 820 million of the original 2 billion were living in areas that had been freed of the disease, and another 850 million lived in areas where eradication was already in progress.[33] However, only a few years after the control program was initiated, it was observed that in some areas the mosquitos had developed resistance to the DDT and other pesticides on which the program depended. The phenomenon appears to be due to the emergence of mutant insect strains, and the resistance of mosquitoes to DDT may result in reintroduction of malaria to areas from which the disease has been eliminated.[33-35] The tendency of disease organisms to adapt to chemical and biologic agents designed for their control may pose increasingly difficult obstacles to disease control in the future.

Evolution of the towns and cities has created new ecological niches for a wide variety of organisms. Even under the relatively hygienic conditions of modern suburban living, man has involuntarily domesticated such varied species as the robin, housefly, cockroach, and blue jay, among many others that could be cited. Most of these are probably harmless and some, like the songbirds of our backyards and orchards, add to the pleasures of life.

Under conditions of congestion and filth, the domesticated species are less desirable, for they include obnoxious and disease-carrying organisms generally classed as vermin. For some of these organisms, the relationship to man is highly intimate, as is that of lice and fleas that can live on his body surface or the worms that can infest the intestines and liver. The rats, fleas, and lice are of particular relevance because they are carriers of disease. Under overly crowded unsanitary conditions, rats can thrive in enormous numbers. Moreover, like the human inhabitants on whom they depend, rats provide habitation for the fleas and lice that transmit the infectious agents of typhus and plague.

Plague was the most devastating disease in history. It is primarily a disease of rats. Humans are involved only incidentally because they become infected by the fleas that are the vectors of the disease. The disease apparently appeared for the first time in epidemic form in the sixth century during the reign of the Emperor Justinian.[36] After that time wave after wave of the disease spread across the world for 1,000 years. It affected the social and political history of Europe. The disease has been variously known as the "Justinian Plague" (after the emperor), the "Black Death" (from discoloration due to the skin hemorrhages that characterize the disease), and "bubonic plague" (from "bubo," an inflamed lymph node). The worst of the plagues came to Europe in the fourteenth century, when it is estimated that within a few years it took the lives of 25 million people, amounting to 25 percent of the population of Europe. Thereafter, for reasons that are not known, the disease diminished in virulence and no world-wide epidemics occurred until 1894, when an outbreak in China spread wildly and caused millions of deaths in India.[37]

Typhus is another disease that has periodically assailed humanity. The first of the typhus epidemics probably occurred in 430 B.C. during the Peloponnesian wars. The rat is the host for the vector, which is the body louse. As in the case of the parasites that cause malaria and plague, the ecological relationships between the host and vector of typhus require delicate forms of adaptation. The louse becomes infected with the microscopic typhus organism, of the rickettsia type, by feeding on an infected host. The organisms grow in the intestinal wall of the louse, from which they are excreted in feces, which in turn infect the rat or human through skin abrasions. As stated by Zinsser: "To the louse, *we* are the dreaded enemies of death. He leads a relatively harmless life . . . then out of the blue, an epidemic occurs: his host sickens, and the only world he has ever known becomes pestilential and deadly. . . ." The louse dies in 12 to 18 days, and the infected humans harbor the rickettsia for only ten to 14 days. Propagation of the disease thus requires that the infected louse get to a new host before dying. Consequently, some minimum population density is required for a typhus epidemic to propagate.

Malaria, plague, and typhus have in common that they are diseases of such virulence as to have had major consequences on human history. All three diseases depend on delicate ecological relationships that have been facilitated by the human way of living.

ENERGY IN PREINDUSTRIAL SOCIETY

For almost all of prehistory, work was performed only by humans, who obtained barely enough energy from their food to support the basic body metabolism and the exertions of the nomadic life. The situation began to change rapidly with the development of husbandry and agriculture. Increased food production made it possible for an individual to produce more food than he himself could consume, and domesticated animals added to the energy available for work.

Technological advances gradually created the need for additional energy. Cooking and preserving food necessitated wood for fires. Additional energy was required for metallurgical and ceramic processes that provided utensils, weapons, and a variety of artifacts that simplified the chores of daily life. Energy was also needed for purposes of illumination, but the situation did not change markedly until manufactured gas was introduced in London in the nineteenth century. (Although oil and tallow have been available to provide artificial illumination for thousands of years, they were too costly for most people.)

Humans learned to sail in boats that utilized the winds produced by solar energy. The invention of the water wheel and windmill made it possible to mill grain, to operate bellows for metallurgical furnaces, and to pump water. However, the Industrial Revolution could not have been brought about if man had been forced to rely on animate or solar sources of energy. A new source was needed, and it was found in the abundant supplies of coal. Seams of coal occur as surface outcrops in many parts of the world, and it takes little imagination to visualize how early man, having learned to kindle fire, accidentally discovered the combustible properties of coal. It is thought that the earliest uses of coal were in England, as early as 1500. B.C.[38] The demand was so small, however, that the needs could be satisfied from outcrops until the thirteenth century, after which mines were developed.

Coal could not be used for home heating until the chimney was invented.[40] Instead, wood fires were built in pits located in the center of the room, and the smoke found its way out of the room as best it could through open windows. The nuisance of the wood smoke was apparently tolerable, but coal could not be burned in this way because its effluents were too irritating and otherwise noxious for even the hearty inhabitants of pre-Renaissance Europe.[39]

Although Mayhew calls attention to the fact that chimneys were mentioned in a Venetian manuscript as early as 1347, flues and chimneys evidently did not come into use in England until the fifteenth century.[40·] The introduction of chimneys made possible the use of coal for domestic purposes, and by the seventeenth century there were said to be 360,000 chimneys in London. For the next 300 years, the smoke from these chimneys combined with the sea fogs to produce a noxious, gaseous mixture that marked the burgeoning industrial centers of England, Wales, and other countries. Headache and catarrh were attributed to the combustion of coal as early as 1577.[40] In 1661, John Evelyn wrote a now-classic pamphlet "Fumifugium: or the Smoake of London Dissipated," which railed against the deteriorating condition of the atmosphere in that city:

What is all this, but that Hellish and dismall clowd of SEA-COALE? Her inhabitants breathe nothing but an impure and thick Mist, accompanied by a fuliginous and filthy vapor, which renders them obnoxious to a thousand inconveniences, corrupting the Lungs, and disordering the entire habit of their bodies, so that Catarrs, Phthisicks, Coughs and Consumptions, rage more in this one City, than in the whole Earth besides.[41]

In contrast, William Harrison, who was a contemporary of John Evelyn and who chronicled life in England, wrote the following as a comment on the increasing use of fireplace chimneys:

Now we have manie chimnies, and yet our tenderlings complaine of rheumes, catarhs and poses. Then we had none but reredoses (braziers in the centre of the hall), and our heads did never ache. For as the smoke of those daies was supposed to be a sufficient hardning of the timber of the houses, so it was reputed to be a far better medicine to keep the good man and his family.[42]

Thus, we see that disagreement about environmental matters existed even in the seventeenth century!

Evelyn noted that by 1661 it was already difficult to grow plants in London. Subsequently, things must have deteriorated even further because in the 1772 edition of Evelyn's work, B. White noted that it was quite remarkable that crops could be grown at all in London, even in Evelyn's time. Evelyn reported that when Newcastle, the main source of coal, was blockaded by war in 1644 and little coal was available in London, flowers bloomed and trees bore fruit.

The preface to White's 1772 edition makes note of the high mortality in London and attributes this to air pollution. "London destroys beyond what it raises" said White, noting that each year 10,000 people must move from the country to London to make up for the excess deaths. The high death rates in the cities of that period actually were of infectious origin due to the unsanitary conditions under which

Figure 2-1. Lead smelting in the sixteenth century. Metallic lead (*D*) is seen dripping from the crude hearth (*C*) on which the lead ore (presumably lead sulfide) is mixed into burning wood. The molten lead is ladeled from the crucible (*E*) into molds (*F*), from which ingots (*H*) are obtained. Lead poisoning would almost certainly develop under these circumstances. The principal mode of exposure would be inhalation of finely divided lead oxide dust formed at the surface of the molten lead and dispersed into the air by the action of ladeling. The natural draft through the vent (T) would be insufficient to prevent the dust from reaching the breathing zone of the worker. In modern times this operation would be enclosed, and provided with mechanical ventilation as well as filters to remove the lead dust from the ventilation air discharged from the smokestack.

people lived rather than to air pollution. However, this was long before the nature of infection was understood, and it is understandable that White should attribute the high mortality rates to the fact that "constant and unremitting poison is communicated by the foul air. . . ."

THE PREINDUSTRIAL OCCUPATIONAL ENVIRONMENT

There is very little written material about the conditions under which the lower classes lived and worked prior to the nineteenth century. This may be due in part to the low esteem in which manual workers were regarded. This attitude seems to have had its roots in the classical Greek era, as suggested by Hunter's quotation attributed to Socrates:

What are called the mechanical arts carry a social stigma and are rightly dishonoured in our cities. For these arts damage the bodies of those who work at them or who have charge of them, by compelling the workers to a sedentary life and to an indoor life, by compelling them, indeed, in some cases to spend the whole day by the fire. This physical degeneration results also in deterioration of the soul. Furthermore, the workers at these trades simply have not got the time to perform the offices of friendship or of citizenship. Consequently, they are looked upon as bad friends and bad patriots. And in some cities, especially the warlike ones, it is not legal for a citizen to ply a mechanical trade.[43]

The health problems caused by the occupations were neglected by the early physicians. The Hippocratic doctors were taught to study the environment of the patients, so far as the community and home were concerned, but they neglected the occupational environment.

A knowledgeable industrial hygienist can visualize what conditions must have been like in the Middle Ages from texts and illustrations in the classic *De Re Metallica,* published by Georgius Agricola in 1556 (Figure 2–1).[44] This is the earliest comprehensive description of working conditions in mines. One can surmise that the incidence of accidents and disease among the miners of Agricola's time must have been shockingly high by modern standards, despite the fact that the safety record of the present-day mining industry provides little reason for pride. (The primitive working conditions that existed in the mines of England and Wales in the early nineteenth century are illustrated in Figure 3.1.)

By the end of the eighteenth-century major technological advances made a better way of life possible. Mechanical energy could be made available wherever it was needed in the form of the steam engine, and among its earliest applications was the locomotive, which could haul fuel to the great population centers. There the fuel could be used to power other steam engines which drove a variety of newly invented machines used to produce manufactured goods at low prices on a massive scale. The human race had progressed to the threshold of the Industrial Revolution.

NOTES

1. Cloud, Preston E., Jr. "Some Early Evidences of Life and their Implications," in: "Infectious Dieseases: Their Evolution and Eradication" (A. Cockburn, ed.), Charles C. Thomas, Springfield, Ill. (1967).
2. Dobzhansky, Theodosius. "Mankind Emerging," Yale Univ. Press, New Haven, Conn. (1962).
3. Sauer, Carl O. "The Agency of Man on Earth," in: "Man's Role in Changing the Face of the Earth" (William L. Thomas, ed.), Univ. of Chicago Press, Chicago, Ill. (1956).
4. Malcolm, James Peller. "Anecdotes of the Manners and Customs of London from Roman Times to 1700," London (1811).
5. Brothwell, Don. "The Question of Pollution in Earlier and Less Developed Societies," *Proceedings of the Eighth Annual Symposium of the Eugenics Society*, London, 1971 (Peter R. Cox, and John Peel, eds.), Academic Press, New York (1972).
6. Ramazzini, Bernardino. "Diseases of Workers" (trans. by W. C. Wright), Hafner Publications, New York (1964). From Latin text "de Morbis Artificum," 1713.
7. Sigerist, Henry E. "Civilization and Disease," Phoenix Books, Univ. of Chicago Press, Chicago, Ill. (1962).
8. Bryant, Sir Arthur. "The Medieval Foundation of England," Doubleday and Co., New York (1966).
9. Trevelyan, G.M. "English Social History," Spottiswoode, Ballantyne and Co., Ltd., London (1944).
10. Woodham-Smith, Cecil. "The Great Hunger," Harper and Row, New York (1962).
11. Butzer, Karl W. "Environment and Archeology: An Introduction to Pleistocene," Aldine Publishing Co., Chicago, Ill. (1964).
12. Stewart, O. C. "Fire as the First Great Force Employed by Man," in: "Man's Role in Changing the Face of the Earth" (William L. Thomas, ed.), Univ. of Chicago Press, Chicago, Ill. (1956).
13. White, Lynn. "The Historical Roots of Our Ecologic Crisis," *Science* 155: 1203 (1967).
14. Martin, P. S. and H. E. Wright. "Pleistocene Extinctions: the Search for a Cause," Yale Univ. Press, New Haven, Conn. (1967).
15. Marsh, G. P. "The Earth Modified by Human Action," Scribners, New York (1874).

16. Osborn, Fairfield. "Our Plundered Planet," Little, Brown and Co., Boston, Mass. (1948).
17. Jacks, C. V. and R. O. Whyte. "The Rape of the Earth," Faber and Faber, Ltd., London (1937)
18. Darby, H. E. "The Clearing of the Woodland in Europe,"in: "Man's Role in Changing the Face of the Earth" (William L. Thomas, ed.), Univ. of Chicago Press, Chicago, Ill. (1956).
19. Glacken, C. J. "Changing Ideas of the Habitable World," in: "Man's Role in Changing the Face of the Earth" (William L. Thomas, ed.), Univ. of Chicago Press, Chicago, Ill. (1956).
20. "Plato: The Collected Dialogues" (Edith Hamilton and Huntington Cairns, eds.), Bollingen Series LXXI, Pantheon Books, New York (4th printing, 1966).
21. Lowdermilk, Walter C. "Palestine: Land of Promise," 2nd ed., Victor Gollancz, Ltd., London (1946).
22. Talbot, Lee M. "Ecological Aspects of Aid Programs in East Africa, with Particular Reference to Rangelands," in: "Ecology and the Less Developed Countries," *Bulletins from the Ecological Research Committee #13*, Swedish Natural Science Research Council (1971).
23. Cameron, Jenks. "The Development of Government Forest Control in the U.S.," Johns Hopkins Univ. Press, Baltimore, Md. (1928).
24. Sears, Paul B. "Deserts on the March," in: "Population Evolution and Birth Control," 2nd ed., W. H. Freeman and Co., San Francisco (1969).
25. Lilliard, Richard G. "The Great Forest," Alfred Knopf, New York (1947).
26. Craven, Avery Odelle. "Soil Exhaustion as a Factor in the History of Virginia and Maryland, 1606-1860," published 1926. Reprinted by Peter Smith, Gloucester, Mass. (1965).
27. Armelagos, G. J. "Man's Changing Environment" in: "Infectious Diseases: Their Evolution and Eradication" (Aiden Cockburn, ed.), Charles C. Thomas, Springfield, Ill. (1967), Ch. 6.
28. Burnet, MacFarlane. "Natural History of Infectious Disease," Cambridge Univ. Press, London (1966).
29. Cockburn, Aidan. "Paleoepidemiology," in: "Infectious Diseases: Their Evolution and Eradication" (Aidan Cockburn, ed.), Charles C. Thomas, Springfield, Ill. (1967), Ch. 5.
30. May, J. M. "The Ecology of Human Disease," M. D. Publications, New York (1958).
31. Zinsser, Hans. "Rats, Lice and History," Little, Brown and Co., Boston, Mass. (1935).
32. *Encyclopedia Brittanica*, William Benton, Publisher, Chicago, Ill. (1958), Vol. 14, p. 706.
33. World Health Organization. "The Malaria Situation in 1975," *WHO Chronicle* 30: 486-493 (1976).
34. Davidson, G. and A. R. Zahar. "The Practical Implications of Resistance of Malaria Vectors to Insecticides," *Bulletin of the World Health Organization* 49:475-486 (1973).

35. "Conquest of Malaria: the Art of the Feasible," *The Lancet* (April 6, 1974), pp. 607-608.
36. Ziegler, Philip. "The Black Death," Penguin Books, Middlesex, England (1969).
37. Hammon, W. D. "Diseases Transmitted by an Arthropod Vector," in: "Preventive Medicine and Public Health," Maxcy-Roseman, 10th ed., Appleton-Century-Crofts, New York (1973), Ch. 3.
38. Marsh, Arnold. "Smoke: The Problem of Coal and the Atmosphere," Faber and Faber, Ltd., London (1947).
39. Brimblecombe, Peter. "Attitudes and Responses Towards Air Pollution in Medieval England," *Journal of the Air Pollution Control Association* 26:941 (1976).
40. Mayhew, Henry. "London Labour and the London Poor," Charles Griffin and Co., London (n.d., circa 1864).
41. Evelyn, John. "Fumifugium or the Inconvenience of the Aer and Smoake of London Dissipated" (1661). Reprinted in: "The Smoake of London," Maxwell Reprint Co., Fairview Park, Elmsford, N.Y. (1969), with preface to 1772 edition by B. White.
42. Trevelyan, G. M. op. cit., p. 71.
43. Hunter, Donald. "Diseases of Occupations," English Univ. Press, Ltd., London 1975).
44. Agricola, Georgius. "De Re Metallica" (1556) (trans. by Herbert and Lou Hoover). Reprinted by Dover Publications Inc., New York (1950).

CHAPTER 3

The Industrial Revolution

The Industrial Revolution was in full swing in England by the first quarter of the nineteenth century, and hundreds of thousands of people were on the move from the country to the cities.

The migration had been brought about by the agricultural changes that took place in sixteenth-century England, more than 200 years earlier, when landowners began to convert large areas of farmland to sheep pasture. This impoverished many peasant farmers and caused them to migrate to the towns and cities.[1]

During the eighteenth century there were also many innovations that increased the productivity of land and reduced the required manpower. Horses replaced oxen, new cultivating tools were used, and agricultural science was born with the design of controlled field experiments. In addition, major improvements in the quality of livestock were brought about by the increase of interest in what was then the art, rather than the science, of animal breeding.[2] The weight of cattle sold at market in Smithfield, England doubled between 1710 and 1795.[3] Obviously, all these changes did not take place simultaneously throughout England.[2] But for the first time in thousands of years, basic changes in agricultural practices were being made, and these would set the stage for the Industrial Revolution, of which the agricultural revolution was really a part.

The Industrial Revolution arose out of the introduction of methods of mass production that for the most part utilized raw materials such as cotton, wool, flax, and animal hides, that were derived from farms, pastures, and forests. Since the financial resources that built the factories in those early years came in a large measure from agriculture, it was only natural that there should have been a return flow of capital from the factories to stimulate agricultural productivity. The earliest years of the Industrial Revolution thus had marked effects on the productivity of farms and pastures.

Agricultural productivity was also stimulated by the network of roads and canals that were constructed during this period, and which greatly expanded the areas in which produce could be marketed. Prior to the eighteenth century, coaches drawn by teams of six horses moved no faster than at a walking pace, along deeply rutted, muddy roads. The subject of highways in eighteenth-century England became so important that 2000 separate road acts were passed between 1700 and 1790.[3] Although the roads were greatly improved by the late eighteenth century, it still took a full day to travel 100 miles.

There has been much speculation as to why the Industrial Revolution began when it did, and why it developed in England. There is probably no simple answer, and in any case I do not intend to give an interpretation of the possible social, economic, and political causes. Suffice it to say that there were many, not the least of which was the appearance in England of many brilliant minds capable of great scientific, mechanical, and administrative inventiveness. Over a period of about 100 years, England's means for producing goods was transformed from what was basically a cottage industry, (in which the work was done with tools owned by the families) to manufacturing enterprises which required the construction of large buildings, investment in expensive machinery, and the recruitment of the former cottage workers who would now be assembled for their labor under a single roof.

The factory system was pioneered in the textile industry and was the logical outgrowth of the invention of mechanical spinning and weaving equipment that favored large-scale production. The factories enabled the manufacturers to lower the unit price of textiles and as a result, increased demand for the products to such an extent that massive capital investments were made to expand the woolen and cotton industries.

The perfection of the steam engine during the late eighteenth century accelerated the move from rural areas to cities. Factories were no longer tied to the remote river valleys where water power could be found. Since use of the steam engine led promptly to the development of railroads, the energy to drive machines could be obtained in any place to which coal could be hauled. The towns and cities soon replaced the river valleys as the centers of industrial development.

When the country people came to the towns and cities to work in the factories, nothing had been prepared for them. Tens of thousands of people lived under environmental conditions that humans in Western society may not have endured as a normal way of life before or since. In addition, conditions in the workplaces, the homes, and on the streets deteriorated to such an appalling state that contemporary descriptions would seem exaggerated if it were not for the consistency of reports written by health officials, humanitarians, writers, and official governmental investigators in England, the United States, and on the continent.[4,5,6]

Those conditions are best documented in a series of reports, the "Blue Books," released by Parliamentary investigatory commissions. The earliest of the Parliamentary inquiries were concerned with the conditions of child labor, a subject of singular ugliness without which no description of the human environment during the Industrial Revolution would be complete. At the start of the nineteenth century, it was not uncommon for children to be sold into the apprenticeships in the mills and mines when they were only five or six years old, and there is evidence that children as young as four years of age were apprenticed to the master chimney sweepers, who found the diminutive size of the children to be advantageous.[5]

In the early days of the Industrial Revolution, the children accompanied their parents into the mines and factories much as they would have done if the parents worked in the fields. Since it was the custom in rural England for children to help their parents plant and harvest at the earliest possible age, it did not seem unusual at first for wives and children to accompany their husbands and fathers to the mills or mines. Moreover, discrimination on grounds of sex was not the rule of the day in most of the collieries, where women worked underground, did the same work and received the same pay as men.[6]

Employment of children was widely believed to be socially desirable in the early nineteenth century. A Connecticut textile-mill owner in 1808 was granted tax exemption because, according to the state legislature, "he had put the energies of women and children to good use." The mill owner himself was reported to have believed that by giving the children employment, he had rescued them from a life of poverty and crime.[7] Earlier, Noah Webster had made a similar proposal. "The children wander the streets," he said. "It would be best for them and the communities if they were put to work in the factories."[8]

This was a period in history when humans who could not look after their needs were sent to workhouses scattered across the countryside. Orphaned infants, the infirm aged, the cripples, the blind, and the mental defectives would live together in common quarters that came into being out of the well-intentioned but ill-conceived laws that had evolved during the eighteenth century.[9] The conditions in many of the English workhouses were so bad that it was easy to understand how a mill or mine owner could believe factory employment was a superior choice to the workhouse—the only alternative for poor people, children, and adults alike.

The situation with respect to orphan children was not destined to change quickly. In the United States it was not until the end of the nineteenth century that children in need would be treated differently from adults in the poorhouses. Until then, the children were confined not only with people whose incapacity arose from their poverty, but also with the criminals, the insane, and the depraved. It was not until 1874 that New York State took its orphans out of the poorhouses and placed them in separate facilities or arranged for their adoption.[10]

The early reformers asked for relatively little in the way of legislation to limit the employment of children, but the few laws that had been effected were resisted strenuously by the mill and mine owners. A bill failed of passage in England that would have excluded children younger than nine years of age from employment in the mills. Children in the age group between nine and 16 would have been limited to no more than 13 hours per day, including one and a half hours for meals and recreation. A similar bill, the Factory Act of 1833, was finally enacted which excluded children under nine from employment, limited the hours of work to eight hours a day for children nine to 13, and to 12 hours a day for those 14 to 18.

A Parliamentary commission estimated in 1842 that about one-third of the mine workers were less than 13 years of age. The chief employment of the youngest children was as "trappers," who attended the air doors that separated various sections of the tunnels. Among the many vivid descriptions of the pathetic conditions under which the youngsters were employed is the following testimony, typical of many given to an inquiry in 1842:

> The trappers sit . . . behind each door, where they sit with a string in their hands attached to the door, and pull it open the moment they hear the corves (i.e., carriages for conveying the coal) at hand, and the moment it has passed they let the door fall too, which it does of its own weight. If anything impedes the shutting of the door they remove it, or, if unable to do so, run to the nearest man to get him to do it for them. They have nothing else to do; but, as their office must be performed from the repassing of the first to the passing of the last corve during the day, they are in the pit the whole time it is worked, frequently above 12 hours a day. They sit, moreover, in the dark, often with a damp floor to stand on, and exposed necessarily to drafts. It is a most painful thing to contemplate the dull dungeon-like life these little creatures are doomed to spend—a life, for the most

part, passed in solitude, damp, and darkness. They are allowed no light; but sometimes a good-natured collier will bestow a little bit of candle on them as a treat.[6]

Many of the seams were less than two feet in height, and Figure 3–1, which is reproduced from an 1842 Parliamentary commission report, gives a vivid, if somewhat crude, expression to the working conditions that then existed. Children and adults alike worked in the dark. Frequently they were required to use painful leg irons and girdles to drag the loaded carts through the narrow passageways.

The Parliamentary investigations revealed that conditions in the habitations of the poor were no less shocking than they were in the factories and mines. The British Parliament, in 1838, became so concerned with the high death rate from infectious disease in the working-class districts that an inquiry was initiated into the conditions under which the laboring population lived. The 1842 report of the investigation, On an Inquiry into the Sanitary Conditions of the Labouring Population of Great Britain, for the first time set forth in detail the conditions of life in the English slums, the sanitary deficiencies, and their effects on the statistics of life and death.[11]

The driving force behind the inquiry was a lawyer, Edwin Chadwick, who came from a prominent family and who developed an interest in the prevention of infectious disease in the course of studying life insurance. He devoted his life to the subject of public health, and deserves major credit for starting the sanitary movement in the nineteenth century.[12] Chadwick was a disciple of Jeremy Bentham, whose earlier work established the first Ministry of Health and laid the legislative basis for much that was to follow. A wealthy mill owner, Robert Owen, also played an outstanding role, both in the manner in which he operated his mills and through his many writings about the responsibilities of mill owners for the well-being of their employees.[13]

Men like Chadwick and Owen received no formal training to prepare them for the work they were to do. They could not read textbooks dealing with environmental matters, public health, or the prevention of disease because no such body of knowledge had yet been accumulated. In the main, it remained for these humanitarians, of whom Chadwick was one of the most notable, to define the relationships between human health and the human environment. The importance of a wholesome supply of water, adequate housing, clean streets, and proper nutrition was to evolve out of their pragmatic studies.

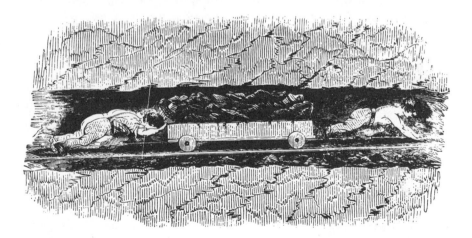

Figure 3-1. Working conditions in the coal mines of England and Wales as illustrated in the report of an 1842 British Parliamentary Commission.[6]

When Chadwick was preparing his reports in 1840, the world was still more than 30 years away from Pasteur's demonstration of the relationship between micoorganisms and disease. The concept of infection, and the relationships that exist between the environment and the biological agents of infectious disease, were not then comprehended. People therefore lived under filthy conditions without understanding the impact of those conditions on health. The major infectious diseases of the period—typhoid, cholera, typhus, yellow fever, malaria, smallpox, and plague—were caused by microorganisms and viruses transmitted by insects and rodents in ways that were not yet understood. Rickets was also endemic in the nineteenth-century cities of Europe, due not only to dietary deficiencies but also to a lack of sunlight. The long hours in the shops and mines and the dark, narrow streets resulted in a high incidence of this bone disease and may account for the grotesque and deformed men and women painted by William Hogarth in his street scenes of the mid-eighteenth century.[14]

If the exact *causal* relationships between environmental factors and disease could not be understood, the *associations* between environmental filth and many diseases could have been recognized by anyone who cared to inquire. Sir Percivall Pott had begun to inquire into the history of men with cancer of the scrotum more than 50 years before Chadwick. He found that chimney sweeping was a common factor, and he correctly deduced that some property of soot was capable of causing cancer.

The investigation made by Pott is one of the earliest examples of the application of epidemiology—the search for the causes of disease. In later years, other medical scientists would use microscopes, sophisticated chemical tests, and highly specialized techniques in their searches, and these "microbe hunters" would ultimately discover the causes of diseases such as malaria, typhoid, and smallpox. But epidemiology can also be a far more empirical art, and as such has been practiced effectively from biblical times until the present.

Sir Percivall was not in a position to identify the specific chemical in soot that caused scrotal cancer among chimney sweepers and we cannot, to this day, describe the exact biochemical mechanisms by which cancer-causing chemicals do their mischief. However, Pott could observe that chimney sweepers were in direct contact with coal soot, and deduced that this was probably an important factor in the cause of the disease. Without knowing the biological mechanism by which the cancer was produced, Pott nevertheless realized that it could be prevented either by scrupulous hygiene or by substituting mechanical methods for chimney cleaning. (Unfortunately, no practical application of this knowledge was made until many years later.

Although machines were available for cleaning chimneys, the master chimney sweepers argued against their use for many decades.)[5]

In the course of his study, Chadwick addressed to medical officers then stationed throughout England a series of questions designed to bring out certain basic information about the conditions under which the working class lived. Chadwick's report includes hundreds of extracts from the letters he received in reply. To change the original text in any way could only detract from the vividness of the firsthand reports. A medical officer reported from Gateshead on the congested living conditions that existed:

> The want of convenient offices in the neighborhood is attended with many very unpleasant circumstances, as it induces the lazy inmates to make use of chamber utensils, which are suffered to remain in the most offensive state for several days, and are then emptied out of windows. The writer had occasion a short time ago to visit a person ill of the cholera; his lodgings were in a room of a miserable house situated in the very filthiest part of Pipewellgate, divided into six apartments, and occupied by different families to the number of 26 persons in all. The room contained three wretched beds with two persons sleeping in each: it measured about 12 feet in length and 7 in breadth, and its greatest height would not admit of a person's standing erect; it received light from a small window, the sash of which was fixed. Two of the number lay ill of the cholera, and the rest appeared afraid of the admission of pure air, having carefully closed up the broken panes with plugs of old linen.[11]

From Glasgow came the following:

> We entered a dirty low passage like a house door, which led from the street through the first house to a square court immediately behind, which court, . . . was occupied entirely as a dung receptacle of the most disgusting kind. Beyond this court the second passage led to a second square court, occupied in the same way by its dunghill; and from this court there was yet a third passage leading to a third court, and third dungheap. There were no privies or drains there, and the dungheaps received all filth which the swarm of wretched inhabitants could give. . . . The interiors of these houses and their inmates corresponded with the exteriors. We saw half-dressed wretches crowding together to be warm; and in one bed, although in the middle of the day, several women were imprisoned under a blanket, because as many others

who had on their backs all the articles of dress that belonged to
the party were then out of doors in the streets.[11]

Some of the larger mines, factories, and mills provided employees
with lodgings that came to be recognized as a major source of
infectious disease. The following report was sent to Chadwick by a
Dr. Mitchel, who investigated conditions in the lodgings associated
with the mines of Durham and Northumberland.

Along one side of the room were three beds, each six feet long by
about four feet and a half wide, the three beds extending the
length of the room; then there were three other beds on the other
side, and at the furthest end was a seventh bed extending from
the one line of beds to the other. Immediately over these seven
beds, and supported on posts, were seven other beds placed
exactly in the same way. Of course the person who slept in each
of the six beds of the upper tier next to the wall could raise his
head only a very little way on account of the roof. Each of these
14 beds was intended for two persons, when only few men were
employed at the mines, but they might be made to receive three
men each, and, in case of need, a boy might lie across at their
feet. There was no opening of any sort to let out the foul air, yet
from 39 to 40 persons might have slept there. . . .[11]

The crowded living conditions were a repetitious theme in the reports
of the medical inspectors:

A large proportion of the cottages in the Union are very miserable places, small and inconvenient, in which it is impossible to
keep up even the common decencies of life. . . . A man, his wife,
and family, consisting in all of 11 individuals, resided in a cottage
containing only two rooms. The man, his wife, and four children,
sometimes five, slept in one of the rooms, and in one bed, some
at the foot, others at the top, one a girl above 14, another a boy
above 12, the rest younger. The other part of the family slept in
one bed in the keeping-room, that is, the room in which their
cooking, washing, and eating were performed. How could it be
otherwise with this family than that they should be sunk into a
most deplorable state of degradation and depravity?[11]

That such living conditions diminished the body as well as the spirit
was shown by Chadwick, who cited French statistics to demonstrate
that during a 90-year period beginning in 1750, the size of men

recruited to the army had diminished from five feet, four inches to five feet, one inch.[11] Observers cited by Chadwick also called attention to the diminishing size of British volunteers, and noted that "though not originally a large race they have become still more diminutive under the noxious influences to which they are subject. . . . They are decayed in their bodies: the whole race of them is rapidly descending to the size of Lilliputian. You could not raise a grenadier company amongst them all."

Only in recent years have the nutritionists and pediatricians discovered that when a child, in its early formative years, is deprived of proper nourishment, its mental ability is damaged. As a result, each generation that lives in poverty handicaps the next by this subtle limitation on intellectual capacity (Chapter 5). This is what Chadwick was referring to 150 years ago when he wrote that

> the teachers of the pauper children at Norwood show that a deteriorated physical condition does in fact greatly increase the difficulty of moral and intellectual cultivation. The intellects of the children of such inferior physical organization are torpid. . . .

Of the many remarkable things about the Chadwick report, not the least was his perception that the cost of public measures for drainage, cleansing, and the supply of wholesome water was far less than the public cost of caring for the diseased poor in the hospitals and asylums of the time. Although he himself was motivated by humanitarianism, his report effectively demonstrated with facts and figures that a program of environmental rehabilitation would pay for itself because the health of the people would improve, family structure would be strengthened, and the thousands of citizens who had been trapped by squalor would begin to enter the mainstream of economic life and contribute to the strength of the community from which they had formerly sapped its resources.

Chadwick inquired of the Assistant Commissioners located in the districts as to whether the sense of social responsibility was found to improve with better living conditions. His representative in Turbey Abbey replied on January 4, 1841:

> I have much pleasure in saying that some cases of the kind have come under my own observation, and I consider that the improvement has arisen a good deal from the parties feeling that they are somewhat raised in the scale of society. The man sees his wife and family more comfortable than formerly; he has a

better cottage and garden: he is stimulated to industry, and as he rises in respectability of station, he *becomes aware* that he has a character to lose. Thus an important point is gained. Having acquired certain advantages, he is anxious to retain and improve them; he strives more to preserve his independence, and becomes a member of benefit, medical, and clothing societies; and frequently, besides this, lays up a certain sum, quarterly or half-yearly, in the savings' bank.

A report from Martson, Stafford, reflected the report quoted above:

> If we follow the agricultural labourer into his miserable dwelling, we shall find it consisting of two rooms only; the day-room, in addition to the family, contains the cooking utensils, the washing apparatus, agricultural implements, and dirty clothes, the windows broken, and stuffed full of rags. In the sleeping apartment, the parents and their children, boys and girls, are indiscriminately mixed, and frequently a lodger sleeping in the same and the only room; generally no window, the openings in the half-thatched roof admit light, and expose the family to every vicissitude of the weather; the liability of the children so situated to contagious maladies frequently plunges the family into the greatest misery. . . . The children are brought up without any regard to decency of behaviour, to habits of foresight, or self-restraint; they make indifferent servants; the girls become mothers of bastards, and return home a burden to their parents, or to the parish, and fill the workhouse. . . .

Although excellent drainage plans had been developed by the Greeks and Romans, good drainage was frequently neglected in the design of early nineteenth-century cities. There was little discipline with respect to the disposal of human excrement or the offal of the butchers and others who prepared food. Putrescible refuse was dumped wherever it happened to be generated because less offensive methods of disposal took more energy than people were prepared to expend. As early as 1657 in New Amsterdam, a minor New World colony later to become New York, the city fathers complained that "within the city of Amsterdam in New Netherland, many burghers and inhabitants throw their rubbish-filled ashes, dead animals, and suchlike things into the public streets to the great inconvenience of the community."[16] Because of the unsanitary practices, the drinking water taken from shallow wells became contaminated by human excrement, which was the cause of the cholera epidemics that took thousands of lives. In a classic investigation during a cholera epidemic in London about 12 years after publication of the Chadwick report, Sir John Snow studied the distribution of cholera cases and found that they were clustered around a public pump located on Broad Street in the Golden Square area. His observation that the outbreak was associated with water from the pump was, like the work of Pott and Chadwick, an early example of the application of epidemiological methods.

The Chadwick report contains detailed descriptions of what the streets were like. For example, in the town of Sterling, open sewers were constructed but there was no water with which to keep them flushed continually. The only available water was collected from rainfall, and this was far from sufficient for flushing sewers. Human excrement and kitchen swill from the town jail were flushed down the main street every two or three days. The slaughterhouse was situated near the top of the town, and the blood from it was allowed to flow down the public streets.

The following is a description of the streets inhabited by the working classes in Manchester:

Many of the streets in which cases of fever are common are so deep in mire, or so full of hollows and heaps of refuse that the vehicle used for conveying the patients to the House of Recovery often cannot be driven along them, and the patients are obliged to be carried to it from considerable distances. Whole streets in these quarters are unpaved and without drains or main-sewers, are worn into deep ruts and holes, in which water constantly stagnates, and are so covered with refuse and excrementitious matter as to be almost impassable from depth of mud, and intolerable from stench. In the narrow lanes, confined courts and alleys, leading from these, similar nuisances exist, if possible, to a still greater extent; and as ventilation is here more obstructed, their effects are still more pernicious. In many of these places are to be seen privies in the most disgusting state of filth, open cesspools, obstructed drains, ditches full of stagnant water, dung-hills, pigsties etc., from which the most abominable odours are emitted. But dwellings perhaps are still more insalubrious in those cottages situated at the backs of the houses fronting the street, the only entrance to which is through some nameless narrow passage, converted generally, as if by common consent, into a receptable for ordure and the most offensive kinds of filth. The doors of these hovels very commonly open upon the un-covered cesspool, which receives the contents of the privy be-

longing to the front house, and all the refuse cast out from it, as if
it had been designedly contrived to render them as loathsome and
unhealthy as possible.[11]

It was recognized at the time that the enormous amounts of dung
were potentially valuable as fertilizers (this was before the development of synthetic fertilizers), but the cost of labor and cartage limited
the general use of these waste products to a distance of about six
miles. One official said in desperation, "I have given away thousands
of loads of night soil: we knew not what to do with it." Nearly 150
years later, we still have no long-range solutions for handling and
disposing of our municipal wastes. The impediments to seemingly
well-conceived plans in the twentieth century are economic in origin,
just as they were in Chadwick's day.

Another not insignificant problem, particularly in the larger cities,
was the accumulation of horse manure. By the middle of the
nineteenth century, the London Board of Health estimated that
200,000 tons were deposited annually in the city streets. Dried particles of manure were an important component of dust dispersed into
the atmosphere by vehicular traffic in dry weather.

What to do about human dung occupied much of Chadwick's
attention, as it did Henry Mayhew, who insisted in 1864 that it was
essential to recycle waste materials: "Now, in nature everything
moves in a circle—perpetually changing, and yet ever-returning to the
point where it started."[15] But he, like Chadwick, recognized the
economic difficulties in returning waste materials to the soil. Even if
dung was used as fertilizer, the practice was not without its problems.
In Edinburgh, the contents of a large proportion of sinks, drains, and
privies of that city were conveyed to covered sewers which entered
into a stream called the Foul Burn, which thus became a large,
uncovered sewer. Nearby farmers diverted the stream so that its
contents could be deposited on their land. About 300 acres of meadow
were so treated, and they did become highly productive, but the
practice was understandably opposed by nearby inhabitants as being
offensive.

From all reports, the early part of the nineteenth century was a
period in the development of Western civilization when men, women,
and children were taxed to the limits of their physical endurance in
their homes, the streets, and their places of employment. Respect for
human life and dignity may have existed within upper-class society,
but rarely could it be encountered for the less privileged. Labor was
treated harshly in the factories, mines, and merchant fleets. The
asylums and workhouses were pestilential, and there was little feeling

of community responsibility for widows, orphans, and the maimed. Both the physical and social environments were harsh for the poor of the period.

Ironically, conditions on board the slave and convict ships were greatly improved in the early nineteenth century. John L. Kennedy indicated the reason why this came about:

> No wilful oversight, still less any oppression, was perhaps imputable to the owners, the captains, or anyone else . . . but somehow or other it happened that fever broke out, and that the mortality during the first voyages was dreadful—sometimes half the passengers were lost . . . but the importance of ventilation was little known at that time, and even the King's ships were ravaged by scurvy, dysentery, and fever. At length, however, the form of contract was altered: instead of the ship-owners being paid per head on the number *embarked*, they were only paid per head *on the number landed alive*; so that the shipowners lost by every person who died on the passage. This form of contract changed the whole face of things: ventilation and other appliances were sedulously attended to; the merchant, at his own cost, provided a medical officer to take charge of the convicts, and the remuneration of that officer was proportioned to the number landed *alive*. The result was that the frightful mortality disappeared, and the voyages have generally been effected with a higher degree of health amongst the passengers, or with less mortality, than would perhaps have occurred amongst the same number of the same class of persons living at large on shore.[11]

Whereas the mortality was formerly as high as 60 percent, the number of deaths dropped to about 1.5 percent under the new arrangement.

This chapter has described, briefly, the kinds of environmental problems with which humans contended as recently as 150 years ago. The conditions described were in part due to the congested living conditions that arose out of the influx of migrants to the towns, primitive working conditions in the mines and factories, the prevailing customs in regard to child labor, and institutional deficiencies that resulted in failure to provide for people in need. The early nineteenth-century environmental problems were not solved all at once, and even now we have a long way to go towards achieving reasonable standards of environmental quality in the homes, workrooms, and streets. When the nineteenth century ended, conditions were still deplorable in cities such as New York and Chicago, but they had already begun to improve, and the worst of today's slums are far better than those of a century ago. The urban poor of modern times usually have an ample supply of fresh water and indoor plumbing that is connected to sewers. The littered streets of today's ghettos cannot be compared to the unpaved, muddy, offal-piled streets and lanes described by Chadwick's investigators. At least today's urban poor are usually innoculated against diphtheria, polio, and other diseases. Typhoid and other waterborne diseases have all but disappeared. Children are no longer so wan and sickly as in the novels of Dickens. There have been many improvements in the way of life of the urban poor, despite the fact that much remains to be done.

NOTES

1. Trevelyan, G. M. "English Social History," Spottiswoode, Ballantyne and Co., Ltd., London (1944), p. 116.
2. Mathias, Peter. "The First Industrial Nation," Methuen and Co., Ltd., London (1969).
3. Trevelyan, G. M. ibid., p. 382.
4. Engels, Friedrich, "The Condition of the Working Class in England" (Blackwell, Oxford, 1845) (trans. and ed. by W. O. Henderson and W. H. Chaloner), Macmillan Co., New York (1958).
5. British Parliamentary Papers: Report of the Minutes of Evidence on the State of Children Employed in the Manufactories of the United Kingdom, Together with a Report of the Employment of Boys in Sweeping Chimneys with Minutes of Evidence and Appendix. House of Commons, 1816 and 1817. Reprinted by Irish Univ. Press, Shannon, Ireland (1968).
6. British Parliamentary Papers: First Report of the Commissioners-Mines. Children's Employment Commission, 1842. Reprinted by Irish Univ. Press, Shannon, Ireland (1968).
7. Kuczynski, Jürgen. "The Rise of the Working Class," McGraw-Hill Book Co., New York (1967).
8. Warfel, Harry R. "Noah Webster: School Master to America," Octagon Books, New York (1966). (First printing, Macmillan Co., New York, 1936.)
9. Longmate, Norman. "The Workhouse," St. Martin's Press, New York (1974).
10. Trattner, W. I. "Homer Folks: Pioneer in Social Welfare," Columbia Univ. Press, New York (1968).
11. Chadwick, Edwin. "Report on the Sanitary Condition of the Labouring Population of Gt. Britain" (1842) (ed. with intro by M. W. Flinn), Edinburgh Univ. Press, Edinburgh (1965).
12. Richardson, B. W. "The Health of Nations: A Review of the Works of Edwin Chadwick," Dawsons of Pall Mall, London (1965), Vols. I and II.

13. Owen, Robert. "A New View of Society & Other Writings," J. M. Dent and Sons, London/Toronto; E. P. Dutton and Co., New York (1927).
14. Loomis, W. F. "Rickets," *Scientific American* 223(6): 76–91 (1970).
15. Mayhew, Henry. "London Labour and the London Poor," Charles Griffin and Co., London (n.d., circa 1864).
16. Berger, Meyer. "New York," Random House, New York (4th printing, 1960). Original copyright, 1953, p. 13.

CHAPTER 4

The Twentieth Century: The Evolution of Environmental Policy

The twentieth century began during a tranquil and optimistic period when the products of technology were beginning to proliferate throughout Europe and the United States. Electric lighting was gradually being adopted in major cities and the nations of the Western world were being introduced to remarkable new agricultural machines, the products of new biological and medical technologies, new developments in mechanical transportation and electrical communication, and major advances in the graphic arts. The existence of relatively stable political systems in Europe, increased literacy among the populations of the advanced nations, and increasing per capita productivity, all seemed to portend a new era of unprecedented opportunity. Yet many of the problems of the living and working environment that had developed during the Industrial Revolution persisted into the new century and, in fact, remain with us to this day. But there was a growing spirit of reform in the early years of the twentieth century, sparked by authors who railed against the living and working conditions of the time, and whose influence gradually took effect.[1-3]

In contrast to the prevailing mood at the start of this century, its last decades have been characterized by apprehension that the new technology, welcomed originally with such optimistic anticipation, has proved to be the cause of many of the seemingly insoluble problems that now face society. A period of uncertainty and confusion has developed that is due in part to the pace of present-day life.

It has been characteristic of this century, in contrast to the past, that many major inventions have moved from the inventor's laboratory to general usage in less than a decade. Mankind waited 2 million years to evolve agriculture and animal husbandry, another 6,000 years

50

to extract small quantities of iron from the earth's crust, and then 2,500 years more to invent the printing press. It took 200 years to develop the steam engine from prototype to practical application. In contrast, technologically advanced countries now produce important inventions in every decade. Electrical energy, the automobile, radio, television, commercial aviation, synthetic fertilizers, pesticides, antibiotics, chemical contraceptives, nuclear energy, transistors, and computers are only some of the important developments of the first half of the twentieth century. Each has had major effects on the human way of life.

The pace has been so rapid that many of the most dramatic midcentury developments were not foreseen 20 or 30 years earlier. The author of a popular post-World War I scientific book wrote that the enormous quantities of energy locked in the nucleus of the atom could be of enormous benefit to mankind, but that "the atom is as much beyond our reach as the moon."[4] In less than half a century, men have not only unleashed the energy of the atom, but have left their footprints on the sands of the moon.

There are people alive in the United States today who began life in the peasant societies of Europe where the way of life had not changed for centuries. The oxcarts, harnessing equipment, and agricultural tools they knew as children were basically those of medieval times. The octogenarians of today were born into a world in which the automobile was a curiosity called the "horseless carriage," and telephones and electric lights were yet to come into general use. The practice of modern medicine was only beginning to develop, and there was no radio, television, or air transport. At the beginning of the twentieth century, the impending changes had hardly been felt in many of the world's rural communities, but by the last quarter of the century, the effects of technology had been experienced everywhere.

The explosive rates at which changes continue to be made in every aspect of human society is another cause of many of the environmental stresses with which we must now deal. The size of the human population, the demand for living space, and the requirements for goods and services are all growing exponentially.

The word "exponential" is so important to the discussions in this book that a brief explanatory digression is justified. Growth can take several forms, the simplest of which takes place at a fixed rate of increase. Using bank interest as an example, simple interest at a rate of 7 percent may be added each year to a deposit of $1,000. The interest is applied to the initial deposit only, not to the accrued interest. This type of growth is known as "linear" and takes the form of the straight line marked A in Figure 4–1. At a rate of 7 percent per

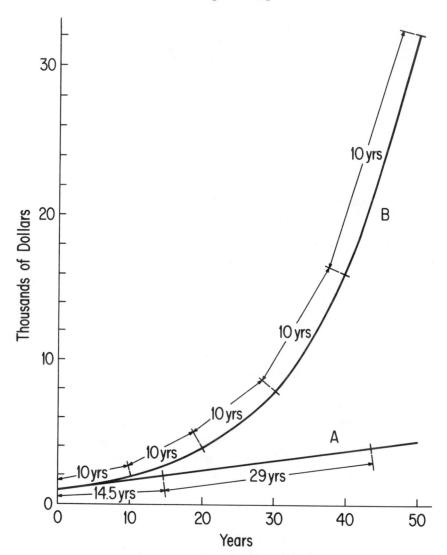

Figure 4-1. The two main forms of growth are illustrated by the difference in simple and compound bank interest. Curve *A* illustrates the manner in which a $1000 deposit will grow over a period of 50 years if the interest is not compounded, at a rate of 7 percent per year. This type of growth is called linear.

In Curve *B*, the interest is also constant at 7 percent per year, but is compounded. This growth, known as exponential (or logarithmic), is characterized by having a constant doubling time (in this case ten years).

annum, $70 will be added each year to the deposit, the size of which at any given time will equal $1,000 + $70 multiplied by the number of years it is in the bank. It is seen from Figure 4–1 that the initial $1,000 deposit will double to $2,000 in 14.5 years. It will then take 29 years for the $2,000 to double to $4,000, and 58 years for that $4,000 to reach $8,000. The rate of growth is constant at $70 per year. The proportionate rate of growth diminishes with time, and the number of years required to double the deposit becomes progressively longer.

The form of growth is very different when the interest is compounded by applying the 7 percent per annum to the total accumulation. This is illustrated by the curve marked B. Instead of the constant rate of increase of curve A, curve B reveals an accelerating rate of increase of the total sum. Whereas the "doubling time" becomes longer in curve A, it is a constant (10 years) in curve B. During a period of 50 years, the money increases from $1,000 to only $4,500 for the case of simple interest, but to more than $30,000 when the interest is compounded.

It is convenient to think of processes similar to accumulating compound interest in terms of "doubling time," that is, the length of time it will take for a given deposit to double in value at a given rate of compound interest. The algebra of these relationships is such that the doubling time in years can be approximated by dividing 70 by the annual increase expressed as a percentage. Thus, a deposit that grows at a compounded rate of 7 percent per year will double in ten years (70 ÷ 7), and will double in 20 years (70 ÷ 3.5) if the interest rate is only 3.5 percent. This type of growth, which is illustrated by curve B in Figure 4–1, is known as "exponential" growth. It can also be called "logarithmic" growth. (The twentieth century has been aptly called "the logarithmic century."[5])

One of the consequences of exponential growth is that, given a short enough doubling time, quantities can increase explosively. An illustration is provided by the school-child query as to which of two options would be preferable as a gift: $1,000 per day during a 31-day month (a total of $31,000) or 1¢ on the first day of the month, with the understanding that the amount would be doubled every day for the duration of the month. It would require remarkable algebraic intuition to avoid choosing the first option, but with access to a calculator or pad of paper, one can readily determine that the second option is much the better of the two. The sum due on the thirty-first day will be about $10 million, and the subtlety of the process is illustrated by the fact that at the end of ten days the gift would be only $5.12, but in 15 days it would have increased to more than $160, and to more than

$5,000 at the end of 20 days, then increasing rather explosively to more than $10 million on the thirty-first day.

Many developments in human affairs have actually become "super" exponential in the twentieth century. Population and economic growth in many cases have not been constant, but have been increasing with time. Also in such cases the doubling times have become progressively shorter. Population and economic growth have increased the demand for goods and services. The demand for goods and services actually is increasing more rapidly than population. This is due to greater per capita demand arising from increasing affluence, particularly in the technologically developed countries of the world. This has caused chain reactions. Thus, the availability of abundant energy raises the standard of living, increases the availability of food, and leads to control over infectious disease. These factors increase the population by reducing the death rate, and greater population results in more demand for food, goods, and services—all of which impact on the environment. (The subject of population growth and its relation to the availability of food, energy, and natural resources is so important to an understanding of contemporary human ecology that it will be treated in detail in the next two chapters.)

The long-range effects of human activities on the environment already were being discussed in the nineteenth century. Among the earliest writers on the subject, G. P. Marsh foresaw in 1874 that human activity would affect the earth's topography to such an extent as to constitute a new geological epoch, "the anthropozoic era."[6]

The finiteness of the human environment was emphasized by Herbert Quick, who published a remarkable popular book in 1913, *The Good Ship Earth*.[7] Regrettably, Quick's views about environmental matters were combined with outrageous bigotry towards Mohammedans and people with black or yellow skin. He used the metaphors "good ship earth" and "terrestrial zeppelin" that preceded others like "spaceship earth,"[8] and "living on a lifeboat"[9] by more than 50 years. Quick emphasized the fact that all living things and the environment on which they depend constitute a single, closed system.

Another perceptive and informative popular book was *Standing Room Only?* published in 1927 by the American sociologist Edward Alsworth Ross, who stressed the rapidly diminishing death rates from infectious disease and the lack of compensatory reductions in birth rates, particularly in the less-developed countries.[10]

These and other concerned writers gradually set the stage for the environmental movement of the 1960s, which occurred when it did, only *in part* because there was a need for it at that particular time. The need had already existed for several decades, and the movement might have been postponed for another few decades had it not been for a series of dramatic events and the emergence of a new and potent personality, the environmental activist. Perhaps the first of these was Linus Pauling, two-time Nobel Prize winner, who led the worldwide assault against nuclear weapons testing in the 1950s. In this he was assisted by Barry Commoner, whose name has been identified with many environmental issues from the mid-1950s to the present time. The nuclear weapons controversy was followed by a wave of concern about pollution, initiated by publication of Rachel Carson's *Silent Spring*, a book of great power that focused on the problems of pesticide contamination.[11] A few years later Paul Ehrlich's book *The Population Bomb* had a marked effect on public comprehension of the global population problem.[12] These and others among the new environmentalists spoke in strident tones, and whether their success was due to the form or substance of their messages is a question that will only be answered from the perspective of the future. For the time being, we can simply conclude that a much-needed environmental awakening was brought into being because of the way the issues were popularized by a relatively small, articulate group of environmental zealots. Some of them may have overdramatized the subjects with which they dealt, and some irritated their colleagues by taking a few liberties with facts. But they did stimulate public interest.

These and others among the new breed of scientist-activist became intensely involved in a series of dramatic issues. The first of these was the debate about testing nuclear weapons in the atmosphere.

THE NUCLEAR-WEAPONS CONTROVERSY

The first test of a nuclear weapon took place in secrecy during World War II in New Mexico, preparatory to the bombings of Hiroshima and Nagasaki in the summer of 1945. The fine particles of radioactive dust created by the explosion drifted across midwestern United States, and enough fallout occurred in Indiana cornfields to cause damage to X-ray film that had been packaged with interleaving paper made from the contaminated corn stalks. A scientific report of this incident was subsequently published in the scientific literature, but attracted little popular attention.[13] It was not until the early 1950s when the United States and the Soviet Union began frequent tests of nuclear weapons in the atmosphere that the subject gradually came to public notice.

The first concerns arose out of tests conducted near Las Vegas, Nevada, beginning in the early 1950s, when scientists at the University of Utah in Salt Lake City first raised questions about possible risks to inhabitants living near the test site, who were known to have been subjected to a number of fallout episodes. Concern was greatly heightened in 1954, when massive fallout from tests by the United States in the mid-Pacific resulted in injuries to inhabitants of a Marshall Island atoll and to 22 men aboard a Japanese fishing vessel. The United States government did not disclose the incident for several weeks, but many of the facts became widely known when the vessel returned to Japan with men suffering from radiation sickness.[14]

The nuclear-weapons testing programs demonstrated not only that mankind has the capability to spread contamination throughout the world, but that certain of the radioactive constituents of fallout can be assimilated by organisms in the human food chain and be passed to humans. For example, strontium-90, a radioactive constituent of fallout similar in chemical properties to calcium, was deposited on grass ingested by cows and ultimately reached the human skeleton via dairy products. Radioactive iodine, another substance in nuclear weapons debris, also found its way to milk and, when consumed by humans, concentrated in the thyroid gland.

Some of the pathways of the radioactive constituents of fallout were more subtle. It was found that cesium-137 was assimilated by Arctic lichens that are consumed by reindeer. As a result, Laplanders and many Alaskan Eskimos, who depend on reindeer as their main source of protein, were exposed to relatively high concentrations of this nuclide. Worldwide concern and fear mounted because it was known that exposure to radioactive substances increases the risk of cancer and genetic mutations.

Whether cancer and genetic mutations would actually occur became the subject of great international debates, and the nonscientific world became aware that scientists were not unanimous in their opinions. They sometimes disagreed violently among themselves. Until the debates concerning fallout, during the late 1950s, scientific disagreements were only rarely revealed outside the conference halls and published literature of the scientific community. But in the case of fallout, leading scientists examined the same body of information and argued publicly about their different conclusions.

Worldwide concern over the dangers of radioactive fallout during the mid-1950s led the United States, the Soviet Union, and the United Kingdom to declare a moratorium, in the fall of 1958, on further weapons tests. In 1961 the Soviet Union broke the moratorium unilaterally and tested about 50 devices. The United States responded, and the two major powers proceeded to test nuclear and thermonuclear bombs on a massive scale. In response to worldwide concern and popular pressure, the United States, the United Kingdom, and the Soviet Union once again signed a nuclear-weapons test-ban agreement in 1963 that has succeeded in eliminating further testing in the open atmosphere by those countries. France, India, and China are not signatories to the agreement and have since conducted tests in the open atmosphere from time to time. To date, these tests have not added significantly to the residues of radioactivity from tests conducted prior to 1963 by the United States and the Soviet Union.

Nuclear-weapons testing alerted the world to the complexities and subtleties of ecology, but the weapons-testing moratorium did not end the controversy. Eventually the controversy moved from nuclear-weapons testing to nuclear power and is far from settled as of this writing. However, the moratorium created a lull in the nuclear controversy. The nuclear-power industry was in its infancy in 1963, attracting little attention from environmentalists. The lull provided the opportunity to set the stage for the next act, which opened in 1963, with publication of Rachel Carson's *Silent Spring*.[11]

SILENT SPRING

Rachel Carson maintained a lifelong interest in biology from early childhood until her premature death in 1964, at the age of 57. Her main interest was scientific writing, and she attained the position of editor-in-chief of the United States Fish and Wildlife Service, a post she held until 1952, when she published the first of her successful popular books, *The Sea Around Us*.[15] This book and its immediate successor, *The Edge of the Sea*,[16] were remarkable for the way in which they aroused popular interest in marine ecology. Her outstanding achievement was *Silent Spring*, a book widely read, highly controversial, and unquestionably the most important single inspiration for the modern environmental movement.

The book was concerned with the use of the organic pesticides, mainly DDT, and their toxic actions on life forms higher than the insects for which the pesticides are intended (see Chapter 9). The title *Silent Spring* refers to the desolation in towns across America where:

Once all life seemed to live in harmony with its surroundings . . . but then a strange blight crept over the area and everything began to change. Some evil spell had settled on the community: mysterious maladies swept the flocks of chickens: the cattle and sheep sickened and died. Everywhere was a shadow of death. The farmers spoke of much illness among their families. In the town the doctors had become more puzzled by new kinds of sickness appearing among their patients. There had been several sudden and unexplained deaths, not only among adults but even among children, who would be stricken suddenly while at play and die within a few hours.

This "Fable for Tomorrow" was a powerful opening scenario which set the tone for the remainder of the book. Carson then went on to say that this town does not actually exist, but that "it might have a thousand counterparts in America or elsewhere in the world." The paragraph had its effect: It communicated a sense of fear that insecticides had the ability to produce widespread sickness and death among humans and their farm animals. This description has been described by the former editor of the British scientific journal, *Nature*, as "a model of calculated exaggeration which has since put its stamp on the literature of the doomsday movement."[17] Rachel Carson was among the first of the modern environmental writers to use literary license to a degree that irritated and even enraged scientists of a more conservative persuasion who repudiated conclusions derived by any process short of those traditionally used by scholars. To many, the end could not justify the means. But the means used by more moderate writers could never have achieved the end, and the time had certainly come to alert the world that DDT and other persistent pesticides could be concentrated by higher life forms and could produce toxic effects on many beneficial insects, birds, and fish. It was imperative to control the emission of toxic chemicals to the environment.

Like radioactive fallout, the use of DDT taught scientists, government officials, and the public that the biological environment was an elaborate network of food pathways, and that a substance introduced into one part of the environment could pass to other parts by devious pathways that were not fully understood.

THE ENVIRONMENTAL MOVEMENT TAKES FORM

Following *Silent Spring*, wave after wave of excitement about environmental matters spread across the United States and other countries. In quick succession, the public soon learned about mercury poisoning, carbon monoxide, smog, cancer-producing substances in foods, and the black-lung disease among Appalachian miners. "The environment" became such a popular subject that newspapers and weekly magazines added articles and special sections dealing with it. Hardly a day has passed since the late 1960s when an environmental article has not been featured prominently in the daily press.

It was during this period that the public and its elected representatives began to understand that the human environment is governed by ecological principles, and that these principles were frequently violated. The human species could no longer conduct its affairs without regard for the environmental effects they caused. Technology could cause such extensive environmental change as to produce irreversible effects that could threaten human well-being. There was clearly a need for a better understanding of the environmental effects of technical innovation. But an unfortunate early by-product of the environmental movement was the wave of hysteria that developed in many countries. Books like *Silent Spring* and *The Population Bomb* were followed by many others with titles and texts that both contributed to and were nourished by the sense of crisis that existed. *Famine, 1975*,[18] *Vanishing Air*,[19] *The Careless Atom*,[20] and *Population Control through Nuclear Pollution*,[21] were a few of the more extreme publications of the period. Some attempted to follow the precedents set by Rachel Carson, but with less effect. *Vanishing Air*, with the apparent authenticity of a serious book about air pollution, begins: "It promised to be a rather pleasant Friday for December: the sun shone brightly over Manhattan." It then describes a period of atmospheric stagnation that resulted in such severe air pollution as to cause rising death rates and increased hospital admissions by the following Wednesday. After the episode is vividly described, leading the reader to believe it is an actual description of an air-pollution emergency, the chapter continues: "What you have just read has not yet happened. . . ."[19] If this style of writing was not deliberately designed to deceive the reader, it certainly had that effect. Most of the literature produced for popular reading in the 1960s made no effort to be objective.

After expressing concern over radioactive fallout, persistent pesticides, and other forms of pollution, the environmental movement began to take an interest in such matters as nonreturnable bottles, roadside litter, the threatened extinction of the blue whale, zero population growth, and the need to recycle waste materials. By the late 1960s, the subject of environment was highly ramified, and had achieved remarkable popular support. The movement in the United States reached a climax on a nationwide "Earth Day," on April 5, 1970. Campuses throughout the country came to a halt for lectures and seminars, and rallies were held in many communities. The intensity of popular interest gradually lessened after "Earth Day." Many of the newspapers and weekly news magazines now no longer include an environmental section on a regular basis, and there are fewer popular lectures on the subject. But an enormous amount of residual interest nevertheless remains.

It is estimated that by 1973 there were 5,000 environmental organizations in the United States. Some of the older organizations, such as the National Wildlife Federation and the Audubon Society, were originally organized to stimulate interest in the preservation of wildlife, but their interests were extended to pollution and other environmental matters in the 1960s. The Sierra Club, formed in 1892 by the naturalist John Muir, is the oldest of the environmental groups and is also among the most aggressive. A number of new and relatively well-financed organizations emerged, including the Natural Resources Defense Council, Environmental Action, and Friends of the Earth. The Natural Resources Defense Council and the Environmental Defense Fund, among others, have concentrated largely on use of the courts to achieve their environmental goals.[22]

PROGRESS TOWARDS AIR-POLLUTION CONTROL

The atmospheres of many industrial cities in Europe and the United States have been smoke-laden for two centuries, but this was tolerated as a relatively minor nuisance since smoking chimneys had long signified a comfortable hearth, economic prosperity, and full employment. This traditional complacency was seriously shaken in 1930 when several days of meteorological stagnation in the Meuse River valley of Belgium caused high levels of air pollution from coke ovens, blast furnaces, sulfuric acid plants, and zinc smelters. [23] By the end of the week, 60 people had died as the result of respiratory complaints.

A similar episode took place in 1948 in Donora, a city in the heavily industrialized Monongahela valley of western Pennsylvania. The meteorological circumstances were similar to those associated with the episode in the Meuse valley, and the symptoms were about the same. Of about 14,000 people who lived in the valley, 43 percent became ill and 20 deaths occurred.[24]

In both the Meuse valley and Donora episodes, the effects were caused by a combination of irritant gases and particulates released from many sources, rather than by any one identifiable toxic substance from a single known source. In contrast, two episodes were reported in 1949 and 1950, in which community disease could be ascribed to specific substances from known sources. In 1949, cases of chronic beryllium poisoning were reported in the vicinity of a plant extracting beryllium compounds in Ohio.[25] These cases had resulted from the emission of beryllium over a period of many years and were similar to the chronic form of beryllium poisoning that had already been reported in beryllium workers. In 1950, exposure to hydrogen sulfide in the vicinity of a Mexican plant designed to recover sulfur from natural gas resulted in 22 deaths.[26]

The air-pollution incident that attracted the most attention occurred in December, 1952 in London, when most of Britain was covered by a foggy mass of stagnant air that resulted in an estimated 3,500 to 4,000 deaths during a one-week period. Following this episode, a search of past meteorological records in the London area was conducted to identify similar periods of stagnation. Mortality records during those periods were then studied, and these indicated that excess deaths had indeed occurred on several occasions as far back as 1873. However, none of these episodes had been as severe as the one in 1952.[27]

The most persistent regional air-pollution problem in the United States has been in the Los Angeles basin, where annoying smogs began to attract attention in the 1940s. It was originally thought that these smogs could be controlled by prohibiting the widespread California practice of incinerating backyard refuse. Despite ordinances to discourage this practice, smog continued to worsen. It was soon shown that the problem of Los Angeles smog was due to automobile emissions and that it resulted basically from the interaction of sunlight with unburned hydrocarbons and nitrogen dioxide (Chapter 12).

It is commonly thought that air pollution in all cities continued to worsen until the relatively recent intervention of the federal government. This is not true. In some cases, marked improvement occurred many years ago because of changes in fuel-burning practices. Among the first of the cities in the United States to study its air-pollution problems was Chicago. In 1915 it published a classic report that led to electrification of the railroads, one of the major sources of air pollution at the time.[28] Another large city to deal effectively with its air pollution was St. Louis, which prior to 1940 had been subject to such dense winter smogs that street lights and automobile headlights were often required by day. An investigative committee in 1939 made recommendations which were enacted into legislation in 1940. Within five years the appearance of the city had changed for the better. The botanical garden could grow plants that formerly had been injured by air pollution, and a reduction in the incidence of upper-respiratory infections was reported.[29] Pittsburgh, which was one of the most polluted cities in the world prior to World War II, achieved similar success during the early 1950s.

Until recently, air pollution was regarded by the federal government as a relatively local problem with few interstate implications. This was a mistake. First, many air-pollution problems are interstate in extent. Second, the economics of pollution control are such that a state with lax standards and enforcement procedures may find it is easier to attract industry than a state with stricter standards. It is certainly not in the national interest that states with less-stringent pollution controls should enjoy a competitive economic advantage over their more conscientious neighbors.

In 1962 President John F. Kennedy requested the Congress to pass a bill, introduced two years earlier, that would authorize the Surgeon General to hold hearings on specified interstate air pollution.[22] This resulted in the Clean Air Act of 1963, which also authorized the Department of Health, Education and Welfare (HEW) to hold exploratory conferences and hearings and made funds available to assist local government in establishing air-pollution control departments.

In the early 1960s, the automobile began to attract attention as the principal source of air pollution. (See Chapter 12.) The first Motor Vehicle Air Pollution Control Act was passed in 1965, and it authorized the secretary of HEW to establish emission standards applicable to new cars. The state of California had already developed such standards, intended to be applied to the 1967 models, and in 1965 it was the intention of HEW to apply the California standards nationally by 1968.

A further step was taken with the Air Quality Act of 1967. This established air-quality regions and laid the groundwork for the establishment of state standards, subject to the approval of HEW. Expenditures for research on the effects of air pollution were greatly increased, and additional assistance to local agencies was provided.

By the end of the decade, there was a great public interest in air pollution, and there was widespread impatience with the slow progress that had been made thus far. These were, in fact, years of substantial accomplishments. Air-pollution control laws had been passed by many states and cities. Standards were being established. Local agencies were being organized and staffs were being hired. Instruments were being perfected so that monitoring networks could gather information on the concentrations of the various pollutants to which people were being exposed. New York City, which reorganized its air-pollution control activities in 1966, had already reduced its sulfur dioxide concentrations by more than 60 percent in early 1970.[30] To professional air-pollution-control officials, all of this represented substantial progress, but the public was impatient, and the subject was receiving such extensive attention in the public press that politicians vying for public office began to outdo each other in their statements about what should be accomplished.

The authorizations contained in the Clean Air Act of 1967 were due to expire in 1970, and a new law was needed. This led to the Clean

Air Act of 1970—overwhelmingly passed by Congress—which, among other stipulations required that the 1970 emissions be reduced by 90 percent no later than 1975, a timetable that has since been delayed. It was an expensive decision, the cost effectiveness of which remains debatable. (See Chapter 12.)

FEDERAL WATER-POLLUTION CONTROL LEGISLATION

Prior to 1948, water-pollution control was the responsibility of the state governments. A federal oil-pollution act was passed in 1924 to control the dumping of oil from oceangoing vessels, but it proved to have little effect. Although many bills were introduced in the Congress during the period from 1935 to 1940, nothing meaningful came out of these efforts, and the subject of federal water-pollution control was then deferred during the wartime years.

It was not until 1948 that the first federal water-pollution-control law was passed. It was a weak law designed to provide loans to municipalities for the construction of sewage-treatment plants, but funds for the loans were never provided. Federal water-pollution-control activities were supervised mainly by the United States Public Health Service (USPHS), an agency of the Department of Health, Education and Welfare (HEW). The USPHS drew increasing criticism for being too conservative and for being concerned with human health to the exclusion of wildlife. With mounting public interest in environmental matters, Congress found it much easier to pass the Water Pollution Control Act of 1965. The act relieved the Public Health Service of its responsibilities for water-pollution control by transferring the functions to a new Federal Water Pollution Control Administration (FWPCA), which was organized as a separate unit within HEW. It also required the states to set standards. If they failed to do so, or if the standards were not acceptable to HEW, then that agency could step in. For the first time this gave the federal government a significant, though still inadequate, role in the control of water pollution.

In 1966, another bill, the Clean Water Restoration Act, was passed. After only a few months, Congress became dissatisfied with HEW's handling of water-pollution-control matters, and transferred FWPCA to the Department of Interior. The 1966 act authorized, for the first time, substantial amounts of money to assist local communities to build sewers and waste-treatment plants. In 1968, $450 million was authorized. This was to rise to $1.25 billion in 1971, with the intent that the federal share of construction costs should gradually rise to 55 percent. The mood of the country had changed. Whereas in previous decades the "buck" was repeatedly passed back to the states, the Congress now accepted the country's mandate and gave the federal government a vigorous and dominant role in water-pollution control.

A number of water-pollution incidents had stimulated interest in the subject. Among the most spectacular was the 1967 grounding of the 970-foot tanker, *Torrey Canyon*, carrying crude oil from the Persian Gulf to England. More than 100,000 tons of oil were released, causing major contamination of the English shoreline. Two years later, in early 1969, an oil-drilling rig in the Santa Barbara channel off the coast of California began to leak about 20,000 gallons of oil per day into the surrounding waters. It was several weeks before the leak could be controlled, and by that time 20 miles of California shoreline had been heavily polluted with oil. This incident increased the clamor for clean water, and early in 1970 President Richard Nixon sent a special message on the environment to Congress, in which he recommended that federally approved standards be required for all industrial and municipal pollution sources. The result was the Water Pollution Control Amendments of 1972. The amendments mandated total elimination of discharges into navigable waters by 1985. The discharges were to be controlled by the best practicable technology by 1976, and by the best available technology by 1981. Despite the report of the President's Council on Environmental Quality that it was not feasible to eliminate all discharges, the bill passed the Senate by 86–0, and the House by 380–14. The bill was then vetoed by the President, but his action was overwhelmingly overriden by both houses.[22]

Environmental engineers do not believe it is possible—or for that matter necessary—to eliminate "all" pollution. At any given time, the allowable discharge can be reduced to any extent short of zero. Discharges can be reduced by from 90 to 99 percent, or more, but there will always be a finite residue that will be measureable if sufficiently sophisticated methods of analysis are used. Standards should be based on the extent to which an effluent can cause harm to either the aquatic ecology or to public health. The law of diminishing returns operates in the environment as it does everywhere, and there comes a point where the discharges will cause no perceptible damage and further reductions can be obtained only at exorbitant cost. However, the concept that some degree of pollution should be "permitted" is abhorrent to many people, and there has been wide public support for the 1972 Water Pollution Control Act amendments. It remains to be seen whether they will survive the test of time.

THE NATIONAL ENVIRONMENTAL POLICY ACT

The National Environmental Policy Act (NEPA) was passed in 1969. It is a landmark bill, unquestionably destined to play a major role in protecting the American environment of the future. Whereas earlier legislation had been concerned exclusively with air and water pollution, the National Environmental Policy Act was directed at the environment in its totality and mandated major organizational and procedural changes.

Prior to NEPA, the environmental activities of the federal government had been fragmentary. Water-pollution control was based in the Department of Interior and air-pollution control in HEW. The radiation hazards of the atomic energy program were the responsibility of the Atomic Energy Commission, but radiation hazards from X-ray machines or the use of radium were the responsibility of the Public Health Service. The U.S. Army Corps of Engineers, relying on the 1899 antidumping legislation which it was charged to enforce, was vying for jurisdiction with the Federal Water Pollution Control Administration, which was based in the Department of Interior. Similar fragmentation existed at the levels of city and state government. New York City, the first governmental entity to recognize the need to consolidate the various environmental functions into a single agency, created its Environmental Protection Administration in March, 1968. That new agency consolidated the municipal responsibilities for control over air and water pollution, solid-waste management, and noise.[30] The federal government was soon to follow that example, and in the National Environmental Policy Act created an Environmental Protection Agency (EPA) that came into being on December 2, 1970. The agency was charged with consolidating *all* federal programs dealing with air pollution, water pollution, solid-waste disposal, pesticides regulation, and environmental radiation.

The year 1970 also saw the establishment of a second important organization, the Council on Environmental Quality. This three-man Council, assisted by a small staff, reported directly to the President and was given responsibility to evaluate the environmental impact of all major actions by federal agencies, and to analyze the national trends with respect to environmental quality.

By far the most important provision of the National Environmental Policy Act is Section 102, which mandates that any agency of the federal government planning to undertake an action "significantly affecting the quality of the human environment" must file an Environmental Impact Statement with the Council on Environmental

Quality. This requirement applies to the effects of new legislation, rules and regulations, and construction being undertaken by the federal agencies; construction being undertaken by contract; and any licenses issued by federal agencies. The Environmental Impact Statement must include a full description of the proposed course of action, the alternatives that have been considered, and the environmental impact of both the proposed course of action and the alternatives. The Environmental Impact Statement is required so that the Council can analyze the extent to which natural resources, including land, will be utilized or disrupted, the effects on aquatic and terrestrial ecology, economic factors, and the effects on human health. Properly interpreted, Environmental Impact Statements can be invaluable to federal decision-making.

The form and complexity of the Environmental Impact Statements varies enormously, depending on the type of action proposed. Some can be relatively trivial, like those related to sewer construction, the building of television-transmission towers, or the improvement of existing highways. However, the Environmental Impact Statements become exceedingly costly and time-consuming when facilities such as nuclear power plants or offshore drilling for oil and gas are proposed.

More than 20 agencies have established procedures for the preparation of Environmental Impact Statements. For many agencies, the NEPA requirement for an Environmental Statement is the first requirement that they incorporate environmental considerations into the decision-making processes. For example, the Interstate Commerce Commission has adopted environmental-impact analysis in the course of approving freight rates that may have the effect of encouraging or discouraging the recycling of waste materials.[31]

Some Environmental Impact Statements are highly technical. To estimate the environmental impact of a power station properly, it is necessary to have available information on the microclimatology of the site, a complete inventory of the indigenous aquatic and terrestrial flora and fauna, and full information about the local hydrology and geology. The impact analysis must also consider effects on nearby historical and archeological sites (if these exist). It may take as long as two years and require the expenditure of several million dollars to prepare the Environmental Impact Statement for a proposed nuclear power plant.

The National Environmental Policy Act raised many questions that generated controversy and required recourse to the courts for adjudication. Intelligently administered, however, this bill can be an indispensable means by which technology can continue to be developed with minimum deleterious impact on the environment. Badly administered, it can lead to widespread dissension and polarization between the "environmentalists" on the one hand, and many branches of government and industry on the other.

OCCUPATIONAL HEALTH AND SAFETY IN THE TWENTIETH CENTURY

Occupational injuries most commonly are the result of falls, burns, dropping objects, or any unanticipated mishaps, often involving machinery (Figure 4–2). Occupational diseases are usually the result of exposure to toxic materials such as lead or mercury. Modern industry exposes its employees to many more toxic substances than in the past, and although we have learned much about how to use them safely, there is still much room for improvement.

The extent to which occupations are responsible for disease is still not known. It is possible to obtain reasonably reliable information about the incidence of certain diseases that are unique to certain occupations, such as silicosis among hard-rock miners, or the rare forms of cancer that occur among workers exposed to vinyl chloride or asbestos. However, there have been too few studies of the health of all industrial workers, and reporting requirements are almost nonexistent. In addition to causing readily identifiable occupational diseases such as those mentioned, the workroom environment may also influence the *general* health of the worker. The effects of occupational factors on heart disease, hypertension, maternal and infant health, and mental health may be far more important than the occupational diseases. There is very little known about the relationship between the job and the general health of the worker.

The fact that workers in any specific occupation have a higher-than-normal death rate does not mean there are causal relationships between the job and the worker's health. Socioeconomic or other factors may predominate. In 1967, the annual death rate among white males was 200 per 100,000 in the United States, but was twice as high among lumbermen, structural iron workers, and boiler makers.[32] The excess mortality could have been due to the higher rate of accidents in the latter industries, or to cultural traits that frequently are characteristic of workers in certain industries and professions. Use of alcohol and tobacco, for instance, is more prevalent in some occupations than in others.

The National Safety Council, a private public-service organization, has been publishing United States statistics on occupational accidents since 1928, and has had limited data since 1912. Its statistics show

Figure 4-2. A deceptively attractive Philadelphia printshop in the early nineteenth century. Note that the machines are driven by unguarded belts. When clothing or limbs were accidentally caught between the pulley and belt, severe injury or death resulted. This hazard can be easily eliminated by enclosing the pulleys with wire mesh guards but this precaution was frequently not taken. Unguarded belt drives continued to be a problem until about 1940, when conversion from centrally powered belt drives to machines driven electrically by individual motors was completed.

that from 1928 to 1975 mortality due to accidents at work decreased by 63 percent, from 15.8 per 100,000 to 5.9 per 100,000.[33]

The death rate from off-the-job accidents is much higher: 31.8 per 100,000 in 1975. This includes off-the-job motor-vehicle accidents, in addition to accidents at home and at play. The National Safety Council reports that in 1912 about 20,000 workers' lives were lost because of accidents on the job. By 1970, when the work force was about 50 percent larger and production was more than nine times greater, there were 14,200 work deaths. The number of occupational deaths *per unit of production* has thus been reduced by 92 percent in a period of about 60 years.

It is now accepted that the price of a manufactured product should include the costs of indemnifying injured employees, as well as the costs of the mechanical safeguards that must be provided to prevent accidents and disease. This was not always so. Until this century, the financial burden of industrial accidents, including loss of wages, medical expenses, and permanent loss of earning ability, was borne entirely by the employee and his family. Industrial practices were influenced by the common law, which held that a worker assumed the ordinary risks of employment when he or she was hired and that the employer was not responsible for accidental injuries unless it could be demonstrated that he was personally negligent. Of course, the employer could not be held guilty of negligence if the accident was due in part to the actions of a fellow employee, or if the injured employee himself contributed to the accident in any way. Hence, it was virtually impossible for an employee to be compensated for injuries sustained on the job. It would be a rare case in which an employee could demonstrate that any negligence was entirely that of his employer without some contribution from either a fellow employee or the injured individual himself.

Today's workmen's compensation laws do not require that negligence be proved in the event of an accident. An employee injured in the course of his work is simply compensated according to a schedule of benefits established by law.

Workmen's compensation legislation was first passed in the United States by the federal government in 1908, and it was applicable only to federal employees. Several states subsequently passed compensation laws (Montana in 1909; New York in 1910; and Kentucky in 1914), but these were repeatedly declared unconstitutional until Wisconsin succeeded in establishing the first permanent law in 1911. It was not until the 1940s that the last of the states adopted compensation laws. Thus, for more than 150 years after the Industrial Revolution, society was satisfied to allow the costs of industrial injuries to be paid by the injured employees. However, it actually was only the personal hardship that was borne by the employees and their families; the financial costs were ultimately paid by the community because the injured workman and his family often became public charges.

With the passage of the workmen's compensation laws there developed various systems of providing insurance for the employers. In the United States, this kind of insurance coverage has been provided mainly by private companies operating according to the laws of the various states. The costs paid by the employer are higher or lower, depending on the accident experience of his company. The insurance system thus provides economic incentives for the employer to reduce the number of accidents.

Workmen's compensation laws were originally drafted to provide compensation to victims of accidents but not to those who suffered from occupational disease. Illinois amended its workmen's compensation laws to provide coverage for occupational disease as early as

1911, but other states were slow to follow suit and the amendments were restrictive in that certain occupational diseases were compensable but others were not. The restrictions have gradually been eliminated, however, and all states now provide for such protection.

If the states were slow to adopt workmen's compensation and occupational-disease legislation, many were even slower in setting standards and developing the required enforcement procedures. As late as 1970, eight states had no programs in occupational health.[34] Four states had no inspection personnel, and the total number of inspectors employed by the 50 states totalled only 1,600, or about one per 50,000 workers. Some states, notably New York, Massachusetts, Illinois, Pennsylvania, and California, had for many years maintained highly competent staffs of physicians, engineers, and chemists who studied the causes of occupational disease and accidents and also undertook the required inspection and enforcement functions. However, the programs varied considerably from state to state in the size and quality of the effort.

The need for uniform standards and federal intervention to assure that all states were meeting at least minimum requirements of enforcement had been recognized for many years, and the federal government actually became involved with health and safety matters as early as 1890, when legislation authorized the establishment of Federal standards for coal mines. However, enforcement was generally considered to be ineffective.

Despite the fact that the federal and state governments have been lax in assuring industrial safety, substantial progress was being made towards reducing the number of industrial accidents and cases of occupational disease. The statistics already cited for accident frequencies are evidence that there has been improvement over the years. The safety records improved, in part, because modernization of industrial plants eliminated many arduous and dangerous manual operations for which mechanization proved safer and more efficient. However, most of the improvement was due to worker education and to a greater sense of responsibility on the part of management.

The unions could have had a stronger influence on occupational health and safety over the years, but until comparatively recently their goal has been one of achieving higher wages and fringe benefits. For this reason, although most union contracts have included health and safety clauses for many years, organized labor was, generally speaking, not an effective force in improving the record of occupational health and safety. However, this changed substantially in the 1960s when, having largely achieved their goals for wages and fringe benefits, the unions began to put more emphasis on working conditions.

The concern of the unions for safe working conditions coincided with the development of environmentalism and widespread media interest that publicized industrial health and safety to an unprecedented degree. As a result, many Congressmen who had for years been attempting to involve the federal government in occupational safety suddenly found the climate changing in their favor, and by late 1969 Congress drafted the needed legislation. The Occupational Safety and Health Act of 1970 was signed by the President late in that year. The act gave the federal government, through the Department of Labor, responsibility for regulating health and safety matters for 57 million employees in 4.1 million establishments. A new post of Assistant Secretary of Labor for Occupational Safety and Health was created by the act and the Assistant Secretary was assigned direct responsibility for the organization required to promulgate standards, and develop enforcement procedures.

The Occupational Safety and Health Act is landmark legislation, but implementation of the act has not been a simple matter. Well-trained industrial hygienists and safety engineers are all too few in the United States, and it will take many years to train physicians, engineers, nurses, and chemists in sufficient numbers to meet the inspection schedules and provide the required highly specialized laboratory services.

There is a dearth of information on which to base standards for the maximum permissible concentration of toxic substances in workroom air, and the act wisely provided for establishment within the Department of Health, Education and Welfare of the National Institute of Occupational Health and Safety, which will be the research arm of the Department of Labor. However, the required information cannot be developed overnight, and balance will be required between the need to push on towards the objective of greater safety, and the caution dictated by the insufficiency of information.

Very few of the states require cases of occupational disease to be reported to central registries. It is a relatively simple matter to obtain information about deaths or disabilities from industrial accidents because these are reported through the workmen's compensation system to insurance companies or to state agencies. It is far more difficult to obtain information about disabilities from occupational diseases because, as noted earlier, not all disability of occupational origin can be readily identified, particularly if it occurred long after exposure has ceased. If a 60-year-old man working in a shipyard develops a partial loss of hearing, is it due to the normal aging process or is it because he was consistently exposed to noise over a period of years? If a heavy smoker exposed to a known carcinogen develops lung cancer, to what extent did occupational factors contribute to his condition? There is very little known about the developmental defects that may appear in the offspring of mothers exposed to toxic chemicals.

Meaningful statistics should become available in the future as a result of the requirement by the Occupational Health and Safety Administration (OSHA) that reports of any disability be centrally recorded. Estimates of death and disability have until recently been based more on speculation than hard facts. In contrast to estimates that occupational diseases cause 100,000 deaths per year in the United States,[35] only 600 deaths were known to have occurred during the last half of 1971.[32] The estimate that occupational diseases cause 100,000 deaths per year first appeared in the 1971 *President's Report on Occupational Safety and Health*.[36] It rests on studies in England and Wales in 1950 that revealed excess mortality in certain occupations. The excess may have been due to socioeconomic associations with specific occupations rather than to occupational risk. The British data were adjusted to the size of the population of the United States and the estimate of 100,000 deaths due to "occupational disease" was thus made. It continues to be a widely quoted figure, but estimates of injuries based on studies in the United States that were published in subsequent presidential reports (1972 and 1973) are very much lower. The records of the New York State Department of Labor indicate that the incidence of occupational disease dropped from 74 per 100,000 workers in 1950 to 24 per 100,000 in 1970, a decrease of about 65 percent in 20 years. Of 1,524 cases receiving compensation in 1970, many were for skin rashes, hernias, and strained ligaments and muscles.[37]

One of the most important provisions of the Occupational Safety and Health Act is that the U.S. Secretary of Labor can implement the law's requirements through the states, which are then required to develop plans that must be approved by OSHA. An ironic development is that many states (including those that had highly effective occupational health and safety programs) elected not to make the changes required by OSHA. Rather, they chose to permit the federal government to take over their responsibilities, mainly because they did not have the financial resources needed to implement the program required by OSHA. When these states (New York, New Jersey, Massachusetts, and Illinois, for example) terminated their effective programs and dismissed their staffs, the federal government found that *it* did not have the means by which to implement its own requirements. Thus, six years after the passage of OSHA, the more-advanced states, which for decades had maintained programs for minimizing occupational diseases and accidents, have not made the progress they would have made if OSHA had never been enacted.

In the years immediately ahead, no dramatic changes in the statistics of occupational health and safety can be expected. Although we must learn more about the relationships between occupational factors and disability and disease, progress cannot be made overnight. Except for relatively rare situations in which changes can be made from information available for years, we must await the results of painstaking epidemiological studies. It helps to be patient (though not complacent), and to remember that great progress has been made in the field of occupational health and safety over a period of many years. Most workers today are safer on the job than off.

NOTES

1 Sinclair, Upton. "The Jungle," reprint of 1906 edition, New American Library, New York (1973).
2. Riis, Jacob. "How the Other Half Lives," reprint of 1890 edition, MSS Information Corp., New York (1972).
3. Riis, Jacob. "The Battle with the Slums," reprint of 1902 edition, MSS Information Corp., New York (1972).
4. Slosson, Edwin E. "Creative Chemistry," Garden City Publishing Co., Garden City, N.Y. (1919).
5. Lapp, R. E. "The Logarithmic Century," Prentice-Hall, Inc., Englewood Cliffs, N.J (1973).
6. Marsh, G. P. "The Earth Modified by Human Action," Scribners, New York (1874).
7. Quick, Herbert. "The Good Ship Earth," Bobbs-Merril Co., Indianapolis, Ind. (1913).
8. Boulding, K. "The Economics of the Coming Spaceship Earth," in: "Environmental Quality in a Growing Economy" (H. Jarrett, ed.), Johns Hopkins Univ. Press, Baltimore, Md. (1966).
9. Hardin, Garrett. "Living on a Lifeboat," *Bioscience* 24: 10 (1974).
10. Ross, Edward Alsworth. "Standing Room Only," The Century Co., New York (1927).
11. Carson, Rachel. "Silent Spring," Houghton Mifflin Co., Boston, Mass. (1962).
12. Ehrlich, Paul R. "The Population Bomb," Ballantine Books, New York (1968).
13. Webb, J. H. "The Fogging of Photographic Film by Radioactive Contaminants in Cardboard Packaging Materials," *Phys. Rev.* 76:375 (1949).
14. Eisenbud, M. "Environmental Radioactivity," 2nd ed., Academic Press, New York (1973).

15. Carson, Rachel. "The Sea Around Us" (1951), rev. ed., Oxford Univ. Press, New York (1961).
16. Carson, Rachel. "The Edge of the Sea," Houghton Mifflin, Boston, Mass. (1955).
17. Maddox, John. "The Doomsday Syndrome," McGraw-Hill Book Co., 1st paperback edition, New York (1973).
18. Paddock, W. and P. Paddock. "Famine, 1975: America's Decision, Who Will Survive?" Little, Brown and Co., Boston, Mass. (1975).
19. Esposito, John C. "Vanishing Air," The Ralph Nader Study Group Report on Air Pollution, Grossman Publishing Co., New York (1970).
20. Novick, Sheldon, "The Careless Atom," Houghton Mifflin Co., Boston, Mass. (1969).
21. Tamplin, A. R. and J. W. Gofman. "Population Control through Nuclear Pollution," Nelson-Hall, Chicago, Ill. (1969).
22. Davies, J. C. and B. S. Davies. "The Politics of Pollution," 2nd ed., Pegasus, Bobbs-Merrill Co., Inc., Indianapolis, Ind. (1975).
23. Roholm, Kaj. "The Fog Disaster in the Meuse Valley, 1930: A Fluorine Intoxication," *J. of Ind. Hyg. & Toxic.* 19: 126 (1937).
24. Schrenk, H. H., H. Heimann, G. D. Clayton, W. M. Gafafer and H. Wexler. "Air Pollution in Donora, Pennsylvania. Epidemiology of the Unusual Smog Episode of October 1948." Preliminary Report. *Public Health Bulletin* 306, 1949.
25. Eisenbud, M., R. C. Wanta, C. Dustan, L. T. Steadman, W. B. Harris and B. S. Wolf. "Non-Occupational Berylliosis," *J. Ind. Hyg. & Toxic.* 31:282 (1949).
26. McCabe, L. C. and G. D. Clayton. "Air Pollution by Hydrogen Sulfide in Poza Rica, Mexico: An Evaluation of the Incident of Nov. 24, 1950," *Arch. Ind. Hyg. & Occ. Med.* 6:199 (1952).
27. Goldsmith, J. R. "Effects of Air Pollution on Humans," in: "Air Pollution," 2nd ed. (A. C. Stern, ed.), Academic Press, New York (1967).
28. Chicago Association of Commerce. "The Chicago Report on Smoke Abatement: A Landmark Survey of the Technology and History of Air Pollution Control" (appraisal by Benjamin Linsky). Reprinted by Maxwell Reprint Co., Elmsford, N.Y. (1971). (Orginally published by Chicago Association of Commerce in 1915.)
29. Mallette, Frederick S. "A New Frontier: Air-Pollution Control," *Proceedings of the Institution of Mechanical Engineers* 168 (2): 595–628 (1954). Institute of Mechanical Engineers, Westminster, England.
30. Eisenbud, Merril. "Environmental Protection in the City of New York," *Science* 170: 706 (1970).
31. Council on Environmental Quality. "Environmental Quality," Second Annual Report of the Council on Environmental Quality, U.S. Government Printing Office, Washington, D.C. (August, 1971).
32. Ashford, N. A. "Crisis in the Workplace: Occupational Disease and Injury," MIT Press, Cambridge, Mass. (1976).
33. National Safety Council. "Accident Facts: 1976 Edition," Chicago, Ill. (1976).

34. Bureau of National Affairs, Inc. "The Job Safety and Health Act of 1970," Washington, D.C. (1971).
35. National Institute for Occupational Safety and Health. Personal Communication from Joseph Gustin, Statistician, to Jacqueline Messite, NIOSH Regional Consultant, *Occupational Medicine* (June 22, 1977).
36. *The President's Report on Occupational Safety and Health* (1971).
37. Messite, J. Personal communication (1977).

PART II

People, Energy, and Resources
Can We Keep Them in Balance?

CHAPTER 5

Population Growth, Food, Energy, and Raw Materials

The subjects of population size, food supply, natural resources, and energy are intimately interrelated and have fundamental effects on the relationships between human beings and their environment. Population size determines the demands for food and, with the standard of living, the need for natural resources and energy. Conversely, the size of the human population has historically been controlled by the amount of available food which, in recent times, has in turn been influenced by the availability of energy and natural resources. Additionally, the by-products of energy production and the wastes from the use of natural resources pollute the environment.

There are, of course, other determinants of human well-being, such as the institutional mechanisms by which society regulates itself, but proper balance of population size in relation to food, energy, and natural resources is basic to a sound relationship with the environment.

Population growth is currently in a state of imbalance with the supply of food, energy, and natural resources in many countries of the world.[1] As a result during the last quarter of this century, and the first part of the twenty-first, there will be tragic contrasts between countries that have and have not achieved population stability. There are signs that the rate of the population growth is being stabilized, at least for the time being, in countries such as Japan, mainland China, the United States, Canada, and others in Western Europe. Regrettably, populations continue to grow alarmingly in many of the less-developed countries, and it may already be too late to avoid mass starvation to an extent unprecedented in history.

79

POPULATION GROWTH

The seriousness of the problem of population growth can most easily be appreciated by considering the fact that the time it takes for the population to double has become progressively shorter during the past 20 centuries. The global population is estimated to have been only 250 million at the start of the Christian era, and it took 1,650 years for the population to double in size to 500 million. It then took only 150 years to double again, and reached 1 billion persons by 1800. The 2-billion mark was reached about 1900, and by 1975 the world's population was already close to 4 billion persons, with a projected doubling time of only 35 years. If present trends continue, the world's population will reach 10 billion by the year 2025. The spectacular prospective increase in the human population is illustrated in Figure 5-1. Such a rate of growth cannot continue for many more decades without imposing an enormous amount of suffering on human beings. One may argue that with planning the earth can comfortably support a population of 5, 10, 20, 50, or even 100 billion. However, even if one could accept a limit of 100 billion, the limit will be reached, if current trends continue, in about 200 years.

It is axiomatic that population stability is achieved only when births and deaths are in balance. However, the relationship between the

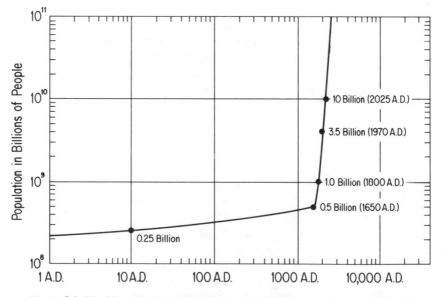

Figure 5-1. World population growth during the Christian era. If the present rate of growth is sustained, the population will reach 100 billion by about 2500 A.D.

rates of births and deaths has varied widely throughout history. The average family size has been known to be as high as ten or as low as two. Life expectancy at birth has varied even in recent times from 20 years to about 75 years, depending on the level of sanitation, nutrition, and the availability of health services. Family size depends in part on economic factors, and is also influenced by religious and cultural mores. In the United States, as recently as 1950, the average number of children in one Mennonite sect was 10.6.[2] There is evidence that reproductive ability is strongly influenced by nutrition.[3]

The influence of births and deaths on the rates of population increase in recent years is shown in Table 5-1. It can be seen that the annual rate of population growth during the period between 1970 and 1975 varied from .6 percent per year in Europe to 2.7 percent per year in Latin America, and 2.6 percent per year in South Asia.[4] The rate of population increase in the underdeveloped countries is about three times that in the more developed countries. This becomes all the more alarming when one considers that among the 4 billion people now alive, three-fourths, or about 3 billion, live in the less-developed countries where the time it takes to double the population is in the range of 25 to 50 years—compared to 100 years or more in some developed countries.

The number of births required to maintain population equilibrium is of course dependent on the death rate. If all newborn babies survived

TABLE 5–1

Birth Rates, Death Rates and Rates of Population Increase, 1970–1975, in the World and Major Areas (Annual Rates per 1000 Population)[4]

Area	Birth rate 1970–1975	Death rate 1970–1975	Rate of population increase (%) 1970–1975	Population Doubling Time (yrs.)
World total	31.8	12.8	1.9	37
More-developed regions	17.2	9.2	0.8	88
Less-developed regions	37.8	14.4	2.3	30
Europe	16.1	10.4	0.6	140
Soviet Union	17.8	7.9	1.0	70
Northern America	16.5	9.3	0.7	100
Oceania	24.7	9.4	1.5	47
South Asia	42.7	16.8	3.0	23
East Asia	26.0	9.8	1.6	44
Africa	46.5	20.0	2.7	26
Latin America	36.7	9.2	2.8	25

through the procreative period and all married, the replacement rate would average two children per family. However, some die prematurely, and others do not procreate. As a result, the replacement rates are actually slightly greater than two children per family.

The population equilibrium can be disturbed in a major way by relatively slight shifts in the ratio of births to deaths. An average of 2.25 children per family in contemporary England would achieve population equilibrium. If the average number of children were three per family, the ratio of the population of one generation to the preceding one would be 3/2.25 = 1.34, an increase of 34 percent. This seemingly small difference in the average family size would cause the population to double in about 70 years.

The historical trends of the rates of births and deaths in the developed countries since the mid-eighteenth century is shown in Figure 5-2, in which it is seen that the birth rate tends to reflect changes in the death rate, with a lag of several years.[5] (Conspicuous in Figure 5-2 are the two peaks in death rates during the two world wars, followed by the two so-called postwar "baby booms.")

The average number of children per woman in the United States has declined dramatically since 1800, as is shown in Figure 5-3.[6] The number dropped from seven in 1800 to a little more than two during

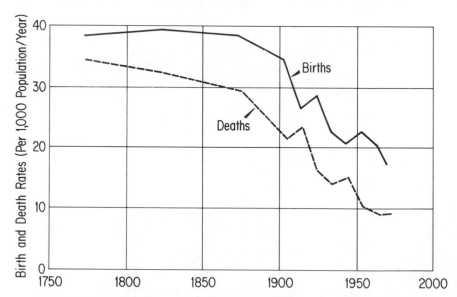

Figure 5-2. Historical downtrend of birth and death rates are shown for the developed countries of the world. The rise in death rates during the two world wars was followed by short-lived peaks in death rates.[6]

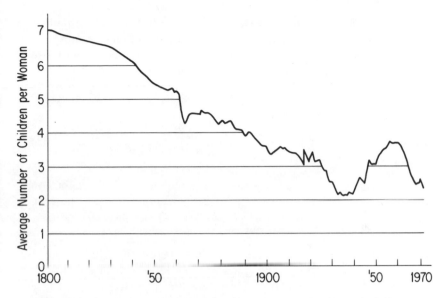

Figure 5-3. Average number of children per woman in the United States, 1800–1970.[a]

the period spanned by the Great Depression and World War II. The average family size then increased to nearly four before dropping again to slightly more than two.

Some of the major statistics that describe the differences in the population of the United States in 1900 compared to 1970 are shown in Table 5-2. The population increased from 76 million to 205 million, and life expectancy has been extended from 47 to 70 years. The birth rate decreased from 32 to 18 per thousand, and the death rate from 17 to nine per thousand. Most important, the annual population growth rate during that period has decreased from 2.3 to 1.1 percent.[6]

Although the mean family size in the United States is very close to the replacement rate of 2.1 children per family, a large number of young adults is now attaining parenthood because of the high birth rates of the post-World War II period. For this reason, if the average replacement rate should be even as low as two children per family, the population of the United States would continue to increase until it reached an equilibrium of approximately 300 million people in about 100 years. Should the average family produce three children, the population would increase more than four times, to nearly 1 billion persons in the next century (Figure 5-4).[6]

In the past, the size of the family could only be controlled by relatively crude forms of contraception, or by abstinence, which in

TABLE 5–2
Demographic Summary of the Twentieth-Century United States[6]

	Around 1900	Around 1970
Population	76 million	205 million
Life expectancy	47 years	70 years
Median age	23 years	28 years
Births per 1,000 population	32	18
Deaths per 1,000 population	17	9
Immigrants per 1,000 population	8	2
Annual growth	1.75 million	2.25 million
Growth rate	2.3 percent	1.1 percent

most countries took the form of late marriages or strict taboos on premarital sexual activity. The spacing of children was also influenced by habits with respect to breast feeding or by restrictions on intercourse following the birth of a child.[2]

The forces that cause fertility to change with changes in the death rate in the manner illustrated in Figure 5-2 are poorly understood. However, lowered mortality is certainly associated with higher material and educational standards which also tend to result in lowered fertility. ("Human fertility" as used here involves more than the biological ability to procreate. Human fertility is modified by social factors such as sexual taboos, customs with respect to age of marriage, and practices concerning contraception and abortion.) In this regard, Chadwick called attention to the history of birth rates and life expectancy in Geneva, where registers of marriages, births, and deaths were begun in 1549. The average lifetime was only 20 years during the last decade of the seventeenth century, and the average marriage produced more than five children. By the end of the eighteenth century, life expectancy was greater than 32 years, and the average family produced about three children. By about 1840, the marriage produced only 2.75 children, at a time when the life expectancy had increased to 45 years.[7]

Sir Edwin Chadwick, in the classic report to which many references were made in Chapter 3, reported a number of relevant observations in the town of Limerick, where the average family produced five children. If the child was nursed by the mother, there was generally an interval of about two years between the birth of one child and that of the next. However, if the child died before weaning, the interval would be much shorter so that, according to Chadwick, the net effect was to stabilize the number of children per marriage at five.[7]

The life expectancy of a population (the average number of years that a newborn will live) is a statistic that is widely used as an index of a country's health. In the more-developed nations, life expectancy at birth is in the range of 70 to 73 years, a bit higher than the biblical allotment of "three score and ten."

Increased life expectancy in the developed countries is due mainly to control of the infectious diseases, particularly those of early childhood. At the present time, in the United States and other developed countries, the death rates among the younger population are so low that significant further improvement in life expectancy can only be made by extending the lives of those over 50 years of age. This is in marked contrast to the undeveloped countries, where childhood mortality continues to be high. During the period between 1965 and 1969, infant mortality in the less-developed regions of the world was 140 per 1,000 live births compared to 27 per 1,000 in the developed countries.[4]

There is also a clear relationship between the level of economic development and life expectancy among the developing countries

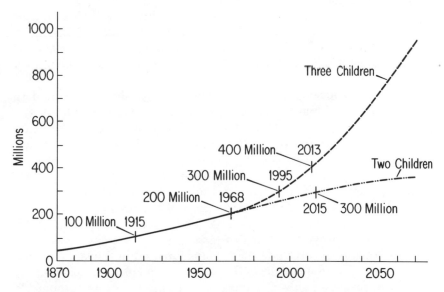

Figure 5-4. Population projections for the United States assuming an average of two and three children per family. At two children per family, the population will continue to grow until the middle of the twenty-first century because of the wave of young adults resulting from the post-World War II baby boom. The population would level off at less than 400 million by 2150. At three children per family the population will then have reached 1 billion and will still be increasing rapidly.[6]

(Figure 5-5). In countries where the per capita gross national product (a useful index of the level of development) is less than about $150 per year, the inhabitants have life expectancies of less than 45 years, whereas life expectancies in countries in relatively modest stages of development are seen to approach those of the technologically developed nations.[8]

Is greater per capita productivity the cause or the effect of increased longevity? In all probability it is both. Along with economic development come improved systems of communication, higher standards of hygiene and education, and better systems for administering health services. The more advanced the country, the more readily can it take advantage of modern developments in public health and agriculture. The vaccines, pesticides, malaria suppressants, and antibiotics that have been produced in the developed countries since World War II have been made readily available to the undeveloped nations, and dramatic reductions in the death rates have taken place. A striking example is Ceylon (now Sri Lanka), where life expectancy increased from 42.2 to 51.8 years in a single year (1946-1947) due to eradication of malaria and introduction of European drugs. In that one year, life expectancy at birth in Ceylon increased about as much as it did in Europe in all of the nineteenth century.[2] When a debilitating disease such as malaria is eliminated from a country, the vigor of the population is increased, which makes it possible to improve economic productivity, and achieve a higher level of technological advancement.

However, reduced death rates are not always due to the modern wonder drugs or improved sanitation. The reduction in death rates in the developed countries during the past 300 years has resulted, to a considerable degree, from better nutrition due to agricultural advances. Inventions such as the horse collar and the iron plough, which seem relatively simple by modern standards, had important effects on agricultural productivity during the past 300 years or so, and this greatly benefited the populations of Western Europe and the United States. The quality of cattle and sheep was gradually improved, and improved methods of land management developed. The average weights of cattle and sheep sold at market in England doubled during the eighteenth century.[9] These developments, and many others like them, resulted in lowered rates of mortality and were responsible for the growth of the Western European population, beginning about 1650, and continuing through the nineteenth century. It is believed that by 1816, the progressively lengthening human life span was due to the substantially greater amounts of animal protein being consumed by the population.[10] The gradual reduction in mortality was further

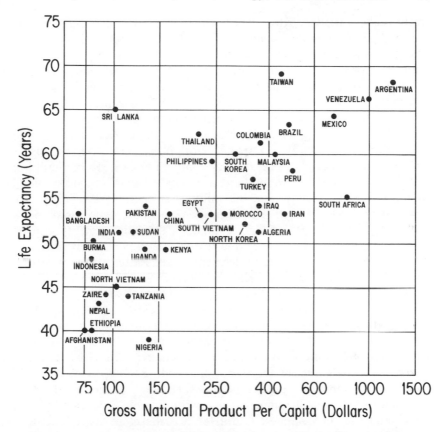

Figure 5-5. Relationship between per capita gross national product and life expectancy in various countries. Note that only countries with a per capita GNP less than about $1,300 per year are shown. In countries with a per capita income greater than about $1,500 per year, the life expectancies are in the range of 70 to 75 years. In 1975 the per capita GNP in Sweden was $6,900 and the life expectancy at birth averaged about 75 years. Per capita GNP in the United States was $6,600 in 1974 and the life expectancy averaged a little more than 71 years (67 for males and 75 for females).[8]

accelerated in the nineteenth century by improved sanitation in the developed countries, and the twentieth century has witnessed spectacular further reductions arising from modern understanding of the nature of infectious diseases, continued improvement in environmental sanitation, and new methods of disease control, such as immunization and use of pesticides.

Modern medicine must certainly be given credit for many of the major advances in the prevention and treatment of disease in relatively recent decades. However, with only a few exceptions such as the smallpox vaccine, medicine had little to offer prior to this century. The microbiological origin of infecious diseases was not understood until Pasteur's work in the latter part of the nineteenth century.

Control over sickness and death is universally desired, and whenever a means is developed for extending life, it is accepted quickly by society. High fertility has also been desired. To some extent this desire has had its origins in religion, such as the biblical command to "be fruitful and multiply." There are, of course, practical reasons as well for the historic desire to have large families. In agricultural societies where land was abundant, a farmer's wealth was determined by the size of his family; and the military strength of a country was determined largely by its population size. Thus, there may be a "natural" reluctance to accept techniques by which fertility can be reduced. As a result, when the death rate of a population is reduced, there is a lag in time before a comparable reduction in the birth rate occurs.[8] And, as a result of this lag, the undeveloped areas of the world are currently experiencing population growth at an average rate of 2.5 percent per year, compared to the developed countries where the growth rate is less than 1 percent, and is continuing to fall. In addition, the methods of modern medicine can be readily adopted by the underdeveloped countries, and the death rates in many of those countries, while still relatively high, are tending to approach those of the more developed countries. However, the reduced death rates have not as yet been accompanied by a comparable reduction in birth rates, and the undeveloped countries are therefore experiencing a greater proportionate rise in population than the developed countries. For example, during the 15-year period between 1960 and 1975, the total world population is estimated to have risen by about one-third, from approximately 3 to 4 billion people. Of this 1-billion increase, nearly 836 million were in the undeveloped areas of the world and 157 million were in the developed areas. As a result of their more rapid growth rate, the less-developed regions are comprising an increasing proportion of the world's total population: 67.4 percent in 1960, and an estimated 71.6 percent in 1975.[4]

The effect of unbridled population growth on land pressure in the undeveloped countries is also becoming increasingly evident. It is estimated that in 1975, the population density in the undeveloped nations was 40 persons per square kilometer. This is more than twice that of the more developed regions. During the 15-year period between 1960 and 1975 it was necessary to accommodate more than 12 additional persons per square kilometer in the undeveloped regions.

The name Thomas Robert Malthus (1766-1834) is prominent in the literature of population growth.[11] He argued that people must consume the products of land to survive, and that there is a natural tendency for a population to increase in size to the limits permitted by the availability of food. Population growth beyond these limits is prevented by war, famine, and pestilence. Malthus urged that the human race need not depend on these cruel counterbalancing forces to control its size, but that it limit population growth by postponing the age of marriage and by adopting strict rules of sexual continence.

The Malthusian theory presented a bleak picture for the future of the human race, but it gave a reasonable analysis for the times in which it was written. What Malthus could not foresee was the enormous effects technology would have on agricultural productivity and family size. For this reason, the arguments have not been valid for a technological era Malthus did not foresee. However, there is no reason to dispute the fact that Malthus was basically correct. If the population of the world continues to expand, it will eventually outrun its supply of food.

Two other major influences on population size not foreseen by Malthus at the dawn of the Industrial Revolution are the development of convenient, effective contraception and a tendency for an increasingly literate population to desire to control family size.

The "pill" and the intrauterine device (IUD), two remarkable methods of contraception, became available during the 1950s.[12] The pill, which is an extremely effective oral contraceptive, is a sophisticated scientific development based on an understanding of reproductive physiology. The intrauterine devices, which are inserted through the cervix, prevent conception by mechanisms that are not clearly understood. The IUD must be inserted by a skilled person, but in most cases it can be left in place indefinitely. Both the oral contraceptive and the IUD have had wide acceptance despite the fact that each can have undesirable side effects in some women.

Vasectomy, a minor but presently irreversible surgical procedure which prevents ejaculation of sperm, is an increasingly popular method of male contraception. Vasectomy is particularly popular among men whose partners have borne the desired number of children. One of the impediments to the wider acceptance of this procedure is the totally erroneous notion that it affects the desire or pleasure of normal sexual activity.

Birth control in developing countries is handicapped by lack of education, cost, and the difficulty of distributing contraceptives and implementing the required educational programs in the rural areas. Some of the Indian states have for some years been attempting to implement a policy of compulsory sterilization of men who have fathered more than three children. In addition to the financial requirements and logistical difficulties of such a program, there are obvious cultural and psychological barriers.

During the 1970s, the prospects seemed favorable for further major developments in the field of biochemical control of fertility. Among the possibilities being widely discussed were "morning after" pills to be taken only following intercourse, oral contraceptives for males, skin implants of contraceptive drugs that would dissolve slowly and last for perhaps two years, and pills that could be taken for a period of two or three days to induce the normal menstrual period at the end of the monthly cycle. Unfortunately, intense interest in effective, economical, and convenient forms of contraception coincided with public pressure to curtail approval of pharmaceuticals that have undesirable side effects. The net effect of the increasing concern about the safety of new drugs caused the pharmaceutical industry to stop trying to develop contraceptives.[13] Thus, some people might accept contraceptives that are less than perfect from the point of view of their undesirable side effects, but the pharmaceutical industry is understandably reluctant to invest the enormous sums required to develop a product, test it for safety and efficacy, and produce the product, only to find that it must be withdrawn from the market because of increasing public pressure for total safety. This problem will be discussed again in subsequent chapters.

The changing attitude towards abortion is a second major development of the past decade. In most countries, abortions have long been illegal, but they have been practiced widely despite their association with censure and medical risk. Illegal abortions have always been a public health problem in the United States because the procedures were usually carried out under conditions conducive to a high incidence of injury and infection. Knowledge about abortions usually reached the public health authorities only because the resulting complications resulted in the need for medical attention. The number of illegal abortions in the United States has been difficult to determine, and estimates have ranged from 200,000 to 1.2 million per year.[13] They have been an important cause of maternal death, which dropped by two-thirds the year after abortions were legalized in New York State in 1970.

By 1970, 15 states had liberalized their abortion laws, and during that year, it was estimated that approximately 200,000 women in the United States obtained legal abortions.[14] The number more than doubled in the following year, to about 500,000. In 1973, the Supreme Court removed the major statutory barriers to abortion, and one year later it was reported that a total of 900,000 legal abortions were performed.[15] This, of course, does not mean that there were 900,000 fewer live births than would have been the case previously, since many of the legal abortions replaced those that would otherwise have been performed illegally or might have been performed for non-residents in those states where the practice had already been legalized.

In countries where infant mortality is low, the number of children per family until recently has usually been determined by two factors: the number of children wanted, and the effectiveness of the contraceptive methods used. By the late 1960s, in the advanced countries, the more general availability of effective contraception and abortion assisted parents in achieving their family goals for the first time in history.[16] In 1961 about 45 percent of the children born were "unplanned" among women who had college educations, compared to 65 percent among women with less than a twelfth-grade education. By 1970, the number of unplanned children born to women with all levels of education was reduced substantially, and the gap between the college-educated (34 percent unplanned) and the women with less than a twelfth-grade education (40 percent unplanned) had narrowed considerably. In other words, unplanned children were born to the less educated at a rate 45 percent greater than to college-educated women in 1961, but at a rate less than 18 percent greater by 1970.

During the decade of the 1960s, the proportion of unplanned births was reduced among blacks and whites, and Catholics and non-Catholics. The largest reduction was among black women. The changes were presumably due to the availability of convenient contraception. Liberalization of the abortion laws did not take place until after 1970 and has undoubtedly resulted in a continuation of this trend.

The evidence from both the developed and undeveloped countries is that the alarming rates of population growth are a by-product of inferior education and socioeconomic status. If the economic and educational levels could be raised in the less developed areas of Africa, Asia, and Latin America, the birth rates would probably be reduced, and the curve of rising population would level off.

FOOD

There are about 32 billion acres of land in the world, of which about 8 billion acres are suitable for cultivation. Another 8 billion acres can be utilized for grazing, but not for raising food crops. The remaining 50 percent of the earth's land area consists of deserts, tundra, and mountains that can be of little use for growing food for the foreseeable future.[17-19] Of the 8 billion acres that were potentially arable in 1969, about 3.5 billion acres were under cultivation, and more than 4 billion acres were being used for grazing. The unused land is not uniformly distributed among the regions of the world. Eighty-three percent of the arable land in Asia is already under cultivation, compared to only 11 percent in South America.

Famine has been a way of life in many parts of the world from prehistoric times until the present. The weather has always been subject to great variations, and a frost, drought, or hot spell can reduce the yields of crops or livestock to such an extent as to cause famine if means are not available to store previous surpluses or to transport food from other geographical areas in which surpluses exist.

War and pestilence have also been classic causes of mass starvation. Biblical literature describes the results of the fierce onslaughts of locusts on the harvests of the Middle East. Throughout history, wars have prevented seed from being planted, have devastated fields, have prevented the harvest, and have caused destruction of stored crops. Most of the famines of history were relatively localized, but some affected large areas of Europe, Asia, and Africa and caused millions of deaths.

Although the historians of the past recorded the dramatic famines that occurred from time to time, they did not have the means to evaluate the effects of chronic dietary insufficiency on the vitality of the populations, on their intellectual capacities, on their resistance to disease, or on other aspects of their physical and mental health. The first worldwide food surveys were not made until after World War II, but such surveys have been conducted periodically since then by the Food and Agricultural Organization of the United Nations and by the U. S. Department of Agriculture.[20]

The increased rate of agricultural and industrial innovation, that began in the nineteenth century and has continued up to the present time, has had profound effects on the social structure of both the developed and less developed countries. In the United States at the start of the nineteenth century, 95 percent of the population lived on farms where they raised food for themselves and the remaining 5 percent of the population. Today, only 9 percent of the population of the United States are farmers, but these are able to raise sufficient food to feed the remaining 91 percent, and supply a very considerable export market.

Although only 50 percent of the arable land in the world is farmed, its productivity, in many areas, has been greatly enhanced by the use of fertilizers. Of the nutrients added to the soil in fertilizers, nitrogen is one of the most important. It is present copiously in the atmosphere in elemental form, but it cannot be used directly by plants, and then by animals, including man. The gaseous nitrogen must be converted to an available form, and this is done naturally by microorganisms living free in soil and water, or in association with certain plants, such as beans, peas, clover, and alfalfa.

Prior to the nineteenth century, soil nitrogen could only be replenished by planting such crops or by adding animal and vegetable wastes to the soil, such as Chilean nitrate deposits (guano) that had accumulated over the years from the droppings of great colonies of sea birds. Commercial production of synthetic nitrogen fertilizers by production of ammonia from atmospheric nitrogen did not become possible until 1913.

Phosphorous is another nutrient element essential to crop growth, and since the middle of the nineteenth century it has been derived from phosphate minerals that are mined and converted into soluble form. Potassium, another essential plant nutrient, has for the past century been supplied primarily as potassium sulfate, which is produced chemically from mineral deposits. The great increase in agricultural productivity during the twentieth century has been due to the use of these mineral fertilizers, combined with a better understanding of plant nutrition, modern techniques of tillage and water conservation, and the use of herbicides and pesticides.

Pesticides have been particularly important both to agriculture and to disease control since the first quarter of this century. It is estimated that nearly 1,000 pesticidal chemicals have been formulated into over 60,000 commercial products that are available for use in agricultural production, public health, animal health, and forestry.[21] Methods for protecting agricultural products from damage by pests are a necessary part of the strategy for producing crops, livestock, and forest products, for pests are destructive at all stages of plant growth. They also destroy food while it is in storage, and during transportation. There are at least 10,000 species of destructive insects in the United States, and plant diseases can be caused by as many as 160 bacterial, 250 viral, and 8,000 fungal species. During the years 1963 and 1964, insects and rodents destroyed grain in India to the extent of 13 million tons, which would have been sufficient to supply a daily loaf of bread to 77 million families for one year. In 19 corn-producing states in America, the grain destroyed by birds in one year would have been sufficient to feed more than 300,000 hogs.[22]

One of the earliest modern methods of protecting crops against plant diseases was to select varieties that were disease-resistant. This has been a remarkably successful technique that has enabled farmers to cultivate wheat, corn, soybeans, and rice which are resistant to fungi, insects, and other infestations which were formerly highly destructive. However, the chemical methods of pest control have been the most successful by far, and these have greatly increased the productivity of land in many parts of the world since they were first introduced on a large scale immediately after World War II. Regrettably, many of these chemicals cause serious environmental side effects (see Chapter 9). However, the fact remains that the combined output of crops, livestock, and forests would be reduced by at least 25 percent in the United States if it were not for the use of chemical pesticides.[22]

The availability of artificial fertilizers and pesticides since World War II, together with development of high-yield strains of grains and better methods of irrigation, have brought into being the so-called "Green Revolution," a term that came into use in the mid-1960s as a result of spectacular increases in the yields of new strains of wheat. A basic problem in wheat culture had been that the strains then being cultivated tended to grow too tall and topple over as they reached maturity. This problem was solved by crossing these high-yield varieties with dwarfed strains, thereby creating wheat with a shorter and stiffer stalk. In 1965, at a time when there was great concern about the insufficiency of food supplies in tropical countries, the wheat seed, developed in Mexico, was distributed to India and Pakistan with spectacular success. In India, the wheat harvest increased from 12.3 million metric tons in 1967 to about 20 million tons in 1969. The annual yields in Pakistan increased from 4.6 million tons to 8.4 million tons.[23,24] Plant breeders also have succeeded in increasing the yields of rice and corn and have produced strains that were more adaptable to a wide range of latitudes and elevation.[25] Thus, three of the major crops of the world became available in high-yield strains that caused the hard-pressed Asians in the late 1960s to believe they had the means to attain food sufficiency. The term "Green Revolution" was quickly coined and was widely touted by the printed and electronic media throughout the world. There was reason to be enthusiastic: in addition to the spectacular increases in wheat yields, rice production in the Philippines was increased by 12.4 percent per year from 1967 to 1972, and by 80 percent in Colombia.[25] But the Green Revolution could not control the weather, and in 1972 severe drought in the Soviet Union, China, Australia, Africa, and Southeast Asia caused a sharp drop in global grain production. The Green Revolution was denounced by some who claimed that the high productivity from 1967 to 1972 was due to good weather, a criticism not supported by the facts at the time, and subsequently disproven by the steady increase in yields of rice and wheat in India and Pakistan.[26,27]

However, a more legitimate criticism of the use of the new types of grain in the less-developed countries is that they require investments in irrigation equipment, fertilizer, and pesticides that are beyond the reach of the typical farmer. Moreover, advanced farming methods require consolidation of farms, often leaving the poor farmer worse off than before. This of course is not a new problem. Farmers in the United States were driven off their land by the agricultural technology introduced in the nineteenth and twentieth centuries: but the displaced farmers could move to the tractor factories or to the plants that manufactured chemical pesticides and other agricultural products. The Green Revolution, however, was introduced to the developing countries at a time when they do not have the industrial capacity to absorb the manpower being displaced from the farms.

Before the underdeveloped countries can take advantage of the Green Revolution, they will have to strengthen their economies. This would permit them to invest in the machinery, fertilizer, and pesticides and forestall mass hunger at least for a period of years until socioeconomic development provides the prerequisites for a decline in population and, accordingly, a sufficiency of food.

Hendricks[18] has examined the limits of global food productivity and has concluded that production could be increased as much as 16 times by utilizing all available arable land and by applying the best methods of land management and technological innovations. Also, it is important to recognize that "optimal" agricultural management will require major changes in the food habits of the developed countries of the world. From an energy point of view, it is far more efficient to favor human consumption of grains, rather than meat-producing animals fed on those grains.

It is widely believed that the seas can provide food in quantities that can significantly augment terrestrial agriculture and husbandry. This appears not to be true. The open sea, which covers 90 percent of the ocean surface and nearly 75 percent of the earth's surface, has been described as a "biological desert." The few productive areas are those in which the surface waters are replaced by upwelling nutrient-rich waters from the ocean deeps. These regions total no more than .1 percent of the ocean area, but they produce about 50 percent of the world's fish supply.[28]

Many fisheries are already showing signs of overfishing. From 1950 to 1970, the world's fish catch climbed from 20 to 71 million tons per year, but the catches declined from 1970 to 1974, despite heavy investments in boats and gear.[29] Assuming that it would be possible for the oceans to provide a sustained yield 2.5 times the 1968 catch (150 to 160 million tons), the oceans at present can provide only about 30 percent of the world's protein needs. Although this is not insignificant, it is far from what will be required to keep abreast of the food requirements of the much larger global population that must be fed by the end of the century. Moreover, although food from the oceans is high in protein (averaging about 22 percent) it is low in calories. The oceans could not provide more than 3 percent of human needs for calories by the year 2,000, according to one estimate.[30]

It has been suggested that the productivity of the oceans could be increased by raising fish and shellfish in the nutrient-rich cold waters of the ocean deeps. Such a system will probably not be economically feasible unless undertaken as a by-product of other processes that utilize deep ocean water. One possibility, which will be discussed in Chapter 6, is that the cold waters from the ocean deeps may ultimately provide an inexhaustible supply of energy by condensing low-boiling vapors passing through turbines driven by the heat of surface water.[31]

Enough food was produced between 1951 and 1973 to meet the increase in the number of inhabitants on the earth. Actually, food production doubled, while the population increased by a half. Between 1951 and 1971 more than half of the increased production was achieved by the wealthy countries, which comprise only 30 percent of the world's population (see Figure 5.6).[32] This also was the case until 1973, but per capita consumption increased in developed countries and decreased elsewhere.

THE ROLE OF ENERGY IN AGRICULTURE

The availability of energy is frequently a limiting factor in agricultural production. In primitive agriculture, food is produced by energy derived from solar irradiation of the field in which the plants are grown. The work that goes into the farm is provided by the farmer, his family, and perhaps his draft animals, all of which derive their energy ultimately from solar sources. In contrast, modern agricultural methods require enormous amounts of inanimate energy to power tractors, harvesters, and grain dryers. From the point of view of energy balance, the relatively primitive farmers are far more efficient than those who practice the methods of the Green Revolution. Agricultural systems in a primitive culture provide as many as 20 calories of energy from the soil for each calorie of energy invested. In contrast, advanced systems of agriculture, such as the feedlot method of raising and fattening cattle, invest up to ten calories for every calorie returned. Modern agricultural methods require substantial energy subsidies to increase the yields per acre. Some of the modern methods of obtaining food require an investment of energy more than

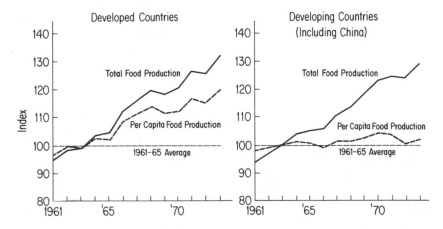

Figure 5-6. Total and per capita food production in developed and developing countries between 1961 and 1973. Total food production is increasing comparably in both the developed and developing countries, but the high birth rates in the developing countries have limited the increase in per capita production.[32]

200 times that of primitive methods.[33] The additional energy can be justifed on economic grounds because more food is grown per acre of land and it can be produced by fewer people.

The energy obtained from food also varies widely due to the type of food being cultivated. Fruits and vegetables return barely one unit of energy for each unit of cultural energy, compared to corn, sorghum, and sugarcane, which return about five units per unit invested. *When high-protein crops like alfalfa are converted to beef or pork, there is a tenfold loss in the protein yield per unit of cultural energy.* However, most people prefer to eat meat, and they are not likely to change their dietary customs, especially those living in countries that have achieved a high standard of living, of which high protein consumption is a central factor.

The solar energy that impinges on farms in the United States varies by a factor of nearly 2, from 4,400 megacalories per acre per year in upstate New York, Vermont, and Oregon to about 7,300 megacalories per acre per year in the Southwest.[34] For example, this is roughly 1,000 times the energy required to produce a corn crop, but only a small fraction of the incident solar energy is converted to food. Most of the light energy is degraded into heat or reflected back into the atmosphere. In addition, the cloudless regions that receive the maximum amount of sunshine are also the arid regions of the world. Here the advantage of a higher input of solar energy is more than offset by the need to provide energy for irrigation. In Israel, an arid region where the land is intensively cultivated, 39 percent of the energy input to the agricultural system is for irrigation, compared to 7 percent for the United States.[35]

The enormous increase in corn yields that has been achieved in the United States since World War II has required substantial amounts of energy. A hybrid variety of corn was introduced in 1945, and by 1970 it had increased the yield per acre by 240 percent, with a requirement for only 60 percent more labor. However, this high productivity required that the amount of nitrogen provided be increased from 7 pounds per acre to 31 pounds, and potassium from 5 to 60 pounds.

We must use 940,000 kilocalories to manufacture this fertilizer. However, we also must use 797,000 kilocalories to refine and distribute the gasoline needed, and 420,000 kilocalories to manufacture the farm machinery. In addition, 310,000 kilocalories of electricity are expended, per acre. Of course, even more energy is used to manufacture herbicides, and insecticides.[36] As a result, 2.8 calories of food are now produced in the United States for every calorie of energy supplied. However, before sophisticated methods of farming were introduced in 1945, 3.7 calories of food were produced. (See Figure 5.7.) This loss can be tolerated because abundant energy has been available in the United States. Of course, this lavish use of energy is far beyond the reach of the developing nations.

To feed a world population of 4 billion at this level of energy expenditure would require the equivalent of 448 billion gallons of petroleum per year. We will see, in the next chapter, that this is an enormous amount of energy that could not long be supplied by the present petroleum reserves. To produce food on this scale would allow no use of oil for any purposes other than food production.[36]

It is not an easy matter to devise methods by which the agricultural requirements for inanimate energy can be reduced. Some of the seemingly easy solutions are impractical, such as the suggestion that farmers use manure for fertilizer. The manure production in the United States totals 1.7 billion tons per year, of which 50 percent is produced in the confines of the feedlots where it is a major source of pollution. Thirty percent of the manure could fertilize all the land used for growing corn. However, the economics of fertilizer use are such that the energy required to transport the manure can be justified only if the cornfields are within one mile of the source of the manure. The use of manure would replace the need for the equivalent of about 40 gallons of gasoline per acre, but this would be offset by the need for equipment, fuel, and human energy.[34] Since it is not practical to relocate the cornfields and feedlots, we must anticipate the continued

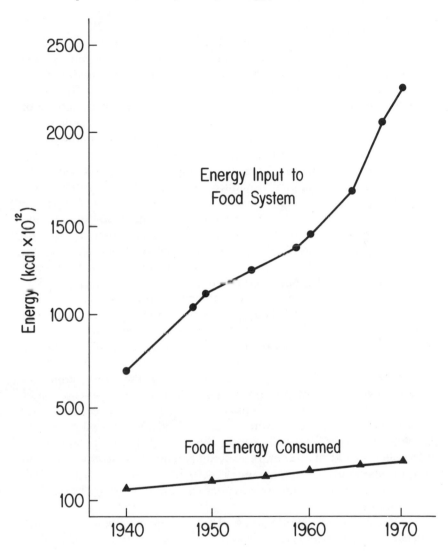

Figure 5-7. Comparison of energy inputs and outputs of the food system of the United States between 1940 and 1970. Food energy consumed has increased only slightly compared to the much greater input of energy.[33]

need to expend nonsolar energy to produce artificial fertilizers, while at the same time unused manure on farms poses a problem of contamination of nearby waterways.

SOME FUNDAMENTALS OF NUTRITION AND MALNUTRITION

A balanced diet that provides all of the constituents needed to meet a person's energy requirements, and the raw materials for body building and replacement, must include carbohydrates, proteins, fats, vitamins, and minerals. Carbohydrates tend to be the main source of energy in undeveloped countries, where they supply 90 percent of the energy requirements, compared to 40 percent in the technologically advanced countries. In agricultural societies, the carbohydrates are consumed primarily in the form of starchy foods, such as potatoes and cassava. Industrialized populations obtain their carbohydrates primarily from sugar.

The amount of fat in the diet varies from country to country and tends to increase with the level of economic development.[37] In the poorer countries, the total amount of energy derived from fat is about 15 percent, but in the more prosperous countries may be as high as 45 percent. (Not all fat is obtained directly as fats from food. The body is able to convert carbohydrates and proteins to fat as well.)

The proteins provide the body with energy and important materials from which it is built and continually repaired. The quantity of protein in the diet is a good measure of the economic health of a country. Fish, beef, milk products, and soy beans are foods that have the highest protein content. Soy beans are by far the most important source of vegetable proteins and, along with fish, have played a major role in supplying the protein requirements of the Japanese. A proper human diet should provide 10 to 15 percent of the required energy in the form of protein.[38]

The diet should also contain essential vitamins and minerals, but except under unusual circumstances, sufficient quantities of these will be obtained from a diet having a proper mixture of carbohydrates, fats and proteins.

Any discussion of the adequacy of food supplies should distinguish between famines (which cause mass starvation) and the more subtle effect of chronic malnutrition. The most frequent cause of famine in modern times is widespread failure of crops due to drought or flooding, causes that are particularly important in tropical and sub-tropical countries subject to wide variations in rainfall. (In temperate climates the famines are more likely to result from war, social disturbances, or crop failures due to blight.) The large populations of India and China have been particularly susceptible to mass starvation: about 10 million Bengalis died from famine in 1769 and 1770, and nearly as many people died in North China a century later. One of the best-documented famines was in Ireland in the mid-nineteenth century, at which time the country depended almost totally on the potato for its food.[39] The potato crops failed year after year because of a blight, and the country endured severe hardships, resulting in an undocumented but large number of deaths and mass migration to North America and other parts of the world.

Although the frequency of famines does not seem to have diminished in recent decades, their severity has been greatly lessened by the ability to forecast crop failures, and by improved systems of storing and transporting food. The human suffering can be greatly relieved if there is a will to do so, both within the affected country and in the rest of the world.

Incredibly, the management of famine relief is still in a primitive state. Despite the fact that there have been more than 50 famines since World War II, there has been a tendency to deal with them as though the "phenomenon were a totally unexpected crisis to be handled in an improvisational manner with whatever personnel and supplies might be contributed on an ad hoc basis. It is as if mankind automatically obliterates the memory of famines as soon as they are over."[40]

One measure of the ability of the world to compensate for crop failures is the reserve of stored grain that can be shipped to hungry nations, if needed. This reserve has gradually diminished from an amount sufficient to feed the world for 95 days in 1961 to an amount sufficient for only 26 days in 1974. The diminution in reserves has been a gradual one during this period. Whether this is the start of a dangerous long-range trend cannot as yet be stated with certainty, but it is clear that the supply-demand equation is precariously balanced.[29]

Chronic malnutrition is more insidious in its effects than famine and may have a greater impact on humanity than even great famines might. This is particularly true of chronic protein deficiency encountered in many tropical and subtropical countries (it also may have a considerable effect on poor people in developed countries as well).

When children are deprived of sufficient protein, their growth is stunted; they become susceptible to disease; and they are also subject to mental retardation. The effect of nutrition on stature is well-documented. The armor worn by athletic knights of the Middle Ages would not fit a 12-year-old boy today, as any visitor to the museums can see. Between 1910 and 1970, the average heights of school boys in Glasgow increased by more than six inches and their average weights by 20 pounds.[38] Asiatics may be slight of build because of genetic factors, but undernutrition is certainly an important factor. In Japan, where the post-World War II economy has greatly improved compared to former times, the weight of 15-year-old children has increased by three kilograms per decade since 1945.[41]

Among the children of the world, two diseases caused by chronic malnutrition are common. These are marasmus, which means "to waste," and kwashiorkor, which is said to take its name from the local dialect in Ghana, where it was first identified, and means literally "the sickness older child gets when next baby is born."[38] Children with kwashiorkor are stunted in growth, have swollen legs and bellies, are subject to infections, and are generally unhappy and apathetic. In many cases the children are wasted, as in nutritional marasmus. Children with kwashiorkor who are given an adequate diet can recover, but in many cases will be permanently cripped, both mentally and physically. Kwashiorkor is a common disease in many countries of the world.

The level of intellectual ability to which a person can develop depends in part on anatomical characteristics of the brain that are determined early in life and which can be affected seriously by dietary protein deficiency. Children who were severely undernourished pre-natally or in early infancy have smaller-than-average brain size and fewer brain cells than normal. This leads to inferior learning ability and to reduced mental performance.[37,40]

Studies of the effects of malnutrition on human mental ability have been complicated because a complex of socioenvironmental factors also influence mental ability. For one thing, undernourished children frequently lived in deprived circumstances that could affect their mental development in many ways. During the past decade there have been many investigations that confirm the relationship of protein deficiency and mental retardation.[42-44] However, there is not full agreement as to whether the effect on mental development is revers-ible if the child is subsequently placed on an adequate diet. A study of undernourished Korean orphans adopted by families in the United States concluded that almost full mental recovery occurred if the children were adopted and properly fed before the age of three.[45]

The effect of malnutrition on mental development is particularly significant because it represents a major additional hurdle in the efforts of the disadvantaged to break out of the system of poverty in which they find themselves. In addition to cultural and possibly genetic handicaps which the impoverished child must suffer, that child must also face the prospect of organic brain damage. This increases the probability that one generation of the underprivileged will pass its legacy of poverty to the next.

THE COMING CRISIS

For more than two centuries there have been unequal races be-tween population growth and food supply in the various parts of the world that are now differentiated into the developed countries and the less-developed countries. The dismal Malthusian forecasts that the human race would eventually be unable to feed its children may yet come true. Although Malthus's pessimism has thus far been proved wrong in the technologically advanced countries, it is possible that even those countries will find themselves subject to Malthusian limits in the centuries ahead. Few of the advanced countries have achieved zero population growth, but the rates of growth have certainly been slowed sufficiently to permit relatively leisurely social adjustments to be made during the next century or two. However, within many of the less-developed countries, the population is growing at rates that are out of control in relation to the supplies of food and other resources available to them.

The less-advanced countries can be subdivided into two groups. There are some, like Brazil and Argentina, where the population is increasing rapidly, but where there are resources in land and raw materials that will make it possible for the population to expand for a considerable period of time without acute distress. If economic development takes place in such countries at a rate greater than the growth in population, both per capita productivity and income will increase and socioeconomic conditions will be improved. The resulting higher levels of education and material wealth can be expected to result in a leveling of population growth. At least, this is to be expected if the experience in Europe, North America, and Japan is applicable. (At present the rates of population growth in these countries are such as to permit a gradual accommodation.) However, if the advantages of greater national productivity are offset by population growth to such an extent that there is no per capita gain in the gross national product, these countries will become subject to Malthusian population stresses in the years to come.

However, not all of the less-developed countries have the low population density and resources of Brazil and Argentina. There are some, like India, Bangladesh, and other South Asian countries, in which the people are already overcrowded and underfed, despite the efforts of their governments to stabilize their burgeoning populations. In some of these countries, there are no known raw materials to be developed, and the population is increasing so rapidly that it does not seem possible, except for the intervention of terrible wars, epidemics, or natural disasters, that the most extreme forms of Malthusian horrors can be avoided. Those countries and their miserable poor are caught in the trap of reduced mortality with little indication of the necessary compensating influence of lowered birth rates. The populations of these countries were already hungry and overcrowded prior to the introduction of life-saving chemicals in the form of antibiotics, antimalarials, vaccines, and pesticides. For a period of time, the increase in population has been accommodated by increased crop yields, but this could do nothing but ameliorate the problem temporarily in the absence of fertility control. The basic principle has been well stated by Hardin: "a food shortage cannot be solved by producing more food. Attempts to do so merely postpone the day when painful decisions must be made, and increase the number of people who will be affected. . . ."[46]

There are contemporary optimists who say mankind will find a way to feed the new masses of the twenty-first century. They emphasize that Malthus was wrong for the reasons given earlier and that somehow humanity will continue to solve its problems. This is a dangerous and unrealistic optimism. There are no longer undiscovered lands waiting to be cultivated. We have learned that advanced agricultural methods can only increase crop yields significantly in countries that have developed the necessary industrial infrastructure. We know that the sea is not a significant future source of food and that it is in fact a biological "desert." And above all, we understand that population stability can only be achieved in literate populations that have achieved a socioeconomic status as yet beyond the reach of poor countries generally, and particularly those that lack natural resources.

The prospects seem even more dismal because countries that have traditionally raised excess food are finding that their own needs for food are increasing and their surpluses are shrinking. It was only a few years ago that the United States government was paying farmers *not* to produce crops. Those days are gone forever: A country with food for export can find many hungry mouths and outstretched hands, although there may not be the means to pay for the food needed so desperately.

When wartime field hospitals are overwhelmed with casualties, they have traditionally resorted to the grim practice of triage, which is designed to use limited medical resources for the most good. Triage divides the casualties into three groups. In the first are those with such minor injuries that their treatment can be deferred. The third category includes those whose wounds are so severe that they cannot be helped. This permits the resources of the hospital to be concentrated on the second group of casualties, those who are so seriously wounded that medical assistance should not be deferred, and for whom there is a reasonable prospect of recovery. Triage involves cruel decisions, but the limited resources of a hospital overwhelmed by battle casualties cannot be dispensed in a way better calculated to do the most good.

Is the world headed for a system of triage in allocating its food? Will the countries be divided into those that can take care of their own needs by a moderate reduction in per capita consumption or by better methods of agricultural management; those that are capable of adjusting their population growth and are showing signs of doing so; and those that are unwilling or unable to stabilize their population and for whom imported food simply increases the problem by raising the population limit imposed by Malthusian principles?

The problems of food allocation now looming on the horizon may one day present the world with unprecedented moral choices. A system of triage applied to food allocation is now almost unthinkable, but such options may not seem unreasonable by the end of this century. It is essential that the developing countries stabilize their populations, but it may take two or three decades to do so, and severe hunger is likely to develop before then. Conditions could quickly become catastrophic, should the world enter a period of unfavorable weather. (See Chapter 14.)

NATURAL RESOURCES

The natural resources on which human existence depends include land, potable water, fuels, wildlife, and the minerals from which many materials are obtained. The need for minerals permeates every aspect of modern life. Electrical energy cannot be distributed without copper or aluminum. The machines of modern society require iron, alloyed for various special purposes with other metals such as molybdenum, chromium, nickel, and vanadium. Chromium, zinc, and titanium are widely used in paints, and silver is a basic ingredient of photographic processes. Mercury, gold, platinum, and other elements serve important uses in the chemical and electronic industries. Metals are required in the refineries, generating stations, and manufacturing plants that produce gasoline, diesel fuels, electricity, fertilizers, and pesticides. Metals are used as catalysts in chemical processes, to distribute water, and they serve hundreds of other purposes for communications, transportation, and manufacturing. The subtlety with which metals enter our everyday life is illustrated by Table 5–3, which lists the 42 chemical elements used in the construction of a single familiar instrument, the modern telephone handset.[47]

Although many of the metals in Table 5–3 play important but relatively unobtrusive roles in modern life, we also need other materials, some of them in large quantities. If the total requirements of the United States for material derived from natural resources and used for

TABLE 5–3

Forty-Two Metals Used in Construction of Telephone Handset[47]

Element	How Used
Aluminum	Metal alloy in dial mechanism, transmitter, and receiver
Antimony	Alloy in dial mechanism
Arsenic	Alloy in dial mechanism
Beryllium	Alloy in dial mechanism
Bismuth	Alloy in dial mechanism
Boron	Touch-Tone dial mechanism
Cadmium	Color in yellow plastic housing
Calcium	In lubricant for moving parts
Carbon	Plastic housing, transmitter steel parts
Chlorine	Wire insulation
Chromium	Color in green plastic housing, metal plating, stainless steel parts
Cobalt	Magnetic material in receiver
Copper	Wires, plating, brass piece parts
Fluorine	Plastic piece parts
Germanium	Transistors in Touch-Tone dial mechanism
Gold	Electrical contacts
Hydrogen	Plastic housing, wire insulation
Indium	Touch-Tone dial mechanism
Iron	Steel, magnetic materials
Krypton	Ringer in Touch-Tone set
Lead	Solder in connections
Lithium	In lubricant for moving parts
Magnesium	Die castings in transmitter, ringer
Manganese	Steel in piece parts
Mercury	Color in red plastic housing
Molybdenum	Magnet in receiver
Nickel	Magnet in receiver, stainless steel parts
Nitrogen	Hardened heat-treated pieces
Oxygen	Plastic housing, wire insulation
Palladium	Electrical contacts
Phosphorus	Steel in piece parts
Platinum	Electric contacts
Silicon	Touch-Tone dial mechanism
Silver	Plating
Sodium	In lubricant for moving parts
Sulfur	Steel in piece parts
Tantalum	Integrated circuit in Trimline set
Tin	Solder in connections, plating
Titanium	Color in white plastic housing
Tungsten	Lights in Princess and key sets
Vanadium	Receiver
Zinc	Brass, die casting in transmitter, ringer

all purposes (food and energy production, roads, commercial and industrial buildings, and industrial processes, among others) are assigned on a per capita basis, each person uses about 20 tons per year![48,49]

Per capita consumption of materials in the United States seems to have leveled off in the past two decades, but has been increasing rapidly in the developing countries. Per capita consumption of steel in India doubled from 1950 to 1970.[48] In the undeveloped countries, where three-fourths of the world's population live, the demand is thus "super exponential," since both population and per capita demand are increasing exponentially.

One of the fundamental dilemmas facing modern society is that the advanced countries have long ago exhausted most of their mineral resources and now depend on the less-developed countries for their supply. The United States has imported copper since 1930, and lead and zinc since about 1920. Rich domestic deposits took care of the national needs of the United States until the increasing population and the greater per capita demands associated with advancing technology caused the needs of the United States for these and other metals to exceed production capacity.[50] In Europe, except for the Soviet Union, essential metals such as chromium, manganese, nickel, cobalt, molybdenum, and tungsten must all be imported from undeveloped countries. By 1970, the United States imported all its requirements for chromium, columbium, titanium, and tin, as well as more than 90 percent of its aluminum, antimony, cobalt, manganese, and platinum.[51]

It is very difficult to project the consequences of the exhaustion of mineral resources. In some cases, substitute materials can be used. Relatively abundant aluminum is already being used instead of copper for cables in electric distribution systems. At one time, the diamonds used for industrial purposes came from nature, but when the demand rose and the price became prohibitive, industry found a way to produce synthetic diamonds at a small fraction of the price of natural stones. To a limited extent, plastic materials are now being used to replace metals for a number of purposes such as water pipes, structural forms, and household appliances. However, plastics are derived from fossil fuels and are thereby involved in the depletion of these resources.

There is another type of substitution that has been called "functional substitution."[47] Examples would include the far more simple miniature electronic calculators that have recently replaced the heavier desk-type mechanical calculators, and the solid-state timepieces that are now competing with the more conventional mechanical or electromechanical clocks and watches. Synthetic adhesives are now used extensively for joining in place of bolts and nuts, and the relatively simple jet engine has replaced the reciprocating propellor engine in aircraft.[52] These are examples of innovations that tend to give equal or superior performance while requiring fewer materials.

Technology can develop new ways of obtaining resources. Aluminum is now extracted from the mineral bauxite by an electrolytic process developed in the last century. According to some estimates, it is possible that the world's bauxite reserves may be exhausted in 50 to 100 years. However, aluminum is one of the most abundant of the earth's metals and is a major constituent of many common rocks and clays. It is not unlikely that the approaching exhaustion of bauxite will stimulate research that will make aluminum available from other sources.

During and shortly after World War II, uranium was obtained from high-grade ores from the Belgian Congo that contained as much as 70 percent uranium. Within a decade, these ores were exhausted, and the United States turned to low-grade deposits in the southwestern United States, where the uranium content was only .3 percent. It has been proposed that the ultimate source of uranium may be granitic rocks, in which the uranium content is only about .01 percent. The price of uranium obtained from such a low-grade source would of course be greater than that for uranium processed from high-grade sources, but this would not raise the price of energy prohibitively. (See Chapter 6.)

The oceans, which are the repository for the enormous quantities of minerals dissolved from the earth's surface by the rivers of the world, are often cited as the ultimate source of raw materials. Some of the elements washed into the oceans, such as magnesium and bromine, are in readily available form and are being extracted from the oceans by processes already developed. Some of the metals precipitate to the bottom as, for example, manganese, which exists in huge quantities in nodules distributed along the ocean floor. However, the manganese is in a siliceous form that reduces its value as a resource for the time being. Other metals have thus far defied the ingenuity of the industrial chemists. Gold is dissolved in the oceans in an amount about 1,000 times greater than the known terrestrial reserves. But the oceans are so vast that the value of gold in seawater is only about 20¢ per million gallons of water. As of 1969, despite many attempts by many organizations to mine the ocean waters for gold, less than .1 milligram had been recovered.[53]

Elements other than gold also are present in seawater, but in too dilute a concentration to be extracted by present technology. To recover $120,000 worth of zinc annually, which is present in a concentration of about ten parts per billion, would require processing 9 billion gallons of sea water per year. This amount is equivalent to the combined annual flows of the Hudson and Delaware rivers.[53]

Because technology moves so rapidly, it is difficult to estimate when the presently known reserves of minerals will become exhausted, what new sources must be developed, or what new substitute minerals must be utilized. For this reason, there is great reluctance to make future projections, although for purposes of formulating public policy they must, of necessity, be made.[54-57] Based on a contemporary set of somewhat pessimistic projections, the known global reserves of many important metals and fuels will be exhausted in less than half a century (if it is assumed that consumption continues to increase exponentially at the present rates of growth). If it is assumed that future exploration will increase the known global reserves by a factor of 5, the extention in time will be increased by a much smaller factor, due to the assumption of continued exponential growth of demand.[58] On the more optimistic side, it is possible to foresee that, given an ample supply of low-cost energy, other materials will be substituted for those now being depleted, although they may be more costly by present standards.[59,60]

From the point of view of public policy, a basic question is whether prospective shortages will be anticipated in sufficient time to take the necessary corrective actions. The question is of global significance and transcends national boundaries. It may only be solved by drastic new international arrangements.[61] International relationships have already begun to adjust. The raw materials required by the developed countries, which were formerly traded in a buyers' market, are now under tight control by the sellers, who in many cases are utilizing their new power to settle long-standing political and economic grievances.[62] The oil-producing nations, which recently have controlled the price of oil under a cartel arrangement, have been followed by the bauxite producers, and by more than one dozen countries that nationalized foreign investments in metal mines in 1974.[63]

It was noted earlier that one of the properties of exponential growth is that a resource can pass, in a very short time, from a condition of great abundance to one of extreme scarcity. This is illustrated in Figure 5-8, which concerns an essential natural resource, arable land.[58] The world's total available arable land is estimated to be 3.2 billion hectares. At present levels of productivity, it is estimated that about .4 hectare is required per person. Based on this per capita requirement, the requirement for arable land is seen in Figure 5-8, extrapolated beyond the year 2000. The rapid rise in this requirement assumes a continuation of current rates of population increase. The availability of arable land diminishes because the increase in population requires that land be taken from agricultural production for purposes such as manufacturing, housing, roads, and recreational areas. Regrettably, land also continues to be removed from the global reserve of arable acreage because of bad land management by primitive as well as more advanced cultures. The land south of the Sahara and in the state of Rajasthan, India is being converted into desert by the same practices of deforestation and overgrazing we described in Chapter 2. In Rajasthan, 300,000 acres are being lost each year because of goat-grazing practices.[64] The curves of available land and required land will cross at about the year 2000, assuming present rates of agricultural productivity. According to this projection, insufficient arable land will be available beyond that time.

Also shown in Figure 5–8 are the curves of required agricultural land, assuming that the present productivity is doubled or quadrupled. Even though agricultural productivity is quadrupled, a shortage of arable land would nevertheless develop before the year 2050.[58]

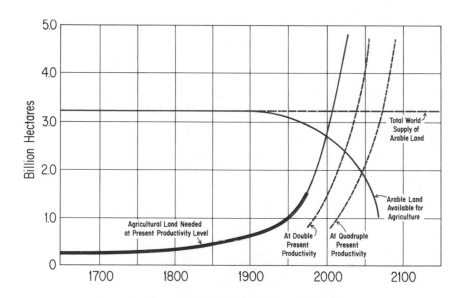

Figure 5-8. Worldwide availability of arable land in relation to needs between 1650 and 2100. According to these projections, from a report for the Club of Rome's Project on the Predicament of Mankind, a shortage of arable land will develop in less than a century even under the most optimistic assumptions.[58]

The interactions of a rapidly increasing human population with the supply and demand of food, energy, and raw materials are so complicated as to defy meaningful quantitative analysis. However, efforts to develop mathematical models that forecast the course of human history do serve to identify the basic characteristics of the interactions. In this regard, the report of the Club of Rome's Project on the Predicament of Mankind utilized the most sophisticated of the models which had been developed at the time.[58] Although it has been widely criticized, it has proven useful for purposes of identifying some of the fundamental characteristics of the system with which we are dealing.[48,63]

Of one thing we can be certain: Exponential growth is inherently a transient condition wherever it is encountered in the natural world, and the exponential growth of many global economic and demographic characteristics can be no exception. The physicist, engineer, and biologist can each cite innumerable examples of exponential growth, but in all cases there are other factors that tend to stabilize the system. Stabilizing influences, for better or worse, must develop: exponential economic and demographic growth cannot long continue.

There is evidence that these feedback loops are already beginning to operate with respect to the populations of the developing countries. When exponential growth involves doubling times of a century or more, there is time to contemplate the consequences of projected increases and take corrective action by deliberate means. However, when the doubling time is of the order of ten to thirty years, it may not be possible to take corrective actions in time to avert severe hardship. In many parts of the world, one or more of the four primary factors that determine man's relationship to his environment appear to be out of control. Population growth is a particularly acute problem in the developing countries, as is food supply. And in both the developed and undeveloped countries, the availability of energy is now showing signs that it may seriously affect the availability of food and natural resources. We will deal more completely with the subject of energy in the next chapter.

NOTES

1. Lee, Everett S. "Population and Scarcity of Food," *Annals of the American Academy of Political and Social Science* 420:1 (July, 1975).
2. Wrigley, E. A. "Population and History," McGraw-Hill Book Co., New York (1969).
3. Frisch, Rose E. "Population, Food Intake, and Fertility," *Science* 199:22 (1978).

4. United Nations. "Concise Report on The World Populations Situation in 1970–1975 and its Long-Range Implications," Department of Economic and Social Affairs, *ST/ESA/SER.A/56*, United Nations, New York (1974).
5. Westoff, Charles F. "The Populations of the Developed Countries," *Scientific American* 231:109 (1974).
6. Commission on Population Growth and the American Future. "Population and the American Future," Report of the Commission (established by the President and Congress in 1969), Signet Books, New York (1972).
7. Chadwick, Edwin. "Report on the Sanitary Condition of the Labouring Population of Gt. Britain" (1842) (edited with introduction by M. W. Flinn), Edinburgh Univ. Press, Edinburgh (1965).
8. Demeny, Paul. "The Populations of Underdeveloped Countries," *Scientific American* 231:149 (1974).
9. Trevelyan, C. M. "English Social History," Spottiswoode, Ballantyne and Co., Ltd., London (1944).
10. British Parliamentary Papers. "Report of the Minutes of Evidence on the State of Children Employed in the Manufactories of the United Kingdom, Together with a Report on the Employment of Boys in Sweeping Chimneys with Minutes of Evidence and Appendix." Reprinted by Irish Univ. Press, Shannon, Ireland (1968). Originally titled "Children Employed in the Manufactories of the United Kingdom," Select Committee Report of Minutes of Evidence (1816) and "Employment of Boys in Sweeping Chimneys," Select Committee Report (1817).
11. Hardin, Garrett (ed.). "Population Evolution and Birth Control," 2nd ed., W.H. Freeman and Co., San Francisco, Cal. (1969).
12. Segal, Sheldon J. "The Physiology of Human Reproduction," *Scientific American* 231:52 (1974).
13. Greep, R. O. "Population Growth, The Environment and Fertility Control," *International Journal of Environmental Studies* 7:51–55 (1974).
14. Sklar, J. and B. Berkov. "Abortion, Illegitimacy, and the American Birth Rate," *Science* 185:909 (1974).
15. Weinstock, Edward, Christopher Tietze, Frederick S. Jaffe and Joy G. Dryfoos. "Legal Abortions in the United States Since the 1973 Supreme Court Decisions," *Family Planning Perspectives* 7(1):23 (1975).
16. Westoff, Charles F. "The Decline of Unplanned Births in the United States," *Science* 191:38 (1976).
17. New York Academy of Sciences. "Food and Nutrition in Health or Disease" (1977).
18. Hendricks, S. B. "Food from the Land," in: "Resources and Man" (Commission on Resources and Man, National Academy of Sciences), W. H. Freeman and Co., San Francisco, Cal. (1969).
19. Revelle, Roger. "The Resources Available for Agriculture," *Scientific American* 235(3):164–178 (1976).
20. Poleman, Thomas T. "World Food: a Perspective," *Science* 188:510 (1975).
21. Randhawa, M. S. "Green Revolution," John Wiley and Sons, New York (1974).

22. Ennis, W. B., W. M. Dowler and W. Klassen. "Crop Protection to Increase Food Supplies," *Science* 188:593 (1975).
23. Wade, Nicholas. "Green Revolution (I): A Just Technology, Often Unjust in Use," *Science* 186:1093 (1974). "Green Revolution (II): Problems of Adapting a Western Technology," *Science* 186:1186 (1974).
24. Johnson, Stanley. "The Green Revolution," Harper and Row, New York (1972).
25. Jennings, Peter R. "Rice Breeding and World Food Production," *Science* 186:1085 (1974).
26. Wortman, Sterling. "Food and Agriculture," *Scientific American* 235(3):30–39 (1976).
27. Jennings, Peter R. "The Amplification of Agricultural Production," *Scientific American* 235(3):180–194 (1976).
28. Ryther, J. H. "Photosynthesis and Fish Production in the Sea," *Science* 166:72 (1969).
29. Brown, Lester R. "By Bread Alone," Praeger Publishers, New York (1974).
30. Ricker, W. E. "Food from the Sea," in: "Resources and Man" (Commission on Resources and Man, National Academy of Sciences), W. H. Freeman and Co., San Francisco, Cal. (1969), p. 87.
31. Othmer, D. F. and O. A. Roels. "Power, Fresh Water, and Food from Cold, Deep Sea Water," *Science* 182:121 (1973).
32. Food and Agriculture Organization of the United Nations. "The State of Food and Agriculture, 1974," Rome (1975).
33. Steinhart, J. S. and C. E. Steinhart. "Energy Use in the U.S. Food System," *Science* 184:307 (1974).
34. Heichel, G. H. "Agricultural Production and Energy Resources," *American Scientist* 64:64–72 (1976).
35. Stanhill, G. "Energy and Agriculture: A National Case Study," Selected Papers on the Environment in Israel, No. 3, Environ. Protection Service, Prime Minister's Office, Hakirya Bldg. 3, Jerusalem, Israel (1975).
36. Pimental, D., L. E. Hurd, A. C. Bellotti, M. J. Forster, I. N. Oka, O. D. Sholes and R. J. Whitman. "Food Production and the Energy Crisis," *Science* 182:443 (1973).
37. Berg, Alan. "The Nutrition Factor," The Brookings Institution, Washington, D. C. (1973).
38. Davidson, Stanley, R. Passmore, J. F. Brock and A. S. Truswell. "Human Nutrition and Dietetics," 6th ed., Churchill Livingstone, Edinburgh, London and New York (1975).
39. Woodham-Smith, Cecil. "The Great Hunger," Harper and Row, New York (1962).
40. Mayer, J. "Management of Famine Relief," in: "Food: Politics, Economics, Nutrition and Research" (P. H. Abelson, ed.), American Association for the Advancement of Science, Washington, D.C. (1975), pp. 79–83.
41. Revelle, Roger. "Food and Population," *Scientific American* 231:161 (1974).

42. Mora, J. O., A. Amezquita, L. Castro, N. Christiansen, J. Clement-Murphy, L. F. Cobos, H. D. Cremier, S. Dragastin, M. F. Elias, D. Franklin, M. G. Herrera, N. Ortiz, F. Pardo, B. de Paredes, C. Ramos, R. Riley, H. Rodriguez, L. Vuori-Christiansen, M. Wagner and F. J. Stare. "Nutrition, Health and Social Factors Related to Intellectual Performance," in: "World Review of Nutrition and Dietetics" (S. Krager, ed.), Basel (1974), Vol. 19.
43. Martin, H. P. "Nutrition: its Relationship to Children's Physical, Mental, and Emotional Development," *Am. J. Clin. Nutrition* 26:766–775 (1973).
44. Lloyd-Still, J. D., I. Hurwitz, P. H. Wolff and H. Shwachman. "Intellectual Development after Severe Malnutrition in Infancy," *Pediatrics* 54(3):306–311 (1974).
45. Winick, Myron and J. A. Brasel. "Early Malnutrition and Subsequent Brain Development," presented at New York Academy of Sciences Symposium, "Food and Nutrition in Health and Disease," Philadelphia, Pa. (December 1–3, 1967).
46. Hardin, Garrett. "Beyond 1967: Can Americans be Well Nourished in a Starving World?" presented at New York Academy of Sciences Symposium, "Food and Nutrition in Health and Disease," Philadelphia, Pa. (December 1–3, 1976).
47. Chynoweth, A. G. "Electronic Materials: Functional Substitutions," *Science* 191:725 (1976).
48. Landsberg, H. H. "Materials: Some Recent Trends and Issues," *Science* 191:637 (1976).
49. Radcliffe, S. V. "World Changes and Chances: Some New Perspectives for Materials," *Science* 191:700 (1976).
50. MacGregor, Ian D. "Natural Distribution of Metals and Some Economic Effects," *Annals of the American Academy of Politican and Social Science* 420:31 (July, 1975).
51. Mesarovic, Mihajlo and Eduard Pestel. "Mankind at the Turning Point: the Second Report to the Club of Rome," E.P. Dutton and Co./Reader's Digest Press, New York (1974).
52. Abelson, P. H. and A. L. Hammond. "The New World of Materials," *Science* 191:633 (1976).
53. Cloud, P. E. "Mineral Resources from the Sea," in: "Resources and Man," National Academy of Sciences Commission on Resources and Man, W. H. Freeman and Co., San Francisco, Cal. (1969).
54. Cameron, Eugene N. (ed.). "The Mineral Position of the United States, 1975–2000," Univ of Wisconsin Press, Madison, Wis. (1973).
55. Landsberg, Hans H., Leonard L. Fischman and Joseph L. Fisher (Resources for the Future, Inc.). "Resources in America's Future: Patterns of Requirements and Availabilities 1960–2000," Johns Hopkins Univ. Press, Baltimore, Md. (1963).
56. "Mineral Facts and Problems," Bureau of Mines *Bulletin #650;* U.S. Government Printing Office, Washington, D.C. (1970).
57. Paley, William S. "Resources for Freedom," President's Materials Policy Commission (1952).

58. Meadows, D. H., D. L. Meadows, J. Randers and W. W. Behrens. "The Limits to Growth: A Report for the Club of Rome's Project on the Predicament of Mankind," Universe Books, New York (1972).
59. Goeller, H. E. and A. M. Weinberg. "The Age of Substitutability," *Science* 191:683–689 (1976).
60. Goeller, H. E. "The Ultimate Mineral Resource Situation: An Optimistic View," *Proceedings of the National Academy of Sciences* 6919:2991-2992 (1972).
61. Aspen Institute for Humanistic Studies. "The Planetary Bargain: Proposals for a New International Economic Order to Meet Human Needs," Report of an International Workshop convened in Aspen, Colo. July 7–August 1, 1975 (policy paper).
62. Grant, James P. "Food, Fertilizer, and the New Global Politics of Resource Scarcity," *Annals of the American Academy of Political and Social Science* 420:11 (July, 1975).
63. Gordon, Lincoln. "Limits to the Growth Debate," *Resources* 52:1–5 (1976).
64. Brooks, R. R. "People vs. Food," *Saturday Review* (September 5, 1970)

CHAPTER 6

Energy Supply and Demand

The food we eat, the homes in which we live, our educational system, our leisure-time activities, and the medical services essential to our health all require access to sources of energy. Energy relieves us of the stressful effects of climate, it enables us to grow food in amounts adequate to fulfill the needs of growing populations, it greatly reduces the constraints of distance on communication and transportation, and it brings to the home of the average person leisure time and relative luxury on a scale unprecedented in history.

In contrast to the way of life in a technologically advanced nation, people in primitive settlements devote most of their time to simple chores that are no longer required where energy has been made available. Bundles of firewood must sometimes be carried on foot for great distances to supply heat for home cooking and heating, and in many villages there are no convenient sources of potable water, which must also be carried on foot for considerable distances. The use of human labor for irrigation of crops is particularly unproductive: a penny's worth of electricity (at 5¢ per kilowatt hour) will provide as much water as can be pumped by hand in a day.[1]

There is a closely coupled relationship between a country's economic development and the energy it uses, as is illustrated in Figure 6-1. The spread in the gross national product between the least-developed and most-developed countries is fiftyfold, but the difference in energy use is even larger. Per capita consumption in the United States is 150 times that in Zaire. However, there is a considerable spread in the energy used per unit of GNP among the developed countries. Sweden uses less than two-thirds as much energy as the United States per unit of GNP (a fact that is cited by some as evidence that energy is used less efficiently in the United States). Although there are undoubtedly opportunities to improve the effi-

116

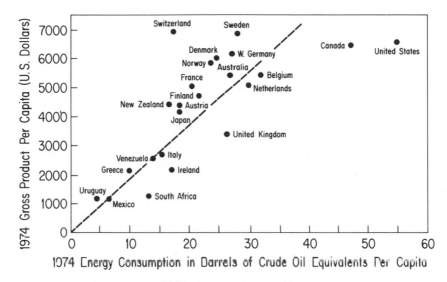

Figure 6-1. Energy consumption per unit of gross national product. Not shown are many undeveloped countries whose GNP is less than a few hundred dollars per year and whose energy production is correspondingly low.[31]

ciency of energy use in the United States, the difference is not due primarily to this factor, but to the different ways in which energy is used. Geographical factors encourage the use of automobiles and trucks in the United States. In addition, the United States produces the kinds of goods that demand large amounts of energy. The differences in energy consumption per dollar of GNP between many states of the United States are comparable to the differences between the United States and Sweden.[2,2a] In fact, the same relationship, in Figure 6–1, between GNP and energy use among the countries of the world can also be shown to exist in the United States when energy consumption is examined in relation to income. The per capita use of energy in 1970 and 1971 was about 3,000 kilowatt hours per year among families of four living in Los Angeles with a gross household income of $4,000 per year, compared to about 13,000 kilowatt hours per year for families of similar size with an income of $30,000 per year.[3]

As recently as 200 years ago, the importance of energy to national well-being was not yet appreciated. The renowned Scottish economist, Adam Smith, did not mention the role of energy when he published his classic study, *An Inquiry into the Nature and Causes of the Wealth of Nations* in 1776. In fact, the subject of energy and its relationship to population, resource development, and food production received surprisingly little general attention prior to the fuel crisis associated with the Mideast war of 1973 and 1974 in the Middle East. The subject was considered prominently in 1957 by Harrison Brown and his associates in an excellent overview of the social and technical problems society would face in *The Next Hundred Years*.[4] M. King Hubbert, of the U.S. Bureau of Mines (whose work we will refer to later in this chapter), has proved remarkably perceptive for several decades in his forecasts of the relationship between long-range supplies and demands for the fossil fuels, and he prepared important studies in 1962 and 1969 that emphasized the relatively fleeting availability of the fossil fuels when viewed in historical perspective.[5,6] However, the subject has more often received less than adequate attention. A study of population growth and its relationship to natural resources, prepared by the National Academy of Sciences as recently as 1971, included 17 excellent papers, none of which dealt with the importance of energy.[7] One of the most widely read of the contemporary texts on human ecology did not deal with energy except as it is a factor in causing environmental pollution.[8] The public has heard much about energy production as a cause of air pollution and water pollution, but there has been a general failure to appreciate the role of energy in relieving human beings of many burdens.

Until comparatively recently, the only energy available to humans was derived from the sun. Muscle power (animate energy) supplied first by people and later by work animals, requires food in which solar energy is stored as a result of the photosynthetic reactions of sunlight with inorganic forms of carbon, hydrogen, and oxygen. Wood, the first fuel that augmented "muscle," is also a form of stored solar energy. The sun is also the source of the energy in coal, petroleum, and natural gas, as well as the power derived from wind and water. Coal, petroleum, and natural gas are the fossil remains of ancient plants. Even water power is ultimately made available by solar energy, for it is produced by rainwater finding its way back to the oceans. Of course, the heat of the sun, is responsible for the creation of rain; it evaporates surface water and convects the vapor to high altitudes from where it falls back to earth as rain or snow. Nuclear and geothermal energy are the only energy sources now available that are not derived from the sun.

HISTORICAL TRENDS

In 1890, when the population of the United States was a little over 63 million, total energy consumption was 7.1 quadrillion British thermal units (7.1 quads). [The Btu. (British thermal unit) is a measure

of energy equal to the amount of heat necessary to raise the temperature of one pound of water by one degree Fahrenheit. It is convenient
to use "quadrillion Btu." (1,000 trillion) as the unit of national energy.
It will hereafter be referred to as the "quad." One quad is
equivalent to the energy consumed by burning 40 million tons of
bituminous coal, or 182 million barrels of fuel oil, or about 100 million
Btu. per capita.] More than 90 percent of this energy came from wood in
1890, and the remainder from coal. By 1973, when the country's population was about 210 million, the total energy consumed was 75 quads, a
per capita annual consumption of 350 million Btu., or about 1 million Btu.
per day. The average American thus now uses about 3.5 times more
energy than an American at the end of the nineteenth century. Since a
human being is capable of doing work at a rate of about 4,000 Btu. per
day, the 1 million Btu. of inanimate energy now available to the
average citizen provides the equivalent of 250 human slaves to lighten
labor, provide food, clothing, transportation, housing, and the
amenities of life. Many of these—the radio, the television, the telephone, wonder drugs, and electric light bulbs could not be provided
by slaves. They are made possible by technology and the ready
availability of energy.

The United States now has 6 percent of the world's population but
consumes about 30 percent of the world's energy.[9] Until recently, the
price of energy was so low, and energy was so readily available, that
it was simply accepted as something that could be had at the flick of a
switch. Prior to the recession of 1974, use of all forms of energy in the
United States was rising at a rate slightly greater than 4 percent per
year, which would cause the demand to double in about 17 years; but
the demand for electricity was growing even more rapidly, and had
doubled every ten years for several decades. Consumption of energy
elsewhere in the world has been increasing at a rate of about 6 percent
per year, which has caused the gap to close slightly. In 1925, the
United States consumed 40 percent of the world's energy[10] compared
to 30 percent at present.

The changing sources of energy from 1850 to 1974 are illustrated in
Figure 6–2.[10] From the figure it can be seen that the rapid rise in the
available energy during the latter nineteenth and early twentieth
centuries was associated primarily with the use of coal; then coal
consumption began to diminish by about 1920 and was largely
supplanted by oil and gas by 1950. Thus, coal accounted for about 80
percent of the energy consumed in the United States in 1910, when
petroleum satisfied only about 10 percent of the demand. By 1974,
petroleum and gas together were providing 76 percent of the energy,
and coal only 18 percent. Until about 1957, the United States was

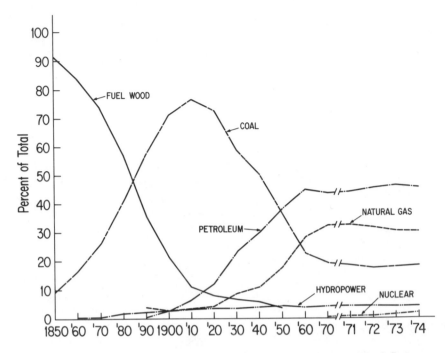

Figure 6-2. Fuels from which energy has been derived in the United States between 1850 and 1974.[10]

self-sufficient with respect to energy as is seen in Figure 6–3.[9] After 1957 consumption exceeded production, and the gap has been widening ever since.

The fuels from which the United States obtained its energy during the period between 1960 and 1974 are shown in Figure 6–4,[10] and the manner in which the energy was used is shown in Figure 6–5.[10] Of about 75 quads consumed in the United States in 1973, approximately 25 were used for transportation, 18 percent for heating buildings, and more than 40 percent by industry.

Nuclear energy became commercially feasible during the 1960s, and is expected to assume an increasingly important role in the future in many countries, although there are uncertainties because of public concerns about safety. (See Chapter 13.) Coal, oil, gas, and uranium will be the only significant sources from which energy can be derived for the balance of this century. Other sources of energy, such as solar and geothermal, have captured the imagination of many environmentalists who see these as clean and inexhaustible sources of power, but we will see later that we do not yet have the technology needed to use

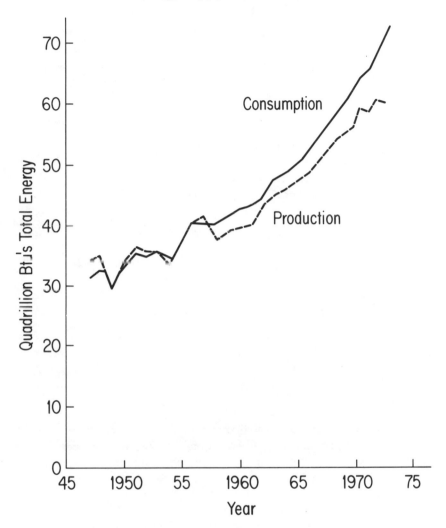

Figure 6-3. Domestic production and consumption of energy in the United States between 1948 and 1973.[9]

these resources on a large scale. Since oil and gas will be increasingly available only from overseas sources, coal and nuclear fuels must be depended upon to provide future needs in order to avoid unacceptable economic and political penalties.

Technology is difficult to project beyond 25 years, and there is considerable risk in attempting to choose among the several foreseeable options that should become available after the turn of the

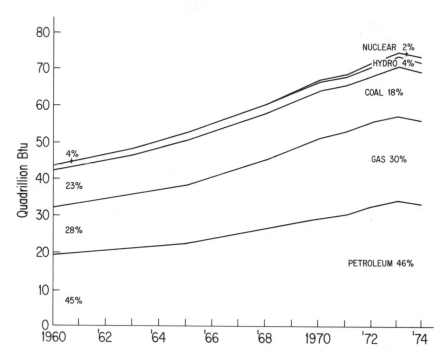

Figure 6-4. Fuels from which the United States has derived energy between 1960 and 1974.[10]

twenty-first century. Controlled thermonuclear power (fusion), solar energy in its many forms, geothermal energy, and the breeder reactor are all viable possibilities, and these may be augmented by new minor sources, such as gas, oil, or alcohol produced microbiologically from organic wastes, or "energy farms."

Our future energy needs must be considered in three time frames. First, there are the short-term actions which can be taken, between now and 2000 to close the gap between supply and demand. In the short term there will be no time to make basic changes either in the sources of energy or the technology by which the fuels are used. During this period, conservation must be emphasized, the coal industry should be vitalized, and maximum use should be made of nuclear energy. Second, we should consider the steps that can be taken by the first decade or two of the twenty-first century, by which time certain advanced technologies that now appear feasible can be put into use. Finally, the long term (50 or more years from now) must be considered when systems of power generation can be developed that are now only in the conceptual stage.

It is essential to bear in mind, as we consider our future options, that while ample new sources of energy are available, they will not be developed overnight; the energy of the future will be far more costly than in the past; and both government and industry may be required to make capital investments for energy development on a scale unprecedented in history.

CONTEMPORARY SOURCES OF ENERGY

Coal. Coal was the predominant source of energy in most parts of the world until the advent of low-cost oil and natural gas in the mid-twentieth century, at which time a rapid transition took place. The railroads converted from coal-burning boilers to diesel engines, and households, commercial buildings, and power plants converted from coal to oil and natural gas.

The environmental problems associated with coal extend from the mine to the place of use, and they are so fundamental to the coal

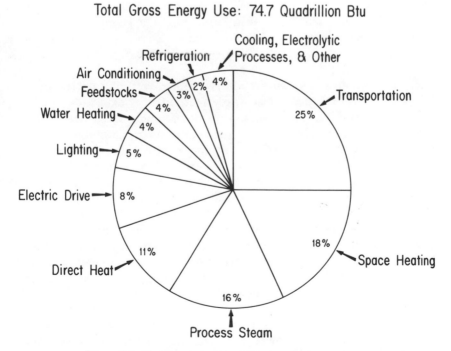

Total Gross Energy Use: 74.7 Quadrillion Btu

Cooling, Electrolytic Processes, & Other

Refrigeration

Air Conditioning

Feedstocks

Water Heating

Lighting

Electric Drive

Direct Heat

Transportation 25%

Space Heating 18%

Process Steam 16%

11%

8%

5%

4%

4%

3%

2%

4%

Figure 6-5. Uses of energy in the United States in 1973.[10]

industry that they must be reviewed in any discussion of the use of coal as a source of energy. These problems include the dangers of mining, land disfigurement, transportation, and air pollution.

Coal mining remains a hazardous occupation despite the fact that there has been steady improvement in the safety record. In 1975, underground coal miners in the United States suffered 35.4 disabling injuries per million manhours worked, which was more than three times the average of disabling injuries in all industries and more than 24 times the rate in automobile manufacturing.[11] In 1973, there were 107 fatalities in the underground mines, compared to more than 1,200 in 1935, a reduction of more than 90 percent. The improvement has been due in part to the fact that coal production has become more mechanized, and fewer miners are employed. Expressed as fatalities per million manhours, the recent rates are about 60 percent lower than in 1935.

A West Virginia coal mine tragedy that caused 78 deaths in 1968 occurred at a time when there was widespread criticism that the federal government was taking little initiative in matters concerning mine safety. The upwelling of concern following that accident caused the Congress to pass the Federal Coal Mine Health and Safety Act of 1969. Health and safety standards were authorized for all coal mines, and provision was made for enforcement of the standards and for accelerated research in the field of coal mine safety. Benefit programs were authorized for miners disabled by lung diseases in the course of their occupation.[12]

This legislation was important and was long overdue. Apart from a high accident rate in the mines, the effects of long exposure to dust and other environmental factors, not all of which have been identified, have been responsible for high mortality rates among underground miners. It is estimated that the death rate among miners during their working years is 60 to 90 percent higher than that of the general population of working males.[13] In addition to higher death rates, the miners are subject to disabling lung diseases that have come to be known as coal workers' pneumoconiosis (CWP), or black-lung disease, which affects about 10 percent of nonsmoking bituminous miners and 26 percent of nonsmoking anthracite miners.[14] In the Pennsylvania anthracite mines, there were five times as many disability awards made for black-lung disease during the period between 1935 and 1969 as for mine accidents.

The unpleasant statistics of morbidity and mortality in the coal mines of the United States represent a major environmental impact of the energy industry, but the Federal Coal Mine and Safety Act of 1969 and, in particular, its 1972 supplements, contain provisions that have proved to be controversial. That is, the 1972 amendments require that benefits be paid to coal miners for disabilities arising from *all* lung diseases, including those such as bronchitis, emphysema, and asthma—all of which occur with high frequency in the general population. The National Academy of Sciences has stated that "the current Black Lung benefits program rests on an unsupportable presumption, namely that all respiratory diseases that may befall a coal miner are due to his occupational exposure." Congress has apparently decided that miners should receive benefits not accorded other workers in hazardous trades. Perhaps this is being done in reaction to the long-neglected conditions under which the miners worked. However, the Academy noted that "there is no need to distort medical knowledge to justify payment of these benefits."[12] There are certainly precedents for according benefits to special groups, such as veterans and farmers. However, the benefits paid to miners under this act exceeded $1 billion in 1973 and would have amounted to $20 billion to $100 billion if applied to workers in other industries.[12] Total disability payments to all employees in the United States totalled $2.4 billion. The 125,000 coal miners, comprising .2 percent of the total work force, thus received 40 percent of all compensation benefits.

Almost all coal in the United States was mined underground until about 1920, and underground mining still accounts for about 50 percent of all coal produced. The coal removed from the mines is separated from the rock, which is then left in great piles on the surface. Runoff of rainwater from these piles is acidic, and has been disastrous to the ecology of streams and rivers draining the underground coal mining areas. The runoff also causes nearby rivers to become turbid and silted. The problem of acid runoff and siltation can be reduced by guiding the water to collection ponds where the solids can settle and the water treated to neutralize its acidity before it is released.

Open strip mining of near-surface deposits has become more popular in more recent years, and now accounts for more than half of the coal mined in the United States. Until recently, litttle effort was made to restore the land after the coal was stripped, and many square miles of land in West Virginia, Ohio, and Illinois have been badly scarred by these activities. It is practical under certain conditions to reclaim the strip-mined area by restoring the original surface topography and replacing the topsoil and vegetation, thus returning the land to approximately the condition in which it originally existed. The feasibility of reclaiming western strip mines is not fully resolved.[9] Many of the deposits are located in semiarid regions in which reestablishment of the native vegetation is a slow process.[15] However, successful revegetation can be accomplished in mixed-grass prairies and forested areas.

After the coal is mined, it is usually prepared for shipment by crushing it to the required size, cleansing it to remove dust and foreign minerals, and drying it with streams of hot air. In coal-washing procedures from 1,500 to 2,000 gallons of water are required per ton of coal processed.[16] This process is particularly noxious to the environment because the waste water from the washing carries high concentrations of suspended solids that must be removed in settling ponds to protect the receiving waterways.

The expense of transporting the coal from the mine to the point of use is one of the largest elements in the cost of coal. A modern 1,000-megawatt, coal-fired electric generating station requires 75 rail-road cars per day, each carrying 100 tons of fuel.

The coal can be transported by barge, truck, or pipeline, but about 70 percent of the coal used in the United States is shipped by rail.[16] The feasibility of transporting coal by pipeline has been demonstrated, and in one installation in Arizona, the coal is pulverized, suspended in water, and transported by pipe as a slurry for 270 miles to an electrical generating plant.[17] Experience with the Arizona pipeline has been satisfactory during its first six years of operation, but the system requires copious quantities of water, which is a disadvantage in arid regions.

Sulfur dioxide (SO_2), a gas with irritating properties, is released to the atmosphere when coal is burned. Sulfur dioxide is converted in the atmosphere to various sulfates, and sulfurous or sulfuric acids. (These oxidized forms of sulfur will be referred to as "the sulfur oxides.") The combustion products of coal also include unburnt carbonaceous particles (soot) and particles of gritty "fly ash," originating from the noncombustible portion of coal which can vary from 2.5 to as high as 32.6 percent in coal mined in the United States.[16] If not controlled, the products of coal burning can be the cause of significant economic loss by soiling, and can also injure human health.

The high sulfur content of coal has discouraged its use on a large scale. Most of the bituminous coals mined in the eastern half of the United States contain more than 3 percent sulfur, but it is necessary to use coal with a sulfur content of 1 percent or less in order to meet clean-air standards in most localities. About half of the sulfur is contained in an inorganic form, and if the coal is finely crushed and cleaned, about 50 percent of this fraction can be removed.[18] The other half of the sulfur is contained in organic form and can only be controlled by equipment designed to remove the sulfur from the combustion products (scrubbers) or by discharging the combustion products from high stacks.

Many of the environmental problems associated with the use of coal would be eliminated if the carbonaceous substances in coal could be gasified or liquified. Liquid fuels were produced from coal in Germany as early as 1930 and were used commercially until 1945. Both gasification and liquification processes have been used successfully in South Africa and Czechoslovakia for many years. By converting the energy-yielding organic compound in coal to liquid form, it would be possible to eliminate many of the environmental problems associated with coal and would provide a substitute for the petroleum products now being imported from abroad in increasing amounts. Coal liquification could provide an important fraction of our energy needs by the end of this century.[19]

It is also possible to convert coal to a gaseous fuel by reacting the carbon contained in it with water at high temperatures, thereby producing methane.[20,21] This concept is not a new one. Coal has been used to produce manufactured gas, a mixture of hydrogen and carbon monoxide, for more than a century. This was done by the "water gas" process developed in the mid-nineteenth century. However, the process was crude and inefficient, and would not be economically viable today. Five major processes for gasification of coal have been designed, but only one, a German invention known as the Lurgi process, has been demonstrated commercially. However, there is little confidence that coal gasification will provide sufficient gas to replace the dwindling supplies of the natural product now being used for domestic and industrial purposes.[22] It is estimated that synthetic gas would cost $4 per million Btu., at which price some geologists believe large quantities of natural gas could be recovered from presently uneconomical sources.[22]

Coal is by far the most abundant of our fossil fuel resources, and there are also great reserves of it in Asia and Europe. M. King Hubbert has projected the use of coal in the United States, assuming that the presently mapped coal reserves are fully utilized (see Figure 6–6).[5] It is predicted that coal production in the United States will peak about 2170, if the rate of production is increased about five times the present rate. Production is expected to diminish thereafter, returning to the present rate sometime around 2400.

Coal reserves in the United States would be fully depleted before the end of the twenty-sixth century. Such forecasts are, of course, very risky but they serve at least to illustrate that there is no shortage of coal for the immediate future.

It would seem as though coal could make up for the depletion that has occurred in our petroleum and gas reserves, but there are many obstacles that will not be solved overnight. With the diminution that has occurred in coal production during recent years, coal-mining

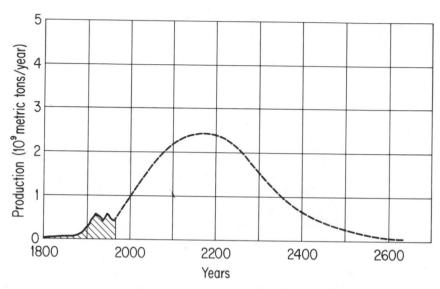

Figure 6-6. Past and projected coal production in the United States.[5]

equipment has become obsolete, the work force has grown smaller, and the railroad equipment necessary to transport the huge amounts of required coal over the rails is not available. Despite these problems, if the United States is to significantly reduce its demands for imported fuels and at the same time meet its future energy needs, it will be necessary to tap the substantial reserves of coal.

Petroleum. Petroleum is currently the most important source of energy in the United States. Petroleum provides more than 95 percent of the fuel used for transportation purposes and 40 percent of the energy consumed for space heating, cooking, and water heating in domestic and commercial buildings. Apart from its value as a source of energy, petroleum serves as source of more than 85 percent of the organic chemicals produced in the United States. It is also a major source of roofing and road materials.

Because of its relationship to the automotive industry, petroleum has played an important part in shaping twentieth-century economic development. The trucks and automobiles powered by petroleum products have freed the merchant and the citizen from the constraints of horse-drawn vehicles, railroads, canals, and harbors. Prior to the Industrial Revolution, towns and cities were built at harbors or along rivers. In the nineteenth century, new cities were developed along the canals and railroads. With the advent of automobiles and the development of road networks, the intercity communities have been able to develop in a pattern independent of the former constraints. Automobiles powered by petroleum have made possible the urban sprawl that is gradually producing gigantic population conglomerates like those along much of the East Coast of the United States.

Crude oil is a complex mixture of more than 1,000 organic compounds that range in properties from light, volatile constituents in gasoline, to the more viscous asphaltic tars. Kerosene was the most important product derived from petroleum during the nineteenth century, when it was widely used for illumination. Demand for the fuel slackened at the turn of the century with the advent of the gas mantle and electric illumination, but the reduced demand for illumination was more than offset by the new demand for gasoline to power the automobile. After World War II there was a rapid shift from coal to oil for home heating, ship propulsion, electricity generation, and railroad locomotives.

In 1973, the United States consumed 6,298 million barrels of petroleum, of which 53 percent were used for transportation, 20 percent by industry, 9 percent for electrical generation, and 20 percent for household and commercial heating (see Figure 6–7).[10] Of the 53 percent consumed by transportation, half is used for automobiles, which accounts for 13 percent of all the energy used in the United States.[23]

In the mid-1950s, at a time when the oil industry in the United States was still more than capable of meeting domestic demands, enormous supplies of oil became available in the Middle East at such low prices that it could be bought in the United States more cheaply than domestic oil could be produced. Oil-import quotas were imposed by the government in 1959 to protect the domestic industry and to avoid too great a reliance on foreign supplies of uncertain dependability. The quotas were maintained until 1973, but despite these restrictions, oil imports tripled between 1970 and 1974, and are continuing to increase.

The period in which we live will be remembered unkindly for the way in which the petroleum resources of the United States were depleted. One estimate of future crude-oil production in the United States is given in Figure 6–8.[6] The figure shows that the petroleum reserves in the United States will be depleted substantially by the end of the century and that 80 percent of the reserves will have been consumed during a period of only 65 years. The data for Figure 6–8 were published by Hubbert in 1969—several years before the oil crisis precipitated by the war in the Middle East in 1973—but attracted little general attention at the time.

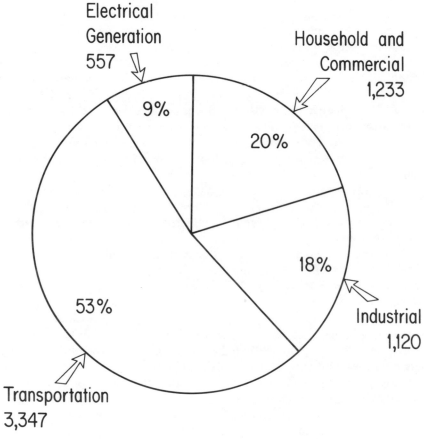

Total Consumption:
6,298 Million Barrels Per Year

Electrical
Generation
557

Household and
Commercial
1,233

9%

20%

18%

53%

Industrial
1,120

Transportation
3,347

Figure 6-7. Petroleum consumption in the United States in 1973.[10]

The production of petroleum in the United States has already peaked, and 42 percent of the petroleum used in 1976 was imported, mainly from the Arab states. Unless new sources of petroleum are found, conservation measures are adopted, or substitute fuels utilized, our dependence on imported crude oil will continue to increase. This dependence could be dangerous, as was demonstrated during the Mideast war in the fall of 1973, when the Arab states cut off shipments of oil to the United States. The embargo lasted only five months, but in the intervening time, the vulnerability to the whims of

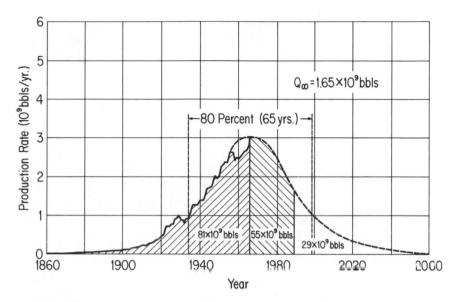

Figure 6-8. Past and projected future crude oil production of the lower 48 states.[6]

foreign oil producers of the United States was made clear by the long lines of automobiles waiting at filling stations and by the reduced availability of fuel oil for space heating. Unemployment rose because many factories were required to curtail production.[25] When the embargo was finally lifted in March of 1974, the Organization of Petroleum Exporting Countries (OPEC) had raised the price of crude oil more than fourfold. The era of cheap energy had come to an end.

The production of petroleum in the United States reached a peak in 1970 at a rate of 11.3 million barrels per day (mbpd). It has since been diminishing gradually, and domestic production in 1977 was 10 mbpd.[24] By 1985, it is assumed that production will have risen slightly to 11 mbpd, including three mbpd from the recently discovered fields on the Alaskan North Slope. At this rate of consumption, and assuming no new discoveries, the known reserves could be fully depleted in slightly less than ten years.[24] The extent to which new discoveries will extend the oil reserves remains to be seen. In any case, there seems little prospect that the petroleum supplies in the United States will be a significant source of energy beyond the end of this century.

The art of forecasting fuel reserves is a difficult one. New technological developments can suddenly convert *resources* (the quantity known to be present in the ground) to *reserves* (the quantity that can be mined economically). Since the price of both oil and gas are regulated by the federal government, the ratio of reserves to resources is dependent on pricing policy. In recent years, the ratio has also been increasingly sensitive to environmental constraints. This has been particularly true in regard to the oil and gas on the Alaska North Slope and the outer continental shelf.

Plans to build a needed pipeline to the ice-free port of Valdez and the Alaskan south coast were delayed for years by the concern of environmentalists over the possible ecological effects of the project. A 1970 court decision required the secretary of the interior to prepare an environmental impact statement prior to issuance of licenses to proceed with the project. The statement was filed in 1972 and the project approved, but environmental organizations succeeded in blocking the start of construction in court because the project would require a right-of-way in excess of the 50-foot width permitted by the 1920 Mineral Leasing Act.[26] It was not until November, 1973, at the height of the crisis created by the Arab embargo on oil shipments to the United States, that legislation was passed specifically authorizing construction of the Alaskan pipeline. When the project reaches full production in the early 1980s, about 3 mbpd will flow to the port of Valdez, from which the oil will be shipped to the lower 48 states. This production will make a significant contribution towards reducing our overseas dependence.[27]

Proposals to drill for offshore oil have also led to disputes over possible environmental effects. Offshore drilling is by no means new to the oil industry: Piers were built to accommodate offshore wells near Santa Barbara, California as early as 1897, and oil drilling rigs located on platforms beyond sight of land were operating by 1947 in the Gulf of Mexico.[28] By 1973, more than 16,000 wells had been drilled off the shore of the United States, of which 10,000 were then still producing 17 percent of all the oil or gas recovered in the United States. A "blowout" that occurred in the Santa Barbara channel in January, 1969 released more than 77,000 barrels of oil, and created havoc to miles of choice California beaches. By late 1977 there had been one dozen offshore drilling accidents that released more than 5,000 barrels of oil, but it has been suggested that the environmental risks of leakage of oil from offshore drilling are actually less than those from tankers importing the same amount from foreign sources.[29]

Natural Gas. Natural gas was first used on a large scale in the United States in 1947 when the first pipeline began to transport gas from the Southwest to the northeastern states. Natural gas soon became popular for home heating, power generation, and industrial heat because it is a clean-burning fuel that can be distributed conveniently and economically. The admirable burning characteristics of natural gas stem from the fact that it is composed of methane and other hydrocarbons that consist only of hydrogen and carbon. Its combustion products are almost entirely carbon dioxide and water, but with small amounts of nitrogen oxides.

The use of natural gas was greatly increased in the mid-1960s with the increased demand for low-sulfur fuels, and by 1970 there were nearly a quarter-million miles of transmission pipelines connected to about 1 million miles of distribution mains. However, the once-abundant reserves of natural gas began to diminish due to the fact that gas was consumed, between 1968 and 1973, at twice the rate it was being discovered.[16]

The effective life of reserves of natural gas in the United States has diminished from 30 years in 1950 to 13 years in 1970, and was only a little more than ten years in 1974.[27] Although drilling for gas was 30 percent greater in 1973 than in 1966, the gas added to the national reserve declined from 13 to 9 trillion cubic feet during that same period of time. Stated another way, the rate of discovery of new gas declined from 536 to 254 million cubic feet per foot of well drilled.[27] There is progressively less gas being discovered per unit of drilling effort. There is every indication that natural gas will be the first of the domestic fossil fuels to be exhausted, and steps are already being taken by the federal government to restrict construction of new industrial plants that depend on the use of this fuel. In some parts of the country, new homes are prohibited from using natural gas for heating.

Hydroelectric Energy. Although most of the major hydroelectric sources were well developed by the mid-twentieth century, this form of energy accounted for only about 4 percent of all consumption in the United States in 1974. It is now conceded that conventional hydroelectric generation, that is, energy obtained from running water or impoundments, will henceforth be limited to small units in the range of three to five megawatts. The total amount of untapped hydroelectric power cannot have an important impact on the nation's future energy needs.

Nuclear Energy. In the mid-1950s, the U.S. Atomic Energy Commission joined forces with industry and constructed a few relatively small power reactors that began operation in the late 1950s and early 1960s. These first power plants were designed to provide operating experience and were not intended to be economically competitive with power plants using fossil fuels. The first nuclear-power plant purchased on the basis of economic competitiveness with fossil fuels was the Oyster Creek Nuclear Power Station at Toms River, New Jersey, ordered in 1963 by the New Jersey Central Power Company. By late 1977, about 275 nuclear-power stations had been operated or were under construction in the United States. Nuclear power was already supplying about 8 percent of the electricity generated in the United States.

The nuclear industry is beset with many problems, of which the most urgent are concerned with safety and will be discussed in Chapter 13. Another problem arises from uncertainty about the future availability of uranium. This factor has a direct influence on the future economics of nuclear power.

The quantity of uranium that can be mined economically in the United States depends on its market price. In 1974, at a price of $8 per pound of uranium oxide (U_3O_8), the estimated reserves in the United States ranged from 227,000 to 450,000 tons. At $30 per pound the reserves increase to 640,000 tons of U_3O_8. In 1977, it was estimated that the installed nuclear capacity in 2000 will be 510,000 megawatts, which will require 1.5 million tons of uranium oxide over the 30-year lifetime of the plants. The known reserves are thus less than the foreseeable demand. However, the reserves may be increased substantially if the price is raised or if further exploration demonstrates the recoverability at present prices of uranium believed to be present. Whether or not the uranium reserves are sufficient to meet future energy needs is crucial to decisions concerning the needs for plutonium recycling and the breeder reactor. These questions will be examined more fully in Chapter 13. The sufficiency of uranium may thus be put to a severe test by the late 1980s. Some estimates suggest that the uranium reserve is less than given above. If so, a serious uranium shortage could develop before the breeder could have a significant effect.[30] Others have estimated the uranium reserves to be sufficient to meet the needs beyond the year 2000, and have recommended on that basis that the breeder not be built.[31] In addition, a demonstration breeder reactor is scheduled for operation in Tennessee by the mid-1980s, but the program has generated massive opposition because it requires the use of plutonium, a fuel that is highly toxic and also susceptible to misuse in the event it is diverted for military or other purposes.

SOURCES OF ENERGY YET TO BE DEVELOPED

Prospective depletion of oil and gas, environmental problems associated with combustion of coal, and uncertainties about the future of nuclear power are bringing into being well-financed research and development programs aimed at introducing abundant and clean new sources of energy. Methods of obtaining energy from the sun, some of them involving ancient principles, are now being reexamined. There is revived interest in heat from geothermal sources, and the advanced countries of the world are cooperating in development of controlled thermonuclear reactors by which inexhaustible supplies of energy could be obtained from the fusion of hydrogen.

Controlled Thermonuclear Fusion. Production of electrical energy by controlled thermonuclear reactions is theoretically possible but is proving difficult to achieve. In *fission* reactors, energy is released when nuclei of heavy elements such as uranium are split. In *fusion* reactors, nuclei of deuterium and tritium, which are heavy forms of the element hydrogen, are combined to form other elements such as helium, and energy is released in the process. The reactions are similar to those that occur within the sun.

Deuterium occurs naturally and is present in seawater in sufficient concentration to provide a source of energy that would be inexhaustible for all practical purposes. Tritium does not occur naturally in significant concentrations, and must be produced artificially by neutron irradiation of lithium, a process that is well developed and has been used for large-scale tritium production for more than two decades.

The fundamental difficulty that stands in the way of a successful controlled thermonuclear reactor is that in order for fusion to occur, it is necessary to impart enormous energy to the nuclei of deuterium and tritium to overcome their mutually repulsive electrostatic forces. This requires that the elements to be fused be heated to the enormously high temperature of about 100,000,000°C. The physics of the controlled thermonuclear reactor requires not only that the deuterium and tritium be raised to so high a temperature, but that they be highly compressed for a sufficiently long time for the fusion reactions to take place. These requirements can be achieved by magnetic containment or laser compression.[20,32,33]

In the magnetic containment system, a mixture of deuterium and tritium is first converted to a plasma, which is a gas heated to such a high temperature that all orbital electrons are stripped, and the gas becomes a hot cloud of atomic nuclei. The plasma is injected into a toroidal chamber, within which it is compressed into a ring by powerful magnetic fields. The combination of heat and pressure creates conditions favorable for fusion of the deuterium and tritium.

Until recently, most of the research effort has involved magnetic containment, but in 1968 the Soviets demonstrated that fusion could be initiated by a laser pulse directed at a small mass of deuterium and tritium. The energy of the laser beam is absorbed so rapidly that the temperature rises to 100 million degrees centigrade in less than one billionth of a second. The heating process is assisted by using several laser beams aimed at the hydrogen pellet from many directions. Since the outer surface of the pellet is being rapidly heated from many directions, it tends to compress in the manner of an implosion, creating even higher temperatures at the center of the pellet. It is said that the implosion technique could reduce the energy required for a practical controlled thermonuclear reactor by a factor of 1,000.[20]

The magnetic confinement technique has been under development since the early 1950s, and the laser approach since the late 1960s. In recent years there has been active collaboration among French, British, Russian and American scientists, with a general recognition of the fact that the technical impediments to successful achievement of a controlled thermonuclear power generator are so formidable as to require the combined efforts of all nations.

It is not now certain that the goal of a controlled thermonuclear reactor can be achieved. There may be inherent technical reasons, now not identified, which will prevent the reactor from ever being built. Even if it is built, the technical problems may make the reactor so costly as to be noncompetitive with other power generators that become available. There seems little probability that the first reactor will be demonstrated before the end of this century and, on the assumption that it will take a decade or more to develop the second-generation plant, large-scale production of thermonuclear power is not likely before about 2025.[34]

Oil Shales. Oil shales are fine-grain sedimentary deposits containing an organic material called kerogen which, when heated, is released in a form that can be converted to a product equivalent to a high-grade crude oil.[16] The oil shales have long been known as a potential source of energy; development of a shale-oil industry was being considered in Appalachia at the time oil was discovered in 1859. Interest developed once again as a result of the petroleum shortage following World War I, but the prospects for oil-shale extraction became economically unattractive with the discovery of the large Texas oil fields. Oil shale is presently being processed commercially in China, Sweden, and Spain, and raw oil shale is burned by power stations in Estonia and West Germany.[16] The recent shortage of crude oil has once again revived interest in the oil shales, and an experimental oil-shale leasing program was approved by the federal government in 1973.

Huge oil-shale deposits are known to exist in the United States. The resource totals about 2,000 billion barrels, the recoverable fraction, or reserve of which will depend on the development of new technology and on the price of crude oil produced by alternate means. (A more speculative estimate places the resource at ten times this figure.[16]) About 90 percent of the identified oil-shale deposits in the United States are located in western Colorado, Utah, and Wyoming, but there are also substantial oil shales in the eastern and central states.

Oil shale can be mined by techniques similar to those used for mining coal, with both surface and underground methods being applicable. It has also been suggested that the oil could be extracted *in situ* by fracturing the oil shale with explosives, introducing heat, and withdrawing the shale oil through wells. If mined by more conventional techniques, the environmental problems will be substantial. The waste rock from a plant producing 100,000 barrels of crude oil per day would cover 40 to 50 acres of land with one foot of debris each day.[19] The quantities of debris are so large that only *in situ* processing may be feasible. Another problem is that large quantities of water are needed for oil-shale processing, and many of the deposits are located in semiarid regions.[35]

Tar Sands. Tar sands are viscous petroleum deposits that permeate porous rocks or sediments. These deposits cannot be extracted easily by present technology, although a number of methods of uncertain economics have been developed in Canada.[16,36] Insofar as the United States is concerned, estimates of the importance of tar sands in the years immediately ahead vary widely. The Department of Interior believes that from 10 to 16 billion barrels of oil can be recovered from tar sands in the next 15 to 30 years, in contrast to estimates made by the Bureau of Mines, which projects recoveries of only 2.5 to 5.5 billion barrels.[16] There are about 550 deposits of tar sands known to be located in 22 states.

Solar Energy. The history books mention that Archimedes built a "burning machine" that used a thousand metal shields to concentrate the rays of the sun on Roman ships and set them afire as they approached the walls of Syracuse more than 2,000 years ago. This legend is not as well documented as are most of Archimedes' other accomplishments, and may or may not be based on fact. However, it does illustrate that humans have long considered ways of using the rays of the sun to provide energy in the form of heat without going through the intermediate photosynthetic cycle by which energy is stored in plants and later released by combustion.

The amount of solar energy that reaches the earth's surface is enormous. It is sufficient to provide the energy needed by the plants and animals of the world for their normal metabolic functions, and it provides the energy for a broad spectrum of atmospheric motions, from the gentle breezes to the great cyclonic storms. Solar energy was obtained from the motion of the winds long before Archimedes' "burning machine," when sails were first invented. It was the winds, also, that powered the Dutch windmills which for centuries have pumped the seawater from behind their dikes. Solar energy has also been used to drive machines, using the power of falling water to turn water wheels and turbines. It has been only in the last two or three decades that the electrical dryer has been substituted for solar energy, in the form of sunlight and fresh breezes, to dry the family wash.

With so abundant a source of free, clean energy, why are we beset with an energy crisis? Why should we waste precious coal and oil that can serve as such valuable raw material for the petrochemical industry? We burn these materials for their heat rather than conserve them for future use in producing pharmaceuticals, polymers, and other organic chemicals. The reasons we have been unable to replace fossil or nuclear fuels with solar energy are the diffuseness with which solar energy reaches earth, its unavailability during periods of cloud cover, the fact that half of the earth is always in darkness, and the inability of solar energy to compete economically with other sources of energy. For the most part these objections will probably be met by the recent infusion of funds into solar-energy research. The federal government spent no money on solar-energy research before 1973, in which year the solar research budget was only $4 million out of a total energy research-and-development budget of $672.2 million. The situation changed dramatically in the following years, and more than $300 million was authorized for solar research in fiscal year 1977.[37-39]

Use of solar energy requires an energy-efficient building design, a subject that overlaps the subject of energy conservation (to be discussed later in this chapter). Buildings can be designed with overhanging roofs that provide shade when the sun is high in the sky during the summer months, yet permit the sun's winter rays to reach the windows and walls during the cold months.

About one-fourth of the nation's energy is used to heat and cool buildings.[40] Much of this energy could be provided by the rays of the sun, using flat plate collectors. These take advantage of the well-known fact that when a dark surface is exposed to the sun, its temperature rises until the rate at which it is reradiating the absorbed energy is equal to the rate at which the energy is being absorbed. The equilibrium temperature will be increased further if the dark surface is covered by a sheet of clear glass, which will not interfere with transmission of the incoming visible light from the sun, but will prevent the escape of infrared radiation. Solar heaters that utilize this principle are widely used throughout the Middle East. In the system shown in Figure 6–9, water is circulated through a shallow space between a dark surface and a glass panel. Water thus heated can be stored in insulated tanks from which it can be drawn after sundown to supply water for washing or heating the building.

If the storage capacity of a solar-heating system is sufficient for a 1-day supply of hot water, the system can produce 50 to 75 percent of the energy required annually to air-condition and heat a building, in addition to providing hot water. It has been estimated that such systems could provide up to 1 percent of the total energy consumed

Figure 6-9. A large heating and cooling system installed in an elementary school in Atlanta. The solar collectors are supported by a wood truss system mounted on steel beams. Reflectors on the back slope of each of the collectors allow them to absorb energy from both the direct and reflected sunlight, so that little energy is wasted. This is a demonstration project for the Department of Energy by Westinghouse Electric Corporation and Georgia Institute of Technology. (*Courtesy of Westinghouse.*)

by buildings in 1985, and from 10 to 35 percent by the year 2000 to 2020. This would be equivalent to about .25 percent of the total national energy requirement in the year 1985 and equivalent to from about 2 to 8 percent of the national demand in the early part of the next century.[41]

There is as yet very little general demand for solar-heating units in most parts of the United States, but they are used in Florida where they provide energy at a cost competitive with electricity.[41] Although such units require no fuel, the capital cost is substantial[40] and they do require maintenance. Space was provided for a collector in a new building in New York City, but the system was not completed due to unfavorable economics. (See Figure 6–10.)[42]

Research during the next few years will certainly improve the performance of panel-type solar heaters and will lower their cost. In addition, they can be expected to become more versatile and, with the use of heat pumps, supply not only heat but refrigeration. Systems that use solar energy for refrigeration are currently designed to operate on the energy of water heated in solar panels to 170°F. The heat is transferred to a low-boiling fluid, the compressed vapors of which are then allowed to expand into a second chamber which is thus cooled. The principle is similar to that used in household refrigerators which operate on gas heat.

Improved commercial solar-heating and cooling units should begin to be available by about 1980. These systems may eventually supply 50 to 75 percent of heating and cooling requirements in homes and commercial buildings, but they must be augmented by electricity supplied from central stations when there is not enough sunlight.[39] A post office in Boulder, Colorado has been selected to be the first Federal building to be retrofitted with a solar heating-and-cooling system,[43] and the Energy Research and Development Administration (ERDA) in 1976 also awarded contracts for installation of solar heating units in 34 nonresidential buildings. This was the first major evidence of progress in the utilization of solar energy in commercial, industrial, educational, medical, and governmental facilities, and was a direct outgrowth of the 1975 legislation that created ERDA (which became part of the Department of Energy in 1977).

The heat produced in rooftop equipment of the type shown in Figure 6–9 cannot achieve the temperatures required for commercial electrical generation. However, following the example of Archimedes, parabolic reflectors can be designed to concentrate the diffuse incident energy. The use of parabolic reflectors is technically feasible, but relatively costly. Because of the importance of a dependable source of sunlight, such units may be practical only in dryer climates, as in the

Figure 6-10. The Citibank headquarters in New York City, showing the roof structure designed to provide solar heating and cooling. The system was not completed because detailed studies proved it to be economically disadvantageous. (*Courtesy of Citibank.*)

southwestern United States. The basic technology is not new. A solar-powered steam engine was built in Arizona in the early 1900s,[44] and a steam engine driven by solar energy was demonstrated at the World's Fair in Paris, in 1878.[41] The first modern solar-powered boiler was built in Egypt in 1912, and utilized a 13,000-square-foot cylindrical trough to focus the sun's rays on a tubular boiler. The steam thus generated supplied power to a 100-horsepower engine,[41] but the experiment was not successful and the project was dropped.[39] In the 1950s, a 1,000-kilowatt (thermal) furnace was built in the French Pyrenees to supply high temperatures (3,300°C) for research purposes.[41] A few other experimental solar-steam generators have been built in other parts of the world.

The principal impediment to further development of high temperature solar energy conversion systems is the high initial cost, which is now four to five times higher than plants that provide power from fossil fuels. Solar-energy systems can operate without fuel, but this advantage will be partially offset by the need for careful, and perhaps expensive, maintenance of the enormous reflecting surfaces. In addition, storage capacity to supply heat during periods of solar insufficiency must be provided. A recent study suggests that economically competitive solar-power plants may be limited to no more than 15 to 20 percent of the electrical generating capacity under optimum conditions, such as those in the southwestern United States.[39]

In addition to the problems of cost and maintenance that solar-energy systems present, they also pose problems that are mechanical and physical. Mechanically, the solar-energy conversion system is not a simple one. The large parabolic reflectors must be driven by clocklike mechanisms to enable them to follow the sun across the sky from morning to night. Physically, a system typical of those now on the drawing boards would utilize an 80-story tower surrounded by solar reflectors covering an area of half a square mile. The enormous extent of such an installation is immediately apparent. A large utility with an installed capacity of 10,000 megawatts would require 100 square miles of solar collectors! This would be possible only on large expanses of desert.

Research during the next few years will certainly improve the performance of panel-type solar-energy systems, and parabolic reflectors may be produced that are less costly.

Another method of utilizing solar energy is the direct conversion of sunlight into electricity using photovoltaic materials of the type that have been used successfully to generate small amounts of energy for spacecraft. (See Figure 6–11.) The principal photovoltaic substances currently in use are crystals of silicon and cadmium-copper sulfide. Single-crystal silicon solar cells have a theoretical conversion efficiency of about 23 percent, but in present practice convert with efficiencies of about 13 percent.[41] The efficiency of the cells will no doubt be improved as the demand for photovoltaic materials increases.

All of these photovoltaic materials are in an early stage of development. They are very costly, and it is not clear how the electricity produced by such devices would be used. The present federal solar-energy research and development program calls for a demonstration

Figure 6-11. An overhead view of the Skylab space station cluster in earth orbit as photographed from the Skylab 4 Command and Service Modules during the final "fly-around" before the return home. The space station is contrasted against a cloud-covered earth. Note the solar shield which was improvised by the second crew of Skylab and which shades the Orbital Workshop in the area from which a micrometeoroid shield has been missing since the cluster was launched on May 14, 1973. The solar panel on the left side was also lost on launch day. (*Courtesy of NASA.*)

of the feasibility of photovoltaic power by the early 1980s, but the first commercial power plant is not likely to be built until after the year 2000. At the present time, the technology for fabrication of large quantities of photovoltaic surfaces does not exist. Incentive for the development of this technology will come, if needed, when the feasibility of the designs for large-scale photovoltaic-power plants and their operation are demonstrated. Should this occur, manufacture of photovoltaic surfaces will become a giant new industry comparable with the largest in the United States. However, enormous technical and economic barriers exist. The costs of photovoltaic conversion surfaces must be reduced by a factor of 50 to 100 for the method to be competitive with coal at today's prices.[43]

The problems of energy storage during night time and periods of cloud cover, as well as the limitations imposed by the large requirements for land, can be solved by an advanced concept in which tens of square kilometers of these photovoltaic surfaces would be fabricated in space station located in synchronous orbit with the earth. The solar energy would be converted to high-frequency radio waves (microwaves) and beamed to the ground. However, several decades of engineering development will be required before we learn how to transmit thousands of megawatts of electric energy by this means. Also, the potential hazards of the microwave-transmission system will require attention, inasmuch as it is known that microwave exposure can be injurious to health. Nevertheless, a space station in synchronous orbit would have the advantage that it could generate electricity for 24 hours per day. The required photovoltaic area would be very much less than for a terrestrial station because the incident solar energy in space is six to ten times greater than at the earth's surface.[41] Photovoltaic systems would release no waste heat at the point of energy production and would produce no polluting gases. Full-scale systems of this kind are not likely to be built before the start of the twenty-first century.

Winds. Wind has been used to translate solar energy to mechanical form for many centuries, but fossil fuels have replaced the winds for propulsion of ships, and the windmill has been replaced by electrically powered machinery in most parts of the world. In both cases the change has taken place because of lower costs, convenience, and better performance. There appears to be no prospect that improved windmill technology will make a significant contribution to our energy requirements in the years immediately ahead, but the strategy developed to meet the energy requirements of the twenty-first century must surely consider the winds for whatever contribution they can make.

The amount of energy generated by conventional windmills is proportional to the third power of wind velocity. That is, if the wind speed is doubled, the energy generated increases eightfold. The most advantageous locations for windmills in the United States are along the coastline of the northeastern states and in the midwestern plain region, where the winds blow at 10 miles per hour or more much of the time. In these parts of the country, power production at a rate of 40 megawatts per square mile would be possible on land set aside for windmills spaced at optimal distances from each other. However, as in the case of solar-heat converters, an economic penalty is inevitable because of the need to construct standby generating capability to supply electricity during periods of low wind speed.

Windmill research and development, like other forms of solar-energy development, did not get under way on a large scale until the energy crisis that followed the 1973 Mideast war. The Energy Research and Development Administration budget for wind-energy conversion systems was about $12 million in the fiscal year 1977. The program covers a broad spectrum of applications from small units designed to provide enough power for heating, refrigeration, and irrigation of small farms to a three-megawatt generator that would utilize a 190-foot rotor, which would be placed on a site with winds of 28 kilometers per hour. The largest windmill built to date has a 125-foot blade built by the National Aeronautics and Space Administration Lewis Research Center at Sandusky, Ohio.[43]

The basic physics of wind-energy systems has been long understood, and with the existing knowledge of aerodynamics, the only remaining problems are in the areas of structural design, economics, and public acceptance. As recently as 1950, about 50,000 small windmills were located on farms in the midwestern United States, but these were gradually replaced by central station electricity with the development of the Rural Electrification Administration.[45] The early windmills were small inconspicuous units, but in the plans for the future, massive 800-foot windmills, straddling the highways at half-mile intervals, are visualized.

Energy from Vegetation. One pound of dried plant material yields about 7,500 Btu. of heat when burned, which is comparable to the energy content of low-grade coal. A ton of dried plant material, properly processed, will yield 1.25 barrels of oil, 1,200 cubic feet of medium quality gas, and 750 pounds of a combustible solid residue.[16] However, photosynthesis is basically an inefficient method for converting energy. The efficiency averages about 1 percent as a year-round average and about 3 percent during the growing season. This efficiency is greatly exceeded by the artificial methods of solar-energy conversion described previously. If one assumes a yield of 10 to 30 tons of plant material per acre per year, the land required for a 1,000-megawatt electric-generating station using plant material as a fuel would be between 250 and 500 square miles! (As noted earlier, a 1,000-megawatt plant utilizing steam produced by solar reflectors would require 10 square miles.)

It seems possible, at first, to dismiss the feasibility of this type of energy production out-of-hand since, if land is sufficiently productive to grow vegetation as a source of energy, it could also be used to grow food. However, there may be large areas of the world in which the local ecology dictates vegetation of types not suitable for food. This would certainly be true in the northern woodlands and possibly in the great rain forests of the equatorial regions.

Plant photosynthesis results in the production of fats, hydrocarbons, proteins, amino acids, and other organic compounds which could be used to produce organic chemicals presently derived from fossil fuels. About 6 percent of our total fuel supply (equivalent to about 700 million barrels of crude oil per year) is used as a raw material for the organic chemical industry. This requirement could be met with materials produced by growing plants.[46] Methanol, which is also known as methyl alcohol or wood alcohol, is a versatile fuel which can be produced easily by the microbial fermentation of wood and municipal and agricultural refuse.[47] In recent tests, a number of unmodified passenger automobiles were operated on a fuel consisting of gasoline to which 5 to 30 percent methanol had been added. It was found that the fuel economy actually increased by 5 to 13 percent and carbon monoxide emissions were substantially reduced. About 7 million bushels of grain spoil each year in the state of Nebraska alone. This waste, properly processed, could yield about 20 million gallons of alcohol. In Nebraska, a 3 percent state-tax credit has been suggested as a means of encouraging production of "gasohol," the name that has been given to the mixture of gasoline and alcohol.[46]

There seem to be a number of possibilities for using photosynthetic products as a source of energy, but for the time being, this subject is also speculative. Like many other "new" ideas, this one has a long history. During the World War II gasoline shortages in Europe and Japan, many automobiles and trucks operated on methanol made from wood chips. Thus, the idea of using methanol for automobile engines is not a new one, but as we have found to be true for other of the nonconventional sources of energy, serious attention has not been given to the subject until the past few years.

Energy from Organic Waste. In 1971 the United States produced the huge total of 880 million tons of solid waste, or about 22 pounds per capita per day. This is not the average amount that each of us puts into our garbage pails. The average city dweller actually creates directly about 3.5 pounds of waste per day, enough to create enormous logistic, economic, and environmental problems for the major metropolitan areas, but this is only the tip of the iceberg.[11] In addition to the paper, plastics, and table scraps that comprise the solid wastes of the household, the United States produces the enormous annual total of 200 million tons of manure, most of which is produced by beef and dairy cattle. To the extent that this manure is produced on the range or in pastures, valuable nutrients are returned to the soil, but much manure is produced in feedlots, where huge numbers of cattle are concentrated for fattening. Feedlot manure cannot usually be used economically for fertilizer, and unless means are provided for proper disposal, the manure becomes a serious source of land and water pollution.

Organic wastes from many sources are available in vast quantities, as shown in Table 6-1.[48] Agriculture is, by far, the largest source. Crop and food wastes alone account for 390 million tons per year. Together with 55 million tons per year from logging and manufacture of wood products, and 200 million tons per year of manure, agriculture and forestry account for nearly three-fourths of all the organic wastes produced in the United States. Actually, not all of these

TABLE 6-1
Quantities of Organic Waste Produced in 1971
(Dry Weight in Million Tons Per Year)[48]

Source	Total Amount Generated	Readily Collectable
Urban refuse [a]	129	71.0 [b]
Manure	200	26.0
Logging and Wood manufacturing	55	5.0
Agricultural crops and food wastes	390	22.6
Industrial wastes	44	5.2
Municipal sewage solids	12	1.5
Miscellaneous	50	5.0
Total	880	136.3

[a] Domestic, municipal, and commercial components of this waste amount to 3.5, 1.2, and 2.3 pounds per capita per day, respectively.
[b] Based on the 100 largest population centers in the United States.

"wastes" are, in fact, wasted. Some manure is returned to pastures as fertilizer. (It only becomes "waste" when it accumulates in great amounts at feedlots remote from the pastures and ranges.) Similarly, the "wastes" left in place after a field is harvested can be plowed back into the soil. However, the agricultural wastes that accumulate at food-processing plants cannot be returned to the soil except at prohibitive cost, but these materials could be used as a source of energy. Lumber mills have obtained steam and electricity from boilers fired by wood scraps for many years, and sugar cane residues (bagasse) have been used as a source of heat at the refineries.

Of the 880 million tons of organic waste produced each year, only 136 million tons (about 15 percent) are readily collectable in a form that can serve as a source of energy. Nevertheless, it has been estimated that if that quantity could be used as a source of energy, it could supply 2 percent of the total energy requirements of the United States. If burned directly as a source of turbine steam, it could produce about 7 percent of the nation's requirements for electricity. Alternatively, there are methods by which organic wastes can be converted into crude oil or gas. If this were done, the wastes would contribute between 3 and 5 percent of the demand for crude oil.[16]

Energy has been derived from burning municipal waste in a few European cities for at least two decades, but the incinerators that are used are too small to satisfy modern requirements for cheap energy. Unfortunately, municipal wastes consist largely of paper, plastic, food scraps, bottles, aluminum cans, iron cans, and a wide variety of items varying from discarded bicycles and bedsteads to mattresses and television receivers. To recover energy from such a heterogeneous mass of material requires first that the waste be shredded by heavy-duty mechanical equipment, then passed through a classifying system that has the capability to separate the ferrous scrap by magnets and other methods. The separation system thus removes some of the noncombustible constituents for possible future recycling and makes the combustible fraction available in a shredded form that can be easily fed into the energy-recovery system.

Until recently, the only means by which municipal wastes could be used as a source of energy involved feeding the wastes into a furnace where they were burned either by themselves or together with oil or gas. These are the systems used in Europe and in a few demonstration power plants in the United States. More recently, attention has been focused on more efficient means for generating energy. Organic waste can be hydrogenated in a manner similar to the hydrogenation of coal to produce a liquid with an energy value comparable to a heavy fuel oil. The feasibility of the process has been demonstrated, and plans are underway for both expansion of the demonstration plant and construction of additional pilot plants that burn animal wastes and wood scrap.[20]

Organic materials can also be "digested" and subjected to anaerobic decomposition to produce methane. The process is basically similar to the "digestion" of sewage sludge in secondary-treatment plants, and is currently being examined for its applicability to solid organic wastes. Another system for obtaining energy from solid waste is by pyrolysis, which is the thermal decomposition of an organic material in the absence of oxygen. Several fuel forms are produced, including a gas with a low-energy content, charcoal, and tarlike oils.[16]

As with other advanced systems of energy production, the role of waste materials in supplying the country's energy needs remains to be seen. The total quantity of collectible solid waste could yield about 2 percent of the nation's total energy requirements, and about 7 percent of the electrical needs. However, it is unlikely that all available wastes can ever be made available as a source of energy, but it might not be unreasonable to assume that 25 percent could eventually be processed, in which case perhaps .5 to 1 percent of the country's energy needs might be supplied in this way. Although this is a small fraction of the total requirement, it would make a contribution to the very significant pool of energy resources that could be assembled if all of the various minor alternatives could be realized. Furthermore, with respect to organic-waste processes, there are by-product advantages: Land-fill requirements would almost completely be eliminated, and it would encourage the recycling of the nonorganic wastes.

Ocean-Thermal Gradients. Another proposal for utilizing solar energy would take advantage of the temperature differences between the warm water at the ocean's surface and the cold water at great depths. The surface temperature of the ocean in semitropical latitudes is about 77° F, and in the ocean depths the temperature is as low as 41° F.[49] This temperature difference could be used to generate electricity in a conventional heat engine using a low-boiling fluid, such as liquid ammonia, that would be heated and vaporized by the warm surface waters passed through very large heat exchangers. The expanding vapor would pass through turbines and would be returned to the liquid state in condensers cooled by water from the ocean deeps. Although the efficiency of this process is only 6.7 percent, it is nonetheless attractive because of the vast quantity of energy that could be obtained. However, the energy must be retrieved from such a dilute reservoir as to require equipment of enormous size and uncertain capital requirements. The technology is presently insufficiently developed to permit any estimate of its feasibility.

It can be expected that utilization of all sources of energy created by the sun will become increasingly attractive as time goes on. The Department of Energy estimates that all the solar-energy systems could contribute as much as 7 percent of the total energy needs of the United States by the year 2000.[43] Since, by one estimate the energy requirements in the year 2000 will be equivalent to the energy available from 26 billion barrels of oil, solar energy could replace the equivalent of 1.8 billion barrels of oil, an amount of energy equivalent to 28 percent of that consumed in the United States in 1975. Some students of the subject go much further and believe that the total requirements for energy can be met in some parts of the world by the use of solar energy.

Geothermal Energy. Geothermal energy is obtained from heat generated in the earth's crust by natural radioactivity and frictional heat from the movement of geological formations. These forces raise the temperature of the deep subterranean rocks, and in some places the heat vents to the surface in the form of steam or hot springs. In other places, the heat accumulates either as steam or hot water in permeable rock and can be released to the surface by drilling. There are also accumulations of heat in dry rocks which can be fractured by hydraulic techniques or by using explosive charges. Water can then be pumped into the crevices thus created and returned to the surface as steam.[16,52] If the temperature and pressure are sufficiently high, the steam can be utilized for driving turbo generators. Lower temperature steam or hot water can be used for domestic or commercial heating.

The feasibility of generating electricity from geothermal steam was demonstrated for the first time in 1904 at Larderello, Italy, and electricity was continuously produced in 1973 from natural dry steam. The capacity of the Larderello generators originally was 12.5 megawatts, but it was expanded to 360 megawatts.[16] Other dry-steam fields have been developed in Japan, where a 20-megawatt plant began operation at Matsukawa in 1963.[53] Commercial production of electricity in the United States was begun in the Geysers area of northern California in 1960. By 1975 about 500 megawatts were being generated there. It is intended that the generating capacity in the Geyser area will be increased by about 110 megawatts a year, and to 2,000 megawatts in 1990.[54] About 7 percent of New Zealand's needs for electricity are provided by geothermal steam, and in Iceland geothermal steam is used by 100,000 people for residential heating and the operation of greenhouses where fruits and vegetables are grown.

Estimates of geothermal energy reserves in the United States vary enormously, depending on one's assumption as to the usability of known accessible heat. Published values vary from 100 to 60,000 megawatts. The situation with respect to resources (geothermal energy that is known to exist but may be unattainable with present technology) is also uncertain, with estimates varying from 400,000 to 148 million megawatts.[16] The U.S. Geological Survey reported in 1975 that geothermal energy could be used to produce 140,000 megawatts from Gulf Coast reserves alone. They estimated that with present technology, and at prices ranging up to twice the 1975 energy cost, a total of about 165,000 megawatts could be obtained.[55]

Estimates of the geothermal capacity that can be installed by 1985 vary from 3,500 to 400,000 megawatts. These estimates, which were made by various groups in 1972 and 1973, simply illustrate the enormous uncertainties that exist concerning the availability of geothermal energy.[16] Most of the estimates agree that geothermal energy can be applied in a major way to meet the electrical demand of the western states and that by 1985 geothermal sources could supply 1 to 2 percent of the requirement for electricity in the United States. The Department of Energy geothermal development program has a goal of 7,500 to 15,000 megawatts by 1985 and 60,000 to 100,000 megawatts by the year 2000.[37,56] Time will tell how realistic these goals are.

The use of geothermal energy at the Geysers presents several problems. For one, the steam comes to the surface at a pressure of about 100 pounds per square inch and a temperature of 205° C. As a result, the turbines operate at such low temperatures and pressures as to be inherently inefficient. Another problem is typical of all geothermal facilities and is concerned with the rate at which the heat will be depleted. In the ideal situation, the heat would be withdrawn no more rapidly than it is produced underground, so that the supply would be inexhaustible. In practice, it is possible that the heat is being withdrawn more rapidly than it is produced, and for that reason geothermal energy may not always be an inexhaustible resource. In addition, geothermal energy has the potential for air and water pollution, which should not come as a surprise to anyone who has been in the vicinity of thermal springs and geysers in Yellowstone Park, Hawaii, or other parts of the world. The waters can be foul-smelling due to sulfurous emissions, and salt deposits are everywhere in evidence. The geothermal waters from the Salton Sea area of California contain six times as much salt as seawater. At Cerro Prieto, Mexico, where the waters are much less mineralized, a 1,000-megawatt generator would produce salt water containing an estimated 12,000 tons of salt per day.[19] In New Zealand, the water from the geothermal wells is known to contain significant concentrations of toxic metals including arsenic and mercury as well as hydrogen sulfide, a highly toxic gas.[57] Some of the wells are also known to emit a substantial amount of radon, a radioactive gas.

At some fields, the possibility must also be considered that removal of large quantities of underground water may cause the land to settle. This latter problem would encourage reinjection of the withdrawn water.

The role that geothermal energy will play in supplying the future needs of the United States will become clearer in the years immediately ahead. As recently as 1973, the total expenditure for geothermal research in the United States was only $4.4 million, but it is expected to reach $88 million in 1978.

Hydrogen as a Synthetic Fuel. Gaseous hydrogen has a number of interesting properties that have encouraged speculation about its use as a fuel. Pound per pound, its energy content is only about one-third that of natural gas, but its lower viscosity allows a threefold increase in pipeline capacity. When compressed into liquid form, its energy content is about 2.7 times greater than that of fuel oil.[20] Hydrogen is a highly explosive gas that was involved in the spectacular disaster to the zeppelin Hindenburg in the 1930s, but techniques for its safe handling have been improved in recent years, and large quantities of hydrogen have been shipped and used safely both in this country and abroad.

Liquid hydrogen would provide a method of storing energy produced by other means. The hydrogen could be produced electrolytically from water, or even by thermal decomposition of water in nuclear reactors operating at high temperatures. Energy would be required to produce the hydrogen, which could then be transported efficiently and used as a source of energy as needed in other places. The principal use of hydrogen might be to operate automobile engines, for which it is admirably suited because it would be a nonpolluting fuel.

DEMAND-SUPPLY PROJECTIONS

Future forecasts of energy supply and demand defy reliable analysis, in part because of the technical complexities, but also because of the unpredictability of human behavior. The shock of the fuel-oil embargo in the winter of 1973 and 1974 seemed to drive home the message that the United States needs a coherent policy that defines the manner in which alternative energy sources will be utilized, assures that the energy is used efficiently, and guarantees a maximum degree of independence of foreign sources of fuels. This was in fact the goal defined by the President of the United States in 1973, when he announced the launching of ill-fated Project Independence, a plan to achieve energy self-sufficiency by 1980. The public soon recovered from the shock of the Arab oil embargo. In addition, there was an economic recession in 1974 and 1975 and two remarkably mild winters and cool summers, which lowered the demand for energy. As a result, the public also forgot that the economic health of the nation was increasingly being threatened by deficient energy resources. Congressional inaction and public apathy delayed progress on a national energy plan through 1976. However, the winter of 1976 and 1977 proved to be as harsh as any in history. Shortages of natural gas developed in many states and resulted in the closing of schools and industrial plants. The unusually cold winter emphasized the nation's vulnerability and encouraged progress towards a national policy. The Federal Department of Energy was created to consolidate all functions that relate to energy, but development of a national energy plan proved fraught with difficult political issues and a policy had not yet evolved by late 1978.

The problem of energy forecasting is difficult enough for the years immediately ahead, but it becomes increasingly complex the further one projects into the future.[50] In 1974, the country consumed about 70 quads of energy. Demand was increasing at a rate of about 3.5 percent per year from 1950 to 1965 and by 4.5 percent per year until 1973.[9] If energy growth continues at this rate, it will be necessary to supply about 180 quads in the year 2000, which would be 2.5 times the 1974 demand. It would also be necessary to develop the mines, refineries, and power plants on such a scale that in less than 25 years the total energy supply would be 2.5 times greater than the capacity developed since the onset of the Industrial Revolution, nearly 200 years ago.

There is, however, every reason to believe that the historical trend will not continue in the United States and that the demand curve will flatten somewhat due to a slowing in the rate of population growth and to the more efficient use of energy. But to what extent? If the number of births per woman remains at about the replacement level of 2.1 children, the population will nevertheless increase to somewhere between 250 and 275 million persons by the end of the century due to the wave of post-World War II babies that have now reached adulthood. Such a population would represent an increase of 15 to 20 percent. But if the number of children per woman should average 2.8, the population would rise to about 300 million, which would be an increase of more than 30 percent. For purposes of our discussion, we will assume that the energy demand in the United States will increase to about 175 quads by the year 2000.

An initial reaction might be that the per capita energy demand should not increase. It is ecologically unsound to be using more and more energy each year. We should learn to use energy more efficiently and diminish our per capita needs. However, the per capita energy demand will probably increase to some extent because of socioeconomic improvement among the poorest third of our population. In 1970, 5 percent of all Americans and more than 20 percent of our nonwhite population lived in substandard housing. One-third of our nation is underemployed. Energy will be needed to provide additional jobs, housing, clothing, and middle-class amenities. This will offset conservation gains, at least to some extent. The demand for energy will also be affected by the general level of the economy, as well as by the price of energy itself. We must also consider the global requirements. Other countries are raising their standards of living and will compete with us increasingly for fuel.

There are also great uncertainties on the supply side of the equation. There are apparently major oil reserves to be found on the outer continental shelf, but these have not been evaluated. There is an enormous supply of energy in the oil shales, but the technology for exploiting these deposits has only begun to be developed. There are also enormous deposits of coal, but there is great uncertainty as to the rate at which the industrial capacity to mine the coal can be developed, and whether we can deal with the environmental problems involved.

Nuclear power, using light-water reactors, has been proved to be an economically viable method of producing electricity. In 1976, nuclear power in the United States supplied more than 8 percent of the total electrical requirement, which was equivalent to nearly 400 million barrels of oil per year.[43] Putting this another way, oil imports would have been about 20 percent greater had it not been for nuclear energy.

Nuclear power is, in fact, the one clean method of producing electricity now available, but it has not found public acceptance and the capital requirements for nuclear plant construction have been increasing so rapidly that they threaten to offset the financial advantages of nuclear power itself.

ENERGY CONSERVATION

For the United States to achieve self-sufficiency in its energy requirements, it must expand the supplies of fossil fuels, increase reliance on nuclear power, and emphasize conservation. During the era of cheap energy now ending, there has been little incentive to conserve. The major areas of energy consumption are the home, commercial structures, industry, and transportation. There are too few homes in which an effort is made to turn off unnecessary illumination or to minimize winter heat losses. There also are too many commercial buildings that were constructed in the 1960s and 1970s with large expanses of energy-wasteful glass, where illumination levels were overdesigned. Some buildings were designed so that windows could not be opened to take advantage of natural ventilation.

It is a relatively simple matter to examine energy usage in a middle-class home and make the judgment that 10 or 20 percent of the energy consumed could be saved by reasonable conservation practices.[58] This, however, does not mean that the demand for energy in the home will be reduced to that extent on a national basis. Once again, we must consider the needs of the less-privileged third of our population. Many of these people live in substandard homes, in which the occupants do not have the means to use energy to the extent that middle-class families do. The demands for energy conservation will require that the more fortunate families learn how to use less energy in the operation of the home. However, much of the energy so conserved will hopefully be used in the years ahead by the less fortunate, whose standard of living will climb above that of the near-poverty level.

Major opportunities for energy conservation in the home can be realized if electrical appliances and mechanical equipment can be designed to last longer. Much of the energy a family needs is consumed in the manufacture of the equipment they buy. In addition, energy is required to provide mechanical service in the course of its lifetime. The energy used per year of operation will thus be reduced if the life of such equipment can be extended, and if it can be built to require a minimum of service. Unfortunately, much of the equipment used in the home is manufactured in such way that service is impractical.

The opportunities for energy conservation in the home are numerous and obvious. Subdued lighting may not only be energy-efficient, but also less tiring on the eyes. People who are fortunate enough to have extra rooms can turn the heat off and keep the doors closed when the rooms are not in use. Air-conditioning can be used less extravagantly. By far, the largest savings can be made by reducing the room temperature to 65° or lower and keeping warm by wearing additional clothing. In one study of the opportunities by which energy could be conserved in the home, it was concluded that families could save the equivalent of nearly 1 million barrels of oil per day by 1980 if they voluntarily decreased their use of energy and if energy standards were established for new buildings and appliances.[59] The study concluded that the annual average growth of residential energy use could be reduced from the present 1.9 percent per year to .4 percent by the end of the century, and that 5 to 10 percent of the energy that is now used by families could be conserved immediately by the simple common-sense measures mentioned above, such as eliminating unnecessary illumination, setting of thermostats lower, eliminating unnecessary air-conditioning, and properly maintaining mechanical equipment. Another study has found that a vigorous conservation program could reduce the increase of residential energy use to nearly zero by the end of the century.[60] Ten to 15 percent of the energy could be conserved during the next two to five years and from 10 to 20 percent over the long term (five to 25 years).[61,62]

Industry uses about 40 percent of the energy consumed in the United States. About 65 percent of this energy is used to supply heat for industrial processes, and about 25 percent for production of electric energy to drive machines. Until recently, the cost of energy has been so low that there has been little incentive for industry to conserve fuels except perhaps in highly energy intensive processes such as in the electrochemical industries. However, it has been estimated that during the next 25 years it should be possible for the industrial sector to reduce its requirements by 30 to 50 percent. Since industry is currently using about 40 percent of all energy consumed in the United States, industrial conservation could result in a saving of 12 to 20 percent of the total consumed in the United States.

Similar savings can be achieved in the commercial sector (including hotels, banks, and commercial and government buildings), which accounts for about 15 percent of the national energy consumption. The energy is used mainly for heating and cooling, but also for illumination, operation of elevators, and office machinery. It is estimated that between 15 to 30 percent of the energy used by the commercial sector could be conserved by the end of the century.[63] This would result in a reduction of from 2 to 5 percent in the national energy use.

About 25 percent of the energy consumed in the United States is expended for transportation purposes of all kinds, and 13 percent is consumed by automobiles. Almost all of this energy is supplied by fossil fuels. Since there are at present only limited applications of electricity for transportation, energy from hydroelectric and nuclear-energy sources could not be applied for transportation in any significant way.

There has been much talk about the need to place less reliance on the automobile, but for the time being the love affair between the citizens of technologically advanced countries and their cars seems bound to continue. By far the greatest savings would be achieved if people would travel less. Encouragement of recreational activities close to home, substitution of electronic telecommunication for business conferences, and locating jobs closer to homes, could achieve enormous fuel economies, apart from the fact that there would be great savings in time, as well as a reduction in the loss of life and limb. Substitution of mass transportation for the individual automobile would also be a major contribution. However, we are dealing here with a basic characteristic of the contemporary way of life that is not likely to be changed easily. This is the age of the private automobile: It gives an individual convenience, status, privacy, and the means to satisfy his wanderlust. As the population moves from the cities to the suburbs, it becomes even harder to reduce dependence on the automobile because mass transportation becomes increasingly difficult to provide.

OPTIONS FOR THE FUTURE

As we look into the future, the year 2000 is a convenient cut-off point, not only because it is the beginning of a new millennium, but because some of the new methods of producing energy will become available by that time, whereas others will not be feasible until a much later date.

Our future energy options, in addition to greater use of coal and nuclear power, include controlled thermonuclear reactions, solar energy for space heating, solar energy conversion to electricity, the breeder reactor, and possibly other sources of energy such as could be derived from temperature gradients in the oceans, geothermal sources, winds, or the use of organic wastes or vegetation. The breeder reactor is sufficiently advanced in its development to assure continued use of nuclear energy, provided the environmental objections can be met. However, it is not likely to be in general use until the end of the century. The future of controlled thermonuclear reactions is somewhat uncertain and, in any case, is not likely to be a major energy producer until well into the twenty-first century. Solar energy may be a major source of energy for heating homes and commercial buildings by the year 2000.

Most of the energy for the rest of this century must be supplied from the conventional sources: coal, oil, gas, and nuclear. However, the total impact of the various new options may begin to be substantial by the end of the century, as can be seen from Table 6–2 which is

TABLE 6-2
Projected Availability of Energy From Certain
New Sources in Year 2000

Source	Energy Production in Year 2000 (Quads)
Oil shale	7.3
Geothermal	4.4
Solar-electric	3.0
Solar heating and cooling	5.9
Waste materials conversion	4.9
Energy from vegetation	1.4
Total	26.9

based on ERDA's projections made in 1976.[43] New sources of energy will account for about 26.9 quads by the year 2000. Assuming that the demand for energy in the United States at that time to be 150 quads, the new sources will account for about 18 percent of the national requirements. Coal gasification and liquification are expected to make a significant additional contribution (14 quads), and oil shales may provide 7.3 quads. The total of geothermal, solar, waste conversion and vegetation conversion may provide nearly 20 quads, or about 13 percent of demand. The advanced systems of energy generation will thus be making a meaningful contribution to our requirements by the end of the century, but the bulk of our needs must continue to be provided by nuclear and fossil fuels. Because of the diminution of domestic oil and gas production, we will rely increasingly on nuclear power and coal.

The oil embargo of late 1973 stimulated a number of studies of our energy needs and the options available for meeting them through the year 2000. One of the most extensive of these studies was that conducted by the Energy Policy Project of the Ford Foundation and the approach taken in that study will be summarized for purposes of illustration.[9] Three alternative assumptions of future energy growth were considered, as shown in Figure 6–12. First, a "historical-growth pattern" scenario assumed that energy demand will grow as it has in the past, reaching a level of a little more than 180 quads in the year 2000. A second, "technical-fix" scenario, assumes growth comparable to historical growth in the level and mix of goods and services, but growth associated with a major national effort to use energy efficiently. The study concluded that, compared to the nearly 200 percent

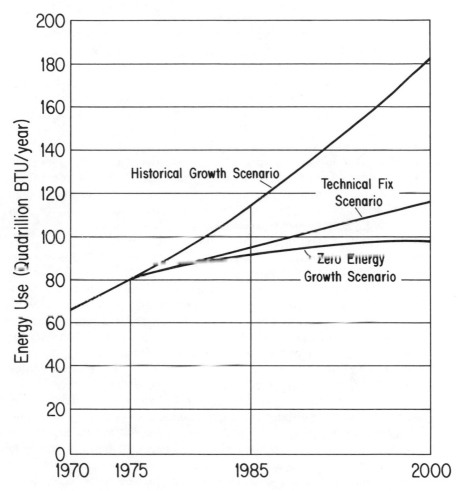

Figure 6-12. Three projected energy consumption patterns to the year 2000.[9]

increase in energy consumption required by the "business-as-usual" historical-growth scenario, the increased energy demand can be reduced to about 45 percent with a determined effort to use energy-saving technologies. The third scenario is that of zero growth, which is not in fact "zero growth": it actually does allow 10 percent more energy per person by the year 2000, at which time national energy consumption would be 100 quads, about 25 percent greater than at present.

The "historical-growth" scenario is probably unrealistic because the price of energy must continue to increase, and this will provide incentives to increase the efficiency with which energy is used. The "technical-fix" scenario, which would require a 45 percent increase in energy supply compared to 120 percent for the "historical-growth" assumption, does not deprive anyone of the benefits of energy. It simply assumes that the energy is used efficiently, and that energy waste is avoided. The "zero-growth" scenario requires that major changes in lifestyle would be required that could only be dictated by a strong central government. Cities would have to be redesigned, and mass-transportation systems installed. Energy-intensive industries would be discouraged. Experience has shown that, with respect to energy, Americans are set in their ways and that the government finds it politically impractical to make major decisions concerning energy policy. Perhaps an energy ethic may develop in the population that would make zero growth possible, but it will grow slowly, and it is not likely to have a significant impact before the end of this century. By the year 2000 the energy requirement for the United States may be somewhat less than the 180 quads estimated on the basis of historical growth, but it will surely be more than the 124 quads estimated in the Ford Foundation's[1] "technical-fix" scenario. In contrast, an independent analysis by the Electric Power Research Institute expects a demand for 150 quads, of which 53 percent would be for electricity.[51]

Most of this chapter has been concerned with the energy needs of the United States. However, the problems of the United States cannot be solved without considering the needs of the world. The United States needs energy to increase the standard of living of its underprivileged, maintain the high standard of the majority, and permit moderate expansion of its population and GNP. But three-fourths of the world's population needs energy in order to raise its standard of living, provide sufficient education, increase agricultural productivity, and develop industries. Only when these goals have been achieved in the developing countries will population stability and agricultural sufficiency be possible. Ample supplies of energy, along with foreign capital and education, will provide the means to bring about the kind of world about which mankind has dreamed for centuries, a world where every newborn child will have an equal opportunity to live no less than three score and ten years, with adequate food, adequate housing, and productive employment.

NOTES

1. Makhijani, Arjun. "Energy and Agriculture in the Third World: A Report to the Energy Policy Project of the Ford Foundation," Ballinger Publishing Co., Cambridge, Mass. (1975).

2. Starr, Chauncey and Stanford Field. "Energy Use Proficiency: The Validity of Interregional Comparisons," Electric Power Research Institute, Palo Alto, Cal. (March, 1977).

2a. Darmstadter, Joel, Joy Dunkerly and Jack Ackerman. "How Industrial Societies Use Energy," The Johns Hopkins Univ. Press, Baltimore, Md. (1977).

3. The Rand Corporation, "The Impact of Electricity Price Increases on Income Groups," Rand Corporation *R-1102-NSF/CSA* (March, 1973).

4. Brown, Harrison, James Bonner and John Weir. "The Next Hundred Years," Weidenfeld and Nicolson, London (1957).

5. Hubbert, M. King. "Energy Resources," Report to the Committee on Natural Resources of NAS-NRC, *Publication 1000-D,* National Academy of Sciences-National Research Council, Washington, D. C. (1962).

6. Hubbert, M. King. "Energy Resources," in: "Resources and Man," W. H. Freeman and Co., San Francisco, Cal. (1969), p. 204.

7. National Academy of Sciences. "Rapid Population Growth: Consequences and Policy Implications," NAS Study Committee of the Office of the Foreign Secretary, Johns Hopkins Univ. Press, Baltimore, Md. (1971).

8. Ehrlich, Paul R. and Anne H. Ehrlich. "Population Resources Environment: Issues in Human Ecology," 2nd ed., W. H. Freeman and Co., San Francisco, Cal. (1972).

9. Ford Foundation. "Exploring Energy Choices: A Preliminary Report," The Ford Foundation Energy Policy Project, Washington, D. C. (1974).

10. U. S. Department of Interior. "Energy Perspectives: A Presentation of Major Energy and Energy-Related Data," Washington, D. C. (1975).

11. National Safety Council. "Accidents Facts," Chicago, Ill. (1974).

12. National Academy of Sciences. "Mineral Resources and the Environment," Supplementary Report: Coal Workers' Pneumoconiosis-Medical Considerations, Some Social Implications. Committee on Mineral Resources and the Environment and Commission on Natural Resources, National Research Council. Washington, D.C. (1976).

13. Enterline, Philip E. "A Review of Mortality Data for American Coal Miners," in: "Coal Workers' Pneumoconiosis" (I.J. Selikoff, M.M. Key and D.H.K. Lee, eds.), *Annals of the New York Academy of Sciences.* 200:260 (1972).

14. Dessauer, P., E. J., Baier, G. M., Crawford and J. A. Beatty. "Development of Patterns of Coal Workers' Pneumoconiosis in Pennsylvania and Its Association with Respiratory Impairment," in: "Coal Workers' Pneumoconiosis" (I. J. Selikoff, M. M. Key and D. H. K. Lee, eds.), *Annals of the New York Academy of Sciences* 200:220 (1972).

15. National Academy of Sciences. "Rehabilitation Potential of Western Coal Lands," Study committee on the Potential for Rehabilitating Lands Surface Mined for Coal in the Western United States, Environmental Studies Board, Ballinger Publishing Co., Cambridge, Mass. (1974).

16. University of Oklahoma. "Energy Alternatives: A Comparative Analysis," Univ. of Oklahoma Science and Public Policy Program, Norman, Oklahoma (1975).

17. Environmental Science and Technology. "How the Coal Slurry Pipeline in Arizona is Working," Vol. 10, No. 12 (November, 1976), pp. 1086–1087.

18. Osborn, E. F. "Coal and the Present Energy Situation," *Science* 183:477 (1974).

19. National Academy of Engineering. "U.S. Energy Prospects: An Engineering Viewpoint," National Academy of Engineering Task Force on Energy, National Academy of Sciences, Washington D. C. (1974).

20. Hammond, A. L., W. D. Metz and T. H. Maugh. "Energy and the Future," American Association for the Advancement of Science, Washington, D. C. (1973).

21. Squires, A. M. "Clean Fuels from Coal Gasification," *Science* 184:340–346 (1974).

22. Hammond, Allen L. "Coal Research (II): Gasification Faces an Uncertain Future," *Science* 193:750–753 (1976).

23. Osborn, E. F. "Coal and The Present Energy Situation," *Science* 183:477 (1974).

24. Office of Energy Policy and Planning. "The National Energy Plan," U.S. Government Printing Office, Washington, D.C. (1977).

25. Council on Environmental Quality. "Sixth Annual Report," U.S. Government Printing Office, Washington, D. C. (1975).

26. Council on Environmental Quality, "Fifth Annual Report," U.S. Government Printing Office (1974), p. 106.

27. Franssen, H. T. "Towards Project Interdependence: Energy in the Coming Decade," prepared for the Joint Committee on Atomic Energy, U. S. Congress. U.S. Government Printing Office, Washington, D.C. (December, 1975).

28. Kash, D. E., I. L. White, K. H. Bergey, M. A. Chartock, M. D. Devine, R. L. Leonard, S. N. Salomon and H. W. Young. "Energy under the Oceans: A Technology Assessment of Outer Continental Shelf Oil and Gas Operations," Univ. of Oklahoma Press, Norman, Okla. (1973).

29. Travers, William B. and Percy R. Luney. "Drilling, Tankers, and Oil Spills on the Atlantic Outer Continental Shelf," *Science* 194:791–795 (1976).

30. Lieberman, M. A. "United States Uranium Resources—An Analysis of Historical Data," *Science* 192:431 (1976).

31. Ford Foundation. "Nuclear Power Issues and Choices," Report of the Nuclear Energy Policy Study Group, Ballinger Publishing Co., Cambridge, Mass. (1977).

32. Post, R. F. and F. L. Ribe. "Fusion Reactors as Future Energy Sources," *Science* 186 (4162):397 (1974).

33. Electric Power Research Institute. "Assessment of Laser-Driven Fusion: Final Report," *EPRI ER-203*, Palo Alto, Cal. (September, 1976).

34. Isaacson, L. K. "Laser-Fusion Program: Summary Report," *EPRI SR-9*, Special Report, Electric Power Research Institute, Palo Alto, Cal. (1975)

35. Metz, W. D. "What Can the Academic Community Do?" *Science* 184:273–278 (1974).

36. Landsberg, H. H. "Low-Cost, Abundant Energy: Paradise Lost?" *Science* 184:247–253 (1974).
37. Joint Committee on Atomic Energy. "ERDA Authorizing Legislation, Fiscal Year 1977." JCAE Hearing, 2nd Session, 94th Congress, on Overall Budget. Part 1, Vol. I: Hearing and Appendices 1 to 11 (partial), U.S. Government Printing Office, Washington, D. C. (1976).
38. U.S. ERDA. "Information from ERDA: Reference Information" (Special Issue), Washington D.C. (May, 1977).
39. Spencer, Dwain F. "Solar Energy," *EPRI Research Progress Report No. FF2*, Electric Power Research Institute, Palo Alto, Ca. (January, 1976).
40. Duffie, J. A. and W. A. Beckman. "Solar Heating and Cooling," *Science* 191(4223:143 (1976).
41. Wolf, M. "Solar Energy Utilization by Physical Methods," *Science* 184(4134):382 (1974).
42. Meyer, J. W. "Why Citicorp Didn't," *Solar Age* (April, 1977), p. 14.
43. U.S. ERDA. "ERDA Selects 34 Solar Demonstration Projects" (76-9), *Weekly Announcements* 2(13):2 (1976).
44. Bos, Piet B. "Solar Realities," *EPRI Journal* 1(1):6 (1976).
45. Wade, Nicholas. "Windmills: The Resurrection of an Ancient Energy Technology," in: "Energy: Use Conservation and Supply" (Philip H. Abelson, ed.), Washington, D.C. (1974), p. 128.
46. Calvin, M. "Solar Energy by Photosynthesis," *Science* 184 (4134):375 (1974).
47. Reed, T. B. and R. M. Lerner. "Methanol: a Versatile Fuel for Immediate Use," *Science* 182:1299 (1973).
48. Anderson, L. L. "Energy Potential from Organic Wastes: A Review of the Quantities and Sources," *Bureau of Mines Information Circular 8549*, U.S. Government Printing Office, Washington, D.C. (1972).
49. Othmer, D. F. and O. A. Roels. "Power, Fresh Water, and Food from Cold, Deep Sea Water," *Science* 182:121 (1973).
50. Starr, Chauncey. "Electricity Needs to the Year 2000." Presented to the Subcommittee on Energy Research Development, and Demonstration House Committee on Science and Technology," Washington, D. C. (February 26, 1976).
51. Starr, Chauncey. "The Year 2000: Energy Enough?" *EPRI Journal* 1(5):6–13 (June, 1976).
52. Ellis, A. J. "Geothermal Systems and Power Development," *American Scientist* 63:510–521 (1975).
53. Robson, G. R. "Geothermal Electricity Production," *Science* 184(4134):371 (1974).
54. "Nuclear Developments," *Electrical World* (October 15, 1977), p. 32.
55. U. S. Department of Interior. "Assessment of Geothermal Resources of the United States—1975," *Geological Survey Circular 726* (D. E. White and D. L. Williams, eds), U.S. Geological Survey, Arlington, Va. (1975).
56. Atomic Industrial Forum, Inc. "Geothermal Energy," *AIF Background Info*, Atomic Industrial Forum, Inc., Washington, D. C. (November, 1975).

57. Axtmann, Robert C. "Environmental Impact of a Geothermal Power Plant," *Science* 187(4179):795 (1975).
58. Darmstadter, J. "Conserving Energy: Prospects and Opportunities in the New York Region," Johns Hopkins Univ. Press, Baltimore, Md. (1975).
59. Rand Corporation. "Energy Use and Conservation in the Residential Sector: A Regional Analysis," *R-1641-NSF,* Santa Monica, Cal. (1975).
60. Hirst, Eric. "Residential Energy Use Alternatives: 1976 to 2000," *Science* 194:1247–1252 (1976).
61. Smith, C. B. "An Energy-Constrained Society," in: "Efficient Energy Use" (C. B. Smith, ed.), Pergamon Press, Elmsford, N.Y. (1976).
62. Fazzolare, R. "Industrial Energy Use," in: "Efficient Energy Use" (C. B. Smith, ed.), Pergamon Press, Elmsford, N.Y. (1976).
63. Taussig, R. T. and C. B. Smith. "Commercial Energy Use," in: "Efficient Energy Use" (C. B. Smith, ed.), Pergamon Press, Elmsford, N.Y. (1976).

PART III

Environmental Contamination:
Some Contemporary Issues

CHAPTER 7

Some General Principles

The next several chapters will be devoted to discussions of certain environmental issues that have received major attention in recent years. Many of the subjects have caused intense debate, with marked polarization of views and disagreements as to facts. The subjects are often complex, and involve scientific, legal, psychological, and political factors that are not always easily separated. While it will not be possible to discuss all contemporary issues, the sample will be sufficiently representative to illustrate many of the basic features that apply generally.

The discussions for much of the remainder of this book will be concerned generally with environmental toxicology, a subject that deals with the sources of toxic substances, the manner in which they reach man after passing through intricate ecological pathways, and their biological effects, with particular emphasis on human health. The problem of environmental pollution has attracted increasing interest in recent years as advancing technology has produced new substances and radiations that affect living beings. The results of this technology have made the subject of environmental health more complex with each passing year. New and potentially toxic chemical substances are being produced at an increasingly rapid rate. One of the first textbooks on the subject of industrial disease, published in the United States in 1914, included only 67 toxic substances.[1] By 1969 a standard reference work listed 17,000.[2] The 1970 Occupational Health and Safety Act required the Department of Health, Education and Welfare (HEW) to compile a complete list of toxic substances used in industry. By 1974, the list contained 13,000 substances, and it is estimated that the number will reach 100,000 when the compilation is complete.[3] A recent list of chemicals suspected of being capable of causing cancer includes 1,500 substances,[4] many of which the public is exposed to in everyday life: pharmaceuticals, food additives, pesticides, dyes, detergents, lubricants, soaps, and the atmospheric and liquid wastes from a myriad of industrial processes.

In addition to toxic substances, the public is increasingly exposed to new physical agents of disease, among which are the ionizing radiations from radioactive substances and X-ray machines, microwaves from radar equipment and electronic ovens, lasers, and ultrasound in medical procedures.

Another factor that adds urgency to the subject is the ability of modern industry to invent a new product, identify its applications, and market it on a worldwide basis in a relatively short period of time. Within a few years after the discovery of DDT, this potent pesticide was being marketed and used throughout the world. Populations everywhere were exposed to it as a contaminant of food, water, and air. This was not so in former times, when new products were produced in relatively modest quantities by modern standards, and were marketed in limited areas, often close to the place of manufacture.

Artificially produced chemicals are often added to food in the course of agricultural practice, food processing, or home cooking. These food additives provide benefits that range from the trivial (a substance that adds color to sugar or preserved fruit) to the important (food additives that prevent food spoilage during transportation and storage, or which diminish spoilage or waste by the actions of insects or rodents). Whether it is socially advantageous to use a given chemical for a specified purpose should depend, as in the case of a pharmaceutical, on a comparison of the benefits and risks. Unfortunately, the benefit-risk equation can rarely be quantified because we know all too little about the risks in many cases.

It will help to understand some of the ramifications of environmental toxicology if we first review some of the physical, biological, and chemical factors that are involved when a contaminant passes from the environment into the human body. Inhalation and ingestion are the principal routes by which pollution affects human health.

THE ATMOSPHERE

Depending on a person's level of activity, an adult will inhale between 25 and 50 pounds of air per day. The importance of air as a vehicle for conveying toxic substances to the body can be appreciated when contrasted with the fact that an adult drinks only about one kilogram (2.2 pounds) of water per day, and eats about the same weight of food.

The earth's atmosphere contains about 78 percent nitrogen, 21 percent oxygen, and .03 percent carbon dioxide. Additionally, there are minor constituents such as hydrogen, argon, neon, and xenon that are present in much lower concentrations. Carbon dioxide is a product of respiration, fermentation, and combustion. (We will see in Chapter 14 that because people burn such vast quantities of coal, oil, and gas, the carbon dioxide content of the atmosphere has been gradually rising, a matter that must be watched carefully because of possible future effects on the world's climate). The atmosphere also contains variable amounts of water vapor, considerable quantities of dusts, both natural and man-made, and gases in the form of vapors, mists, fumes, and smokes.

Under *normal* conditions of temperature and pressure, a true gas exists only in the gaseous form. Oxygen, hydrogen, and nitrogen are such examples. However, sometimes a substance exists in liquid or solid form under normal conditions but is in equilibrium with a vapor phase present in the atmosphere. Water vapor is an example with which everyone is familiar. Many liquids used in industry also exist both in liquid and vapor phases that are potentially harmful to employees exposed to them. These include gasoline, paints, solvents, degreasing agents, alcoholic products, and literally thousands of organic substances used in the chemical manufacturing industry. However, very few of these vapors are introduced to the general atmosphere in sufficient concentration to be of concern to the general population. One important exception is the vapor from gasoline, which interacts with other atmospheric constituents under the influence of sunlight. It produces a class of compounds known as oxidants, some of which are irritating to the eyes, and may cause other deleterious effects. Gasoline vapor is an important precursor of the smog associated with automobile usage.

Liquids, when dispersed as mists, are another class of atmospheric particulates, of which fog, the naturally occurring mist of water, is a well-known example. Until recently, there were few important examples of noxious artifically produced mists to which the general public was exposed as air pollutants. However, aerosol dispensers have become popular in recent years for application of many substances, including household paints, pesticides, and cosmetics. Exposure to artifically produced mists is thus now more commonplace in the home. Although it has not been demonstrated that this exposure is injurious to health, the practice of using "aerosol cans" certainly represents a major new source of artificially produced air pollution. The compressed gas used as the propellant is an inert gas (Freon) that has been accumulating in the atmosphere to such an extent that atmospheric chemists are concerned it may allow increased transmission of ultraviolet light to the earth's surface as a result of its interaction with the protective ozone layer.

Another class of atmospheric particulates are smokes and fumes, produced by condensation of combustion products. Tobacco smoke, the smoke from chimneys, and the visible emissions from automobile tailpipes are examples of this class of pollutants.

Finally, we come to dust, which is the form of airborne particulate produced mainly by mechanical action, such as abrasion or crushing.

The various particulate forms of mists, smokes, fumes, and dust are classed together as aerosols. Once dispersed, the physical and biological behavior of an aerosol is markedly influenced by its particle size. (The unit of measurement is the micron, equal to 1 millionth of a meter, or about one twenty-five thousandth of an inch. A period on this page is about 500 microns in diameter. The smallest particle that can be seen comfortably with the unaided eye is 100 microns.) When a dust particle is sufficiently small, it can become suspended in the atmosphere for periods of time ranging from minutes to months, depending on its density and meteorological factors.

A 10-micron particle of quartz, a common mineral substance from which most beach sands are formed, and an important constituent of the earth's crust, settles through the atmosphere at a rate of about a foot per second, or about 60 feet per minute. Thus, a 10-micron quartz particle will not remain suspended for more than a few minutes unless released to the atmosphere from a very great height. In contrast, the settling rate of a quartz particle having a diameter of about .1 micron is imperceptible, and such a particle would remain suspended indefinitely were it not for the fact that the electrical properties of dusts tend to favor the attachment of the very small particles to larger particles that do settle more rapidly. Dust particles can also be washed out of the atmosphere by rain.

When dusts are first formed, as in a rock-shattering explosion of dynamite, the "particles" first produced will constitute a wide spectrum of sizes ranging from large chunks that fall to the surface immediately, to the more finely divided particles that remain suspended for long periods of time. Persons exposed to the cloud of dust from a few minutes to an hour after it is produced are likely to inhale more large particles, that is particles larger than 5 micrometers in diameter. Such persons are likely to be those employed in the immediate vicinity of the source of dust. Members of the general public, are more likely to be exposed to a dust cloud from which all but the very fine particles will have settled. Because the larger particles tend to settle so quickly, more than 50 percent of the mass of particles suspended in the general atmosphere are usually less than about one micrometer in diameter. However, because the mass of an individual particle increases in proportion to the cube of its diameter, a single 10-micron particle will weigh as much as 1,000 1-micron partricles, or 1 million .1-micron particles. Although half of the mass of particles may be smaller than 1 micron, and the other half larger than 1 micron, the *number* of particles in the fraction weighing less than one micron will be very much larger than the number of particles in the larger fraction.

This has significance from the point of view of the effects of dust on human health. The anatomy of the respiratory tract (Figure 7–1[5]) is such that the region of the lung in which inhaled dust is deposited, and its subsequent behavior within the body, is largely determined by the size of the inhaled particles.[6] When inhaled particles larger than about 5 micrometers are passing through the tortuous passageways of the nose, pharynx, trachea, and upper bronchial region, they tend to be centrifuged by the turbulent motions of the air stream to the mucous-lined walls of these structures. The tracheal and bronchial linings are provided with a layer of cells having hairlike appendages that wave in a coordinated manner so as to escalate the mucous film upwards to the mouth. This lining is the ciliated epithelium, one of the major defenses of the lung against inhaled dust or other foreign objects such as bacteria that land on the surfaces of the upper respiratory tract. In a matter of hours, dust deposited in this way is transported out of the lung with mucous and is secreted into the esophagus and usually swallowed. This is a major protective mechanism by which the upper respiratory tract is cleansed of the deposited particulates.

Particles smaller than 1 or 2 micrometers penetrate deeper into the lung structure and may reach the aveolar region, which is constructed of microscopic air sacs (aveoli) lined with capillaries that transport blood and lymph to and from the lung. The alveoli are designed to facilitate transfer of gases between the blood and air. Although they are individually microscopic in size, they are very numerous and the surface area available for gas exchange is enormous. It is estimated that the total area of the alveolar surfaces within the human lung is about 1,000 square meters. This large surface area permits oxygen to pass readily from the inhaled air to the blood. The carbon dioxide that is produced as a waste substance in the tissues of the body, passes from the blood to the alveolar spaces from which it is eliminated in each expired breath.

Some of the particles that penetrate to the alveolar spaces are removed by a type of white blood cell (phagocyte) that has

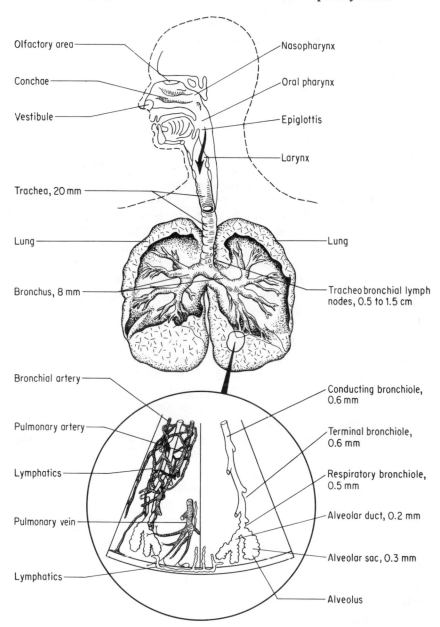

Figure 7-1. The principal anatomical features of the human respiratory tract.[5]

amoeba-like properties that permit it to envelop a dust particle and transport it either to the bronchial tree for mucociliary clearance, or into lympathic capillaries in the alveolar walls. The dust transported to the alveolar capillaries in this way can be concentrated into focal accumulations surrounding the smaller bronchial passageways.

Dust particles that are insoluble or of low solubility can be transported to lymph glands that filter and accumulate them. Dust that is soluble dissolves in the moist linings of the respiratory tract and is absorbed directly into the blood.

The behavior of pollutants in the atmosphere depends, not only on their particle size, but also on their chemical properties and their tendencies to react with other constituents of the atmosphere. When the pollutant is inhaled, its behavior within the body varies, depending on metabolic factors. Thus, many organic compounds accumulate in the liver, while others accumulate in fatty tissue, and calcium-like substances, such as radioactive strontium, are deposited in bone.

THE SOIL-FOOD PATHWAYS

Air pollutants that are not inhaled directly can be absorbed into the human body via several other pathways, as shown in Figure 7–2.[7] The contaminant can deposit on soil, from which it can then be absorbed by crops that are later ingested by humans. The contaminant also can deposit directly on the exposed surfaces of standing crops that are consumed by humans, or it can pass from soil to plants that are eaten by animals which then transmit the contaminant to humans in milk or meat. The air-grass-cow-milk pathway is a particularly important route of human absorption. A cow consumes grass from as much as 1,600 square feet of soil per day, and many trace substances consumed with the forage can be secreted in the milk in relatively concentrated form.

Soils consist of mineral and organic matter, water, and air arranged in a complicated physiochemical system to which plants have marvelously adapted to fulfil their nutritive requirements.[8] Most of the nutrient chemicals required for plant growth are not contained within the soil particles, nor are they dissolved in the soil water. Rather, these nutrients are tightly bound, in atomic form, to the surfaces of the soil particles. As a matter of fact, dissolved nutrients would not remain long in soil but would be leached out by rainfall or flood water were it not for the extraordinary ability of the soil clays to bind many substances in ionic form. The particles contained in a cubic foot of rich soil have a total surface area of more than 500,000 square feet. The nutrients held on the particle surfaces are available to plant roots,

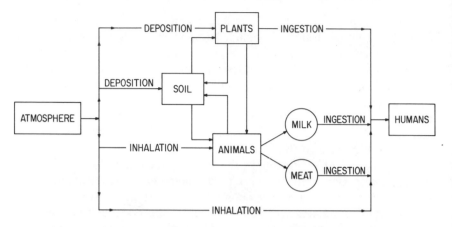

Figure 7-2. Pathways by which radioactive materials released to the atmosphere can affect man.[7]

which are similarly well contrived for providing tiny surfaces that come in contact with individual soil particles. It has been estimated that the roots of a single winter rye plant have a surface area of more than 6,800 square feet!

Although almost every chemical element is present in all soils, only 16 are considered necessary for the growth and reproduction of plants. These are carbon, hydrogen, oxygen, nitrogen, phosphorous, sulfur, potassium, calcium, magnesium, iron, manganese, zinc, copper, molybdenum, boron, and chlorine. The plant derives all these elements from soil except carbon, hydrogen, and oxygen, all of which are supplied by the atmosphere. Some trace elements not essential to plant growth can be readily absorbed, but others are so tightly bound to the clay fractions that their availability is greatly reduced. For example, during the period when nuclear weapons were being tested in the atmosphere, fallout added both strontium-90 and cesium-137 to the soil. Strontium-90, which is chemically similar to calcium, was readily absorbed from the soil by plants and ultimately appeared in the milk of grazing cows. Some strontium-90 also reached cows milk after it had fallen on plants, but the most important pathway to milk was via the roots. Thus, strontium-90 persisted as a contaminant of milk long after nuclear-weapons testing was curtailed, since the soil contained a substantial reservoir of this radioactive substance, which has a 28-year half-life. Cesium-137, is chemically similar to potassium. However, it binds so tightly to soil particles that it is not readily absorbed by the plants. Cesium-137 did pass to the milk of grazing cows during the period when fallout was actually occurring because it was deposited on the plant surfaces, and thereafter consumed by cows.[9] However, after the weapons-testing programs were curtailed, the presence of this radioactive material in milk diminished rapidly, because the reservoir in soil was not available to the plants.

THE PATHWAYS FROM WATER TO HUMANS

Pollutants can be discharged as liquid wastes and reach the human body in a variety of ways.

If the pollutant is introduced to the aquatic environment in particulate form, it can be consumed by small organisms, or become attached to plant surfaces. Pollutants in solution can adsorb on suspended organic and inorganic solids or be assimilated by the plants or other biota. The suspended solids, dead biota, or excreta can settle to the bottom and become part of the organic substrate that supports the community of organisms living in the sediments. The specific route by which a pollutant reaches man depends on the chemical and physical properties of the pollutant and on the particular ecological system into which the pollutant is introduced. (See Figure 7–3.[9]) The figure illustrates diagrammatically two possible pathways by which a contaminant can pass stepwise from water to fish. As the contaminants pass from one level to the next (these are called the trophic levels) the quantity of pollutant per gram of biological substances can either be

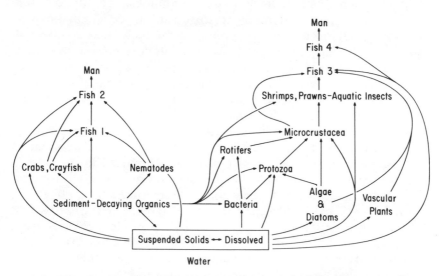

Figure 7-3. Possible routes for the transfer of trace metal through an estuarine ecosystem.

increased or diminished. Mollusks, such as clams and oysters, concentrate many trace substances because they feed by filtering large volumes of water. Thus mollusks grown in fresh water can concentrate the metal manganese by a factor of about 300,000. That is to say, if the water contains one part per *billion* of manganese, the flesh of mollusks at equilibrium with their environment could contain about 300 parts per *million*. However, this is an extreme case, and concentration factors rarely approach so high a level. In some cases, the concentration factor is less than one, which is another way of saying that biological discrimination impedes the passage of the substance up the food chain, as it does against radioactive strontium.[9] Discrimination against this substance takes place at each step from grass to cows milk to human bone because the strontium is progressively diluted by a preference for calcium, an element that is chemically similar to strontium.

SOME BASIC PRINCIPLES OF TOXICOLOGY

The effect of exposure to a toxic substance varies from species to species and also depends on a wide variety of factors, including the size of the dose, the chemical properties of the substance, and the manner in which the dose is administered. Knowledge of the effects of toxic substances usually comes either from human experience or from laboratory experiments. There are inherent difficulties in applying information about the lower animals to humans, because different species may respond differently to toxic substances. Animal experiments are useful to identify the organs of the body that are affected, and the mechanisms involved. However, the dose-response relationships can best be obtained from human experience if such experience is available. There are, of course, ethical considerations that limit the extent to which humans can be knowingly exposed to toxic substances. Unfortunately, there have been all too many tragedies in which exposure occurred either because of ignorance of the effects of the substance, or due to accidents. Yet much valuable information has become available as a result of their being used in the past in industry without precautions. Lead and mercury were used for centuries before studies were undertaken to determine their effects on industrial workers. The harmful effects of radium and asbestos first became known in this century, and the effects of vinyl chloride were discovered only in the past decade. These are only a few examples of toxic or carcinogenic (cancer-causing) substances, the effects of which were first observed among industrial workers.

Toxic effects can be divided initially into those that result from short-term exposure to large doses of a substance (acute exposure) and those due to small doses administered over a long period of time (chronic exposure). For example, inhalation of high doses of certain beryllium compounds for only a few minutes produces an acute lung reaction similar to pneumonia. This reaction usually occurs within a day or two after relatively heavy exposure and, if the patient survives, complete recovery can take place in a matter of weeks. However, exposure over a period of many years to very low concentrations of this metal can produce another type of lung disease that has long-term effects that may prove fatal.[10]

It is important to distinguish between toxic effects that are observed immediately, and those that become apparent only after a long latent period. Carbon monoxide produces effects within minutes after exposure. However, when the exposure is terminated, the chemical is promptly eliminated from the body. There are no delayed reactions so far as is known, unless the exposure is lethal or near lethal, in which case permanent brain damage can result. In contrast, many chemicals may cause no effect at the time of initial exposure, but may produce toxic reactions after many years. Beryllium poisoning has been known to develop decades after workers inhale relatively low concentrations for brief periods of time. Cancers caused by exposure to chemicals may develop 30 or 40 years after exposure. The fact that some toxic and carcinogenic effects may have long latent periods provides one of the most difficult challenges in evaluating the risks from compounds to which humans are exposed.

Another factor that must be considered is *synergism*, a phenomenon in which the biological effect from exposure to two compounds is greater than the additive effects of exposure when the substances are administered separately. It has been observed that the incidence of lung cancer among people who smoke and are exposed to asbestos dust or to the radioactive gas, radon, is much greater than among nonsmokers similarly exposed. Smoking during exposure to the carcinogens increases the susceptibility of the person, and the effect is greater than the additive effects of smoking and exposure to the carcinogen. That is, the substances are synergistic: One intensifies the effect of the other.

STANDARDS OF PERMISSIBLE EXPOSURE

The first standards designed to protect industrial workers were established early in this century. They were established for lead, mercury, arsenic, certain mineral dusts, and organic solvents. These standards have come to be called threshold limit values, and they define the permissible limits for exposure during a 40-hour working week.

When human experience is not available, threshold limit value will be set on the basis of experiments with laboratory animals and to some extent by analogy with chemicals of similar properties and for which human experience is available.

There are a number of reasons why standards of permissible exposure for the general public should be more conservative than standards established to protect industrial workers. Workers are not ordinarily exposed to contaminants for more than 40 hours per week, whereas pollution of community air or water can expose people continuously, seven days a week, 24 hours per day. Moreover, industrial employment does not begin until a person is nearing adulthood and it usually terminates in early old age, whereas it is known that some pollutants have their most pronounced effects on the very young and the very old. Furthermore, the population of industrial workers is less likely to include individuals who have physical disabilities that make them unusually sensitive to pollutants. This is particularly true for air pollutants that affect the cardiorespiratory system. Exposure to air pollutants consisting of acid gases and mists, such as sulfur dioxide or sulfuric acid, for example, tend to act as bronchial constrictants, may have insignificant effects on healthy persons but prove fatal to someone with advanced cardiovascular disease. Additionally, some toxic substances can pass across the placenta to the fetus, which is particularly sensitive to some chemicals.[11] These are among the reasons why the limits of exposure to toxic substances for the general public are usually set by regulatory authorities at between one-tenth and one-hundredth of those permitted for occupationally exposed individuals.

It is important to understand that a toxic effect may occur, yet not be observed. Whether an effect is observed depends on the number of persons exposed, the probability of injury, and the frequency with which such injuries occur spontaneously in the general population. Certain occupational diseases such as silicosis, asbestosis, and manganese poisoning do not occur in the general population. Other diseases such as leukemia, certain liver dysfunctions, and lung cancer do occur in the general population, but in addition are also known to result from exposure to chemicals in the workroom. For example, a number of industrial chemicals are capable of causing lung and bladder cancer, and ionizing radiation can cause bone cancer and leukemia. These kinds of cancers occur in the general population as well.

Let us assume that we have a work population of 2,000 workers (a large number to be exposed to a highly toxic substance in significant concentrations) who are exposed to asbestos dust. The 2,000 workers can be screened by routine physical examinations and a single case of asbestosis can be identified as being of occupational origin, since this disease is only known to occur among asbestos workers. Asbestosis is a pneumoconisosis (dust disease of the lung) caused only by exposure to asbestos dust. It should not be confused with the forms of cancer resulting from exposure to asbestos which may also involve the lung.

If the workers are exposed to radiation or to chemicals capable of causing leukemia the problem becomes more complicated. Leukemia is a blood disease that occurs normally in the general population at a frequency of about 60 cases per year per million persons. Thus, a population of 2,000 workers would be expected to have .12 cases of leukemia per year, or roughly one case every eight years, due to the normal occurrence of the disease. But the statistics of small numbers permit considerable variability from year to year, and the cases would probably vary from zero cases in some years to one, two, or even more cases in other years. Moreover, the "natural" incidence of leukemia is influenced by many ethnic, geographical, and socioeconomic factors that are not well understood, so that the normal incidence for the group could be more or less than the assumed incidence.

If the incidence of leukemia in this group of 2,000 workers should be increased by 50 percent, the expected incidence would increase from .12 cases to .18 cases per year. This is a substantial difference, but the point is that it would not be detectable *above the expected rate* except after an impractically long period of observation.

Statistical difficulties place the industrial epidemiologist in a dilemma. He may not be able to detect an effect, but he cannot be certain there is *no* effect. One possible way out of the dilemma is to consider the risk acceptable if the effect occurs so infrequently as to be undetectable against the normal frequency with which the effect occurs spontaneously in the general population. This does not mean to say that the level of permissible exposure would be deliberately established to allow occasional cases to occur. It does mean that in the absence of information to the contrary, a substance can be judged to be sufficiently safe if the effects of exposure cannot be observed for the reasons given. The only alternative would be to avoid technological innovations that expose employees to chemicals whose *absolute* safety has not been proved. Unfortunately, although this is a desirable goal in the abstract, it is not feasible in practice since no techniques exist for establishing the absolute safety of a chemical. Many substances that have proved to be toxic to humans were given a clean bill of health on the basis of animal experiments.

So much for those effects that cannot be detected because they occur so infrequently as to be lost in the "noise level," the frequency with which the effects develop spontaneously in the general population. What about diseases that do not occur normally in the general population, but are caused by exposure to environmental contaminants? The difficulty of setting standards for such substances depends to a large degree on whether the effects are seen immediately or are long delayed. As noted earlier, the effects of carbon monoxide develop within a matter of minutes. Thus, information can be gathered quickly, and standards of permissible exposure established that will prevent injury from acute exposure.

In contrast, the effects of exposure to some chemicals may not become apparent for many years. This is particularly true of those capable of producing cancer (carcinogens). Vinyl chloride, a chemical used in plastics manufacturing, has been found to be capable of causing a rare type of liver cancer after many years of occupational exposure. Here is a case where animal experimentation initially provided no clue that this rare tumor could be caused by vinyl chloride, and many years passed before the first human case was seen. Thus, when use of a new chemical is contemplated in industry, even extensive animal research beforehand cannot always provide assurances that the substance will be innocuous when humans are exposed. (See Chapter 8.) It may be reasonable to consider a risk acceptable for an industrial worker if the effects occur so infrequently that they cannot be differentiated from those that occur naturally. However, this cannot be followed as a criterion for the general population.

The problem of justifying permissible risk to members of the general public is far more complicated. Above all, we are dealing with a much larger population than the industrial segment. Returning again to leukemia, at a mortality rate of 60 cases per million of population per year, about 13,000 cases of leukemia would be expected to occur each year in the United States. The probability of developing leukemia in any randomly selected individual is thus about one in 16,000 per year. This is much less than the probability of dying in a motor vehicle accident (one in 4,000) and somewhat less than the probability of dying from a fall (one in 10,000). It is about twice the risk of drowning, and one-eighth the risk of dying in an accident of any kind in any given year. Thus, the probability of developing leukemia is sufficiently low that an individual is likely to be less concerned about the risk of developing the disease than about the prospect of dying in an automobile accident or from a fall. Many persons will not likely be greatly concerned if the risk of leukemia is increased by 50 percent. However, the number of persons at risk in the general community is so very large that a 50 percent increase in the incidence of leukemia would be a matter of considerable concern on the part of the public generally, and the health authorities. A 50 percent incease in the annual incidence of leukemia in the United States (an additional 30 cases per million persons) would result in an additional 6500 cases per year. The individual may not be concerned about the added personal risk, but from the point of view of a health official, the matter would warrant prompt application of preventive methods.

Another difference so far as the public is concerned is that the people rarely are given the opportunity to decide whether a risk is acceptable. The risks are usually imposed without the public's consent or knowledge. Moreover, the risk-benefit relationships may not be easily evaluated because neither the risk nor the benefits are always amenable to quantification. In many cases the risk is taken by one person but the benefit accrues to another. If a plant manufactures a product that exposes people in its neighborhood to toxic chemical wastes, the risk may not be offset by any benefit perceptible to the individual. He has no choice but to accept the risk or move to another neighborhood, which would be inconvenient and costly. If the person works in the plant, or uses its products, he may receive adequate benefits in exchange for what *he* perceives as a slight risk. But if he does not work at the plant and does not use its products, the individual may not perceive any benefits. Yet there may be benefits that are not apparent. The plant may pay substantial taxes, or it may ship its products to a second factory in another city, where the product is incorporated in some way into goods that are sold back to the first community.

In contemporary America, there is a widespread intolerance of any risk from air or water pollution or from substances that may become incorporated into food, either deliberately or inadvertently, due to agricultural or manufacturing practices. This attitude presents great practical difficulties in setting standards and administering environmental protection programs. Unfortunately, it is not possible for the environmental toxicologist to accept a target of "zero risk."

The classical relationship between the dose of any noxious agent and the effect it produces is shown in curve *A*, Figure 7–4. The S-shaped curve, which mathematicians call a sigmoid, has certain important characteristics. Along the horizontal axis, there is a certain value below which there is essentially no effect. As the dose is increased above this point, the effects become manifest very slowly at first, but then increase more rapidly until the curve again flattens out because all of the susceptible members of the population have been

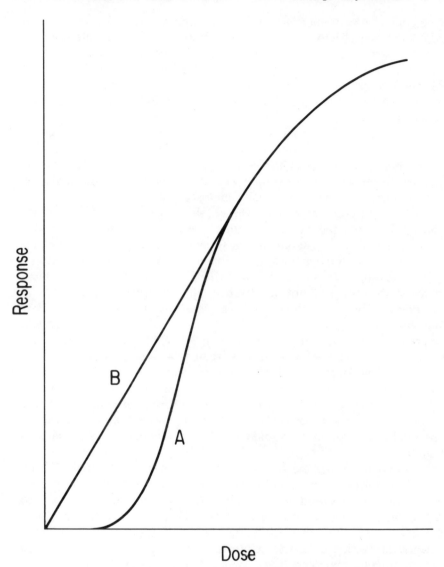

Figure 7-4. Two possible dose-response relationships: *A*, classic sigmoid response, with threshold; *B*, linear response, with no threshold. Other possibilities clearly include curve *A* with no threshold and curve *B* with a threshold. It is also possible that over a limited portion of the two curves, the shapes could be adjusted so that one would be statistically indistinguishable from the other.

affected. According to this curve, there is a threshold dose below which no effect is observed.

In recent years, this classical relationship has been challenged for a variety of reasons. The information on which these relationships are based has been obtained from studies that involve relatively high doses, and either by using experimental animals, or by studying industrial workers. For certain of the radiations, as well as for many chemicals, there is evidence that the relationship between dose and response may in fact, be linear (as is shown in the straight line *B* of Figure 7–4) and that there is no threshold. In other words, the effect is proportional to dose, and any dose, however small, has a finite effect. It is also possible for the S curve to intercept the point of zero effect only at zero dosage. If there would be no threshold, any exposure, however small, would be expected to affect the more sensitive persons in the general population. If there is no threshold, there is no dose that is absolutely safe. The question that must then be answered is not "what dose is 'safe?'" but "how 'safe' is 'safe enough?'"

The shapes of the dose-response curves at low concentrations are perhaps the most perplexing questions facing modern toxicologists. The central problem is that the effects of low doses can be determined only by studying impractically large populations over long periods of time. For example, in most parts of the world, the average person receives a dose of about .1 rad per year due to naturally occurring sources of radioactivity. (The rad is a unit of ionizing-radiation dose. Although natural radioactivity results in a whole-body dose of about .1 rad per year in most parts of the world, the lung receives a higher dose because of the presence in the atmosphere of a naturally occurring chain of radioactive substances derived from radon, a radioactive gas that emanates continuously from the surface of the earth.)

In certain parts of India and Brazil, people are exposed to about .5 rad per year due to the presence of unusually high concentration of naturally occurring radioactive minerals in the soil. It would seem that such localities would offer excellent opportunities to determine if radiation doses at these elevated levels are capable of increasing the incidence of leukemia. It is well established that leukemia is more prevalent among the survivors of Hiroshima and Nagasaki, and an elevated incidence of leukemia has been observed among patients who received heavy doses of X-rays for treatment of a disease known as ankylosing spondylitis. From these and other studies it has been estimated that leukemia develops in an irradiated population at a rate of about 20 cases per million persons per rad. These can develop over a 15- to 20-year period following exposure.

We should bear in mind that this information was obtained at high doses of the order of hundreds of rads. If one assumes that leukemia is produced at a rate proportional to dose, one can calculate that if people are being exposed to .5 rad per year (about five times the dose the average person receives from nature), one year of exposure should produce about ten cases per million persons, distributed over 15 to 20 years. One can then calculate the increased incidence that might be anticipated in a population with a lifetime exposure of .5 rad. It can be shown that in order to demonstrate a 20 percent increase in incidence, it would be necessary to accumulate 1 million person-years of experience. That is, it would be necessary to observe 1 million people for one year, or 100,000 people for ten years.[12] The populations exposed to five times the normal level in Brazil and India are too small to permit studies on this scale, even if the funds could be obtained and if good medical records were available (which they are not).

It should be noted that this approach assumes that the effect (in this case, production of leukemia) is independent of the *rate* at which the dose is delivered. The survivors at Hiroshima or Nagasaki were exposed for a few seconds and the spondylitics were exposed for a matter of minutes. Theoretical reasons, as well as evidence derived from experience with laboratory animals, strongly suggest that the effect per unit of radiation exposure for the types of radiation considered here is somewhat less when the exposure is protracted, as it is in natural radioactivity. In other words, the *rate* at which the dose is delivered has an influence on the effect. If this is so, correspondingly larger populations would be required for study.

NOTES

1. Thompson, W. Gilman. "The Occupational Diseases," D. Appleton and Co., New York and London (1914).
2. Gleason, M. N., R. E. Gosselin, H. C. Hodge and R. P. Smith. "Clinical Toxicology of Commercial Products: Acute Poisoning," 3rd ed., Williams and Wilkins Co., Baltimore, Md. (1969).
3. U. S. Department of Health, Education and Welfare. "Facts of Life and Death," *DHEW Pub. No. (HRA) 74–1222*, U. S. Government Printing Office, Washington, D.C. (1974).
4. Christensen, H. E., T. T. Luginbyhl and B. S. Carroll "Suspected Carcinogens: A Subfile of the NIOSH Toxic Substances List," National Institute for Occupational Safety and Health, U.S. Department of Health, Education and Welfare, U.S. Government Printing Office, Washington, D.C. (June 1975).
5. National Academy of Sciences-National Research Council. "Effects of Inhaled Radioactive Particles," *NAS-NRC Publ. No. 848* (1961).

6. Lee, D. H. K. (ed.). "Environmental Factors in Respiratory Disease," Academic Press, New York (1972).
7. Environmental Protection Agency. "Estimates of Ionizing Radiation Doses in the United States 1960–2000," Report of Special Studies Group, Division of Criteria and Standards, Office of Radiation Programs (1972).
8. U.S. Department of Agriculture. "Soils," Washington, D.C. (1957).
9. Eisenbud, M. "Environmental Radioactivity," 2nd ed., Academic Press, New York (1973).
10. Stokinger, H. E. "Beryllium: Its Industrial Hygiene Aspects," Academic Press, New York (1966).
11. Harada, Masazumi. "Intrauterine Poisoning: Clinical and Epidemiological Studies and Significance of Problem," *Bulletin of the Institute of Constitutional Medicine* XXV (Suppl):1–60 (Mar. 25, 1976).
12. Albert, R. E. and R. E. Shore. "Fundamental Epidemiological Considerations," in: *Proceedings of the International Symposium on Areas of High Natural Radioactivity*, Academia Brasileira de Ciencias, Rio de Janeiro, Brazil (1977), pp. 15–17.

CHAPTER 8

Environmental Cancer

Cancer is a group of many diseases that can attack almost all tissues of the body. All cancers break down the stability of cell populations, and cells begin to proliferate without control. A cancer may exist for years at its place of formation in benign form without doing perceptible harm, but it may eventually become malignant and invade other tissues (metastasize). When this happens, the likelihood for survival diminishes greatly.

It is possible that most people beyond middle age are living with *in situ* cancers of which their physicians are unaware. For example, the annual incidence of known prostate cancer among 70-year-old men is about 200 cases per 100,000, or .2 percent per year. However, routine autopsies of 70-year-old men who died of other causes have revealed microscopic evidence of prostate cancer in 15 to 20 percent of the cases.[1]

The most common cancers are known as carcinomas, and start in the epithelial tissues that line organs such as the lung, breast, skin, and stomach. Another type of cancer is the sarcoma, which develops in connective tissue in bone, the liver, or certain other organs. The several kinds of leukemias are a third form, and involve the mechanisms by which white blood cells are produced. The major cancers can be differentiated, not only as to the organ in which the cancer develops but also with respect to cell type. All together, there are more than 100 kinds of cancer, but most of them are comparatively rare. Three, of the lung, large intestine, and the female breast, account for nearly half of all cancer deaths in the United States.

The causes of the cancers are not fully understood, but there is ample evidence that the majority are caused in some way by environmental factors. The World Health Organization and others have analyzed the epidemiology of cancer and have concluded that 60 to 90 percent of all cancers are due to chemical, physical, and perhaps biological factors in our environment.[2-6] The factors that cause cancer are known as *carcinogens*.

The WHO estimates have been widely popularized in the press and environmental literature, and are frequently misunderstood to imply that 60 to 90 percent of the cancers are due to man-made environmental factors, such as air and water pollution, or food additives. This conclusion is not warranted, and was not suggested by the original authors of the estimates. Many cancers are caused by natural factors, and tobacco smoke has for many years been the leading cause of cancer in men. These facts are frequently ignored by writers who simply say that "60 to 90 percent of cancers are due to environmental factors" in a context that implies that the responsible factors are associated with industrial pollution.[7-12]

Even the professional journals have been known to be misleading. A review article in a publication of the American Chemical Society has said:

Fact:
There can be no cancer without a cancer-causing agent. As many as 90% of all cancers may be caused by environmental factors—a substantial portion of which are chemicals—and these cancers are potentially preventable.[9]

Another typical example of a well-meaning but nevertheless misleading statement is the following, which appeared in the newsletter of a leading environmental organization:

Finally, we must face the fact that the "chemical revolution" of the past fifty years appears to be one of the chief factors behind the rapid rise in the incidence of cancer. And we may only be seeing the tip of the iceberg, because most of the suspected chemical carcinogens did not come into widespread use until after World War II. . . .
But if 8 out of every 10 cancer cases are attributable to environmental factors, then it is obvious that cancer is not, except in a minority of cases, inevitable.[7]

Statements such as the above serve to distort a rational approach to cancer prevention. There is no evidence that the increased use of chemicals during the past 50 years is "one of the chief factors behind the rapid rise in the incidence of cancer." The trends in cancer incidence and mortality are due to many factors, of which the practice of cigarette smoking is the foremost. Other major causes are sunlight, diet, and factors about which we know next to nothing. However, there is every indication that they are of natural origin or are rooted in cultural traits other than those associated with the use of chemicals. The evidence does not suggest that cancer is predominantly due to air and water pollution or to cancer-producing substances which we add to our foods.

It is estimated that cigarette smoking caused the deaths of 81,000 persons from lung cancer in the United States in 1975, and accounted for 22 percent of all cancer deaths among males.[13] There is also epidemiological evidence that other cancers, such as of the stomach, colon, pancreas, breast, and cervix, are due to natural factors related to diet or personal hygiene. While we cannot exclude pollution, food additives, and medicinal drugs as being carcinogenic, the fact remains that epidemiological research has not been able to establish more than a minor influence for these factors up to the present time. Additionally, the suggestion has been made, from time to time, that natural radioactivity is the cause of some cancers, but this is not supported by studies of the relationships between cancer incidence and the levels of natural radioactivity exposure, which is known to vary widely.[14,15] It remains possible that some cancers are due to natural radioactivity, but if so, they are too few to be observed.

One important exception are the cancers of occupational origin. It is well established that cancer-producing chemicals can affect the workers who are exposed to them, and occupational exposure to carcinogenic chemicals may be responsible for 1 to 3 percent of all cancer deaths.[5,6,16]

One of the great concerns of public health officials is that cancers caused by environmental factors may not develop for several decades after exposure. Skin cancers due to the effect of sunlight frequently do not appear until people are in their sixties or seventies. Cancers among chemical workers may not develop for 20 or 30 years. Thus, the effects of a carcinogen introduced into the environment in a manner capable of causing cancer in the general community may not be experienced for several decades, by which time many members of the community will be irreversibly destined to develop the disease. Thus, while the facts may be misrepresented by some, there is good reason to be concerned about the future role of pollution as a cause of cancer.

CANCER TRENDS

In recent decades, cancer has emerged as a major cause of death in the United States, second only to diseases of the heart.[17] This is in contrast to the situation at the beginning of the century, when the cancers were eighth on the list of causes of death. The increasing importance of cancer is both relative and absolute. It has moved higher on the list primarily because the infectious diseases have come under control since 1900, when the Bureau of the Census first reported on cancer mortality rates.[18] In 1900 cancer was responsible for only 3.7 percent of all deaths in the United States compared to 16.5 percent in 1975. However, 11 percent of all deaths were then due to tuberculosis, compared to .4 percent in recent years. Deaths from many other infectious diseases have been similarly reduced. For this reason, diseases that have not come under control have become more prominent on the list of death causes. However, during this period the rate of cancer mortality has increased more than 250 percent, from 64 per 100,000 in 1900 to 162 per 100,000 in 1970.[19] This has been due in part to the older age of the 1970 population, and in part to better medical services that provide more reliable information on causes of death. Public health records in 1900 left much to be desired, and because the population was predominantly rural, a higher percentage of deaths occurred at home under conditions that most assuredly resulted in many cancers going undiagnosed. The importance of this factor is not known quantitatively.

The cancer trends for males and females in the United States from 1930 to 1975 are summarized in Figure 8-1A and 8-1B.[13] Lung cancer in men, a comparatively rare disease in 1930, has increased 20 times, and the upward trend appears to be continuing. Lung cancer among women increased only gradually until the late 1950s, but has recently been increasing more rapidly. The rise in lung cancer among men has been due to cigarette smoking and the increasing use of tobacco among women, beginning around World War II, is now beginning to take its toll. For reasons that are not understood, but which are believed to be linked in some way to dietary factors, stomach cancer has decreased dramatically in both males and females. The death rate from cancer of the uterus has decreased about two-thirds. The death rates from other types of cancer have either remained about the same or increased slightly during this period. Two exceptions are cancers of the pancreas (which we will see later is due, in part, to smoking), and of the colon and rectum. Some of these have increased about twofold, but the rates are so low that they do not contribute in a major way to total cancer mortality.

The net effect of these trends on total cancer mortality is that deaths from cancer have increased among both white and black males, but have decreased slightly among white females between 1935 and 1970 (see Figure 8-2).[17] This is due to the substantial reduction in the incidence of stomach and uterine cancer. The upward trend, beginning about 1965, is due to the increased incidence of lung

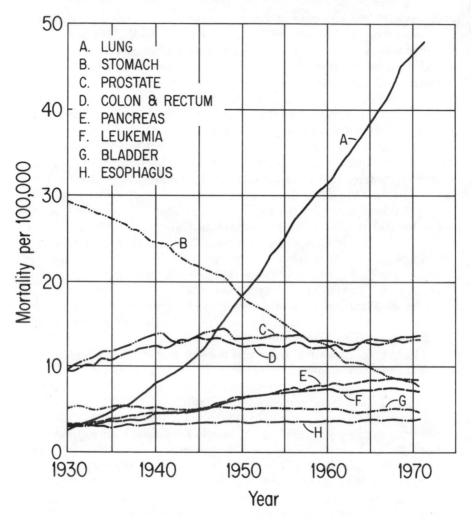

Figure 8-1A. Male cancer death rates in the United States, 1930–1971 (standardized for age in the population of the United States in 1940).[13]

cancer. The increase among men can also be largely attributed to cigarette smoking, but occupational exposure may also be a factor, as we will see later in this chapter.

Until recently, it was widely believed that cancer is an inevitable consequence of old age and that "if we lived long enough, we would all die of cancer." The relationship of cancer to age is, in fact, so important that it explains, in part, why cancer mortality has increased so remarkably in this century. In 1900, the average life expectancy at birth was only 47.3 years, compared to 70.9 years in 1970. Since the death rate from cancer at 70 years of age is about ten times the rate at age 47, it is immediately apparent that the cancer-mortality rate should be greater now than it was at the beginning of the century. It is for this reason that it is necessary to standardize cancer statistics for purposes of comparison from year to year. However, the higher incidence of cancer among the old is not the result of increasing susceptibility, but rather that the cancer-causing agent is given a longer period of time to complete the process of cancer production.[20]

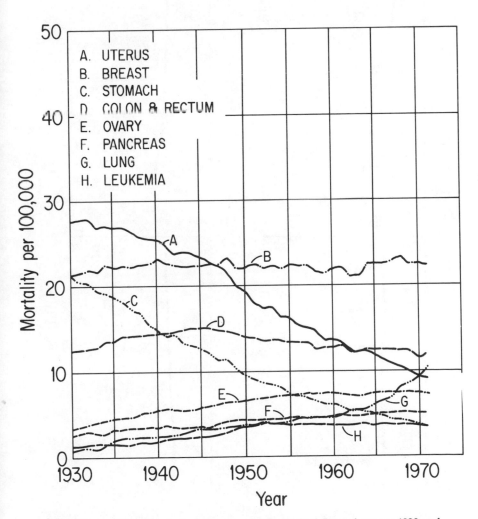

Figure 8-1B. Female cancer death rates in the United States between 1930 and 1971 (standardized for age in the population of the United States in 1940).[13]

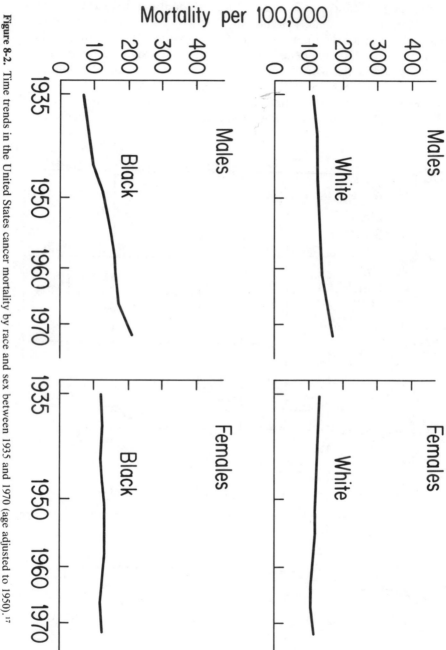

Figure 8-2. Time trends in the United States cancer mortality by race and sex between 1935 and 1970 (age adjusted to 1950).[17]

There are a number of cancers of unknown cause that occur most frequently in children. Leukemia is most common around four years of age among whites, but not among nonwhites. Wilm's tumor, a cancer of the kidney, occurs most frequently in five-year-olds.[21] These and other tumors of early childhood currently cause greater mortality at age five than at age 15. Beyond that age mortality increases, as is shown in Figure 8–3.[22] The reduction in childhood mortality due to cancer is the result of improvements in treating the two major cancers of early life, Wilm's tumor and leukemia.[23,24]

The rate of cancer mortality from all causes increased substantially among adults in the United States during the 30 years prior to 1974, but the higher total figure for cancer deaths appears to be due to the increase in lung cancer, which, in turn, is due primarily to cigarette smoking. This is shown in Figure 8–4,[22] in which cancer mortality among white males in the United States is plotted against age for the periods 1943 and 1944 and 1973 and 1974. The lower curves show age-specific mortality for the United States due to lung cancer for the same two-year periods. The well-known increase during the past 30 years is evident at all ages above about 45 years. When the 30-year increase in lung cancer is subtracted from the curve of total cancer mortality for 1973 and 1974, it is seen that the mortality from cancers other than lung cancer is slightly lower now than it was 30 years ago. Lung cancer among women is not as yet so pronounced as in men, but if the statistics are considered separately from the cancers due to smoking, a more pronounced reduction in cancer mortality can be demonstrated in women than in men, due to the combined effect of lower mortality from both stomach and uterine cancer.

EVIDENCE FOR ENVIRONMENTAL CAUSES OF CANCER

Interest in the environmental causes of cancer was stimulated in the 1940s by knowledge that cigarettes were playing a role in the increasing incidence of lung cancer, and that cancers of the lung, bladder, bone, and blood-forming tissues were sometimes due to occupational factors. There are undoubtedly many carcinogens in the natural environment, but only a few have been identified. It is likely that natural carcinogens explain many of the wide geographical variations that have been observed in the occurrence of cancer. For example, it has been known for a century that chewers of betel nuts, which are valued for their stimulatory effect by people of Southeast Asia, have a high incidence of cancer of the mouth, larynx, and esophagus.[25] A more dramatic example is the high incidence of cancer of the esophagus along the southern shores of the Caspian Sea in Iran.

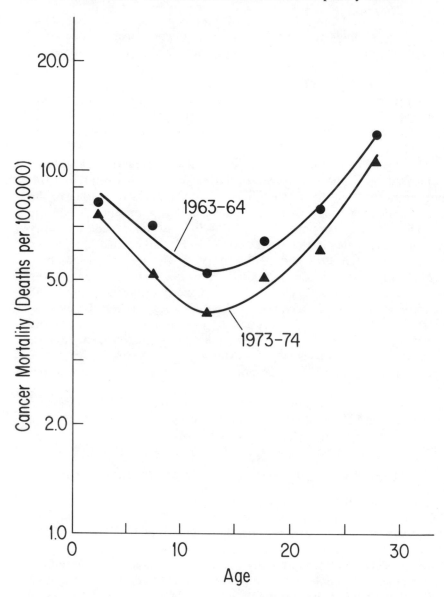

Figure 8-3. Comparison of cancer mortality among young white females in the United States between 1963-1964 and 1973-1974.[22]

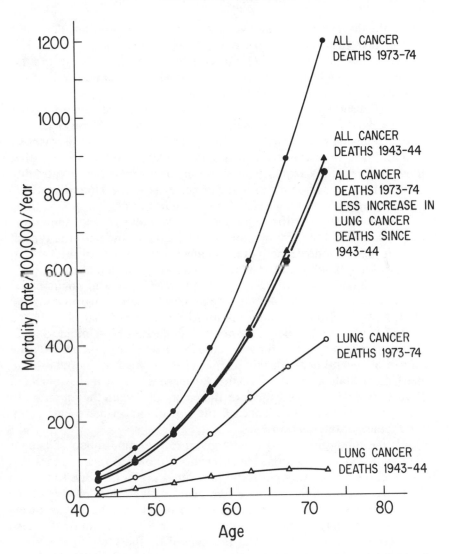

Figure 8-4. Effects of deaths from respiratory tract cancer between 1943-44 and 1973-74 on total cancer mortality among white males in the United States. When the increase in lung cancer is subtracted from the 1973-74 curve of the total cancer mortality, the residual falls slightly below the curve of total cancer mortality in 1943-44.

There the disease occurs among women at a rate more than 125 times greater than the incidence in women living in the United States. The cause of this anomaly is not known, but since the people are far removed from industry, grow their own food locally, and rarely drink or smoke, the cause is believed to be some factor in the soil.[26]

It was observed as recently as 1944 that liver cancer among blacks in the United States was occurring at a rate very much lower than in Africa, showing that African blacks do not have a genetic predisposition to this form of cancer, as had been supposed previously, but that they were probably exposed to some environmental factor (possibly aflotoxins) in their homeland that did not exist in the United States.[27] This finding initiated a number of other studies of migrant populations that tended to support the hypothesis that whereas recent immigrants to a country tend to develop cancers according to the patterns typical of their home country, the pattern changes with time, and the susceptibility of migrants to cancer eventually approaches that of the populations in the country in which they have relocated. A well-documented example is to be found among Japanese who have migrated to the United States. Stomach cancer is more common in Japan than in the United States, but when Japanese migrate to the United States, the frequency with which they develop this disease is reduced, over the course of several generations, to that of the rest of the population in the United States. In addition, the Japanese living in their homeland have a low incidence of cancer of the colon, but when they migrate to the United States the incidence of this type of cancer increases within the lifetime of the migrant.[5]

At one time, great weight was given to the importance of genetic factors in cancer induction, but it has been shown in recent years that many cancers once thought to be hereditary are in fact due to natural and or cultural influences. The one major exception is cancer of the skin, which is by far the most common of all cancers but is rarely fatal, due in a large measure to the ease with which it can be detected and treated. Most skin cancers are caused by the ultraviolet component of sunlight, so that they can, in fact, be classified as being due to an environmental factor. However, there is an important indirect genetic component, since light-skinned people are far more susceptible than those with pigmented skins, and skin pigmentation is determined largely by heredity.[28]

Circumcision also plays a role in the epidemiology of cancer, as indicated by the near-absence of cancer of the penis among Jews and Moslems. The causative factor is believed to be the smegma secreted by the foreskin of the penis. The smegma may also play a role in the development of cancer of the uterine cervix, a type of cancer that is classified as being due to an environmental chemical—but one that occurs naturally, and which can be prevented by circumcision. The incidence of cancer of the cervix correlates inversely with socioeconomic status, suggesting that personal hygiene on the part of the male may also play a role.

In countries like the United States, where the population is heterogeneous in ethnic and religious origin, marked differences in cancer incidences can be observed among the various groups. Seventh-Day Adventists, an evangelical religious sect of which about 100,000 live in California, have a reduced incidence of several cancers. This sect prohibits smoking as well as use of alcohol, and encourages a milk, egg, and vegetarian diet. There is evidence that this diet protects against colon cancer.[30] Mormons, who likewise abstain from tobacco and alcohol, also have a lower incidence of many cancers. The predominantly Mormon state of Utah has a cancer incidence that is three-fourths that of the United States generally and is the lowest of any state.

Some intriguing relationships between cancer and the cultural environment have been shown to exist. For example, the per capita annual consumption of beer in several countries is shown to correlate with the aged-adjusted annual incidence of rectal cancer. (See Figure 8–5.[31]) A similar relationship can be shown when rectal cancer is plotted against per capita beer consumption in the individual states of the United States.[31] One must be very careful, however, in interpreting data of this kind, since the association between rectal cancer and beer consumption may be an association that is linked by other factors. For example, a close correlation can be shown between per capita energy production and cancer of the colon.[32] Similar correlations have been shown between energy and cancer of the prostate, rectum, and lung, but there is no correlation between energy use and cancer of the stomach. There is no explanation for these relationships. In some way, the advanced cultures that produce the most energy tend to adopt dietary patterns, use food additives, or become exposed to environmental pollution so as to influence the incidence of certain cancers. Whatever the causes may be, the data suggest that the "spontaneous" incidences of cancers of the colon, to take one example, must be close to zero, and that almost all such cancers can be attributed to environmental factors.

Tobacco Smoking. The practice of cigarette smoking is almost totally responsible for the increase in cancer mortality in the United States in recent years. Many other countries of the world are now also in the midst of similar lung-cancer epidemics.

The practice of smoking cigarettes in the United States and in much

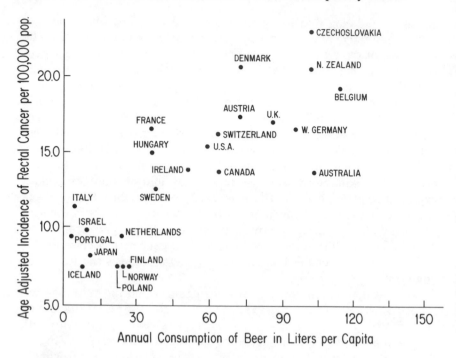

Figure 8-5. Relationship between per capita beer consumption in 1970 and estimated average annual age-adjusted incidence rates between 1960 and 1972 for rectal cancer among males in 24 countries. This association must be interpreted with caution. A similar correlation does not hold for distilled alcoholic beverages.[31]

of Western Europe increased rapidly beginning about 1920, and in late 1938 German and United States investigators observed independently that lung cancers constituted 10 to 15 percent of all carcinomas and that almost all lung-cancer patients had a history of heavy smoking.[33,34] The increasing trend was first seen in males, among whom the practice of smoking increased rapidly in the years during and after World War I. The women of that period were not yet indulging in some of the pleasures that their liberated daughters would enjoy in the years ahead, and lung cancer did not begin to increase significantly among women until the 1950s.

The rise in both male and female lung cancers since 1900, and the relationship between cigarette consumption and lung cancer, is illustrated in Figure 8–6.[1] The lung-cancer incidence among men has increased about 15 times since 1930, during which time there has been proportionately a much smaller increase among women. However, the rate of increase for women has changed rapidly since the early 1960s, reflecting their greater use of cigarettes since World War II.[35]

The studies linking cigarette smoking to lung cancer have been thorough, and the findings are unequivocal. By 1950, the relationship between lung cancer and smoking had been demonstrated in large-scale studies in which the smoking habits of patients with lung cancer were investigated. It was clearly demonstrated that those members of the population who developed lung cancer smoked more heavily than those who did not develop lung cancer.[36] These studies were followed by a number of prospective investigations, the most conclusive of which was investigation by the American Cancer Society of more than a million men and women. Their smoking habits were carefully documented, and they were followed for about five years.[37] A surprising finding was that, in addition to markedly higher rates of cancer of the lung and other portions of the respiratory tract, there was significantly higher mortality due to leukemia and cancers of the bladder,

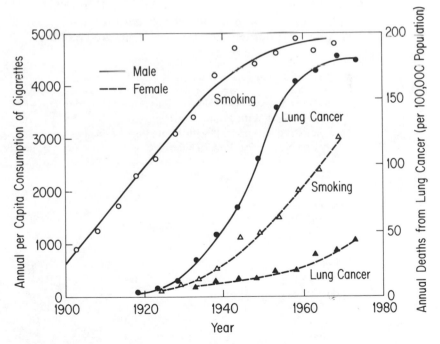

Figure 8-6. Relationships between cigarette consumption and lung cancer mortality for men and women. Tobacco smoking became popular among women a generation later than among men, and the rise in lung cancer was similarly delayed.[1]

kidney, pancreas, liver, and stomach. Mortality from all cancers among smokers was more than twice as high as among nonsmokers. The essential conclusions of the American Cancer Society study are presented in Table 8–1.

Additional studies by the American Cancer Society showed remarkably consistent relationships between the cancer mortality and the number of cigarettes smoked per day, the age when smoking began, and the extent to which the smoke is inhaled (Table 8–2).[25] A man who smokes 40 or more cigarettes per day is subject to about 19 times the risk of dying of lung cancer, compared to a man who has never smoked regularly.

More recent studies have revealed associations between cigarette smoking and diseases other than cancer. In addition to 62,000 tobacco-related cancer deaths, it was estimated that in 1967, smoking caused 129,000 deaths due to arteriosclerotic heart disease and other circulatory impairments, with an additional 30,000 deaths due to bronchitis and emphysema, stomach and duodenal ulcers, cirrhosis of the liver, and influenza, and pneumonia. *Cigarette smoking in 1967 was estimated to account for 18 percent of all male cancer deaths and 27 percent of all deaths due to arteriosclerotic heart diseases.* Stated in another way, if it were not for the effects of cigarette smoking, 3.7

TABLE 8–1[37]

Ratio of Cancer Mortality Among Men
Ages 45–64 With History of Cigarette
Smoking, Compared with Men Who
Never Smoked Regularly

Type of Cancer	Ratio
Lung	7.8
Mouth and pharynx	9.9
Larynx	6.1
Esophagus	4.2
Bladder	2.0
Kidney	1.4
Pancreas	2.7
Liver	2.8
Stomach	1.4
Leukemia	1.4
All cancers	2.2

(Hammond, 1966.)[37]

TABLE 8–2[25]

Lung Cancer Mortality Ratios Among Men
Aged 35–84 Years, According to Current
Number of Cigarettes Smoked per Day,
Degree of Inhalation, and Age Began Smoking

Current no. cigarettes/day	Ratio
1–9	4.62
10–19	8.62
20–39	14.69
40+	18.77
Degree of inhalation	
None	8.00
Slight	8.92
Moderate	13.08
Deep	17.00
Age began smoking	
25+	4.08
20–24	10.08
15–19	14.69
<15	16.77
Never smoked regularly	1.00

(Hammond, 1975.)[25]

million Americans who died from various tobacco-related causes might have been alive in 1973.[38]

The risks due to tobacco smoking are further increased by the fact that there are marked synergisms between tobacco smoke and a number of chemicals known to be carcinogenic. We will see that lung cancer is rare among nonsmoking uranium miners and asbestos workers, but occurs among smokers in those trades at an incidence far higher than can be explained by the combined effects of the cigarettes and carcinogenic dust to which the men are exposed in the course of their work. Cigarette smoking may also increase the carcinogenic action of air pollution.

The risk of developing lung cancer is reduced substantially when a person stops smoking and, even among heavy smokers, the risk is decreased to that of nonsmokers ten years after cessation.[36]

By 1964, the evidence for a causal relationship between cigarette smoking and lung cancer was so overwhelming that the United States Public Health Service issued a Surgeon General's Report on Smoking and Health.[39] Since that time the public has had an opportunity to become well aware of the relationship of cigarette smoking and lung cancer. However, because of the emphasis on lung cancer, the public is not sufficiently aware that cigarette smoking also increases the risk of dying from other cancers, emphysema, and diseases of the cardiovascular system.

Considering the amount of public attention that has been given to other cancer-causing agents, the absence of a national crusade against tobacco is both disappointing and surprising. However, early in 1978, the Secretary of Health, Education, and Welfare announced an educational campaign against tobacco use. It remains to be seen if the federal investment for this purpose will be commensurate with the urgency of the problem.

In recent years, the tobacco manufacturers have been required to place the following on each package of cigarettes: "Warning: The Surgeon General has determined that cigarette smoking is dangerous to your health." Cigarette advertising on the electronic media has been banned and the American Cancer Society and other public health agencies have been increasing the size of their antitobacco advertising budgets. However, the amount of money spent by the tobacco companies to advertise the qualities of their product in the printed media continues to be enormous and is far greater than the money available to inform the public about the harmful effects of smoking.

It has been found that as of 1970 about 60 percent of all smokers in the United States had tried to stop smoking and that 26 percent of the men and 17 percent of the women succeeded in stopping for a significant period of time.[40] However, tobacco is a powerfully addictive drug, a fact to which any habitual smoker who has tried to break the habit can testify. Some people succeed in quitting on their own, whereas others have sought hypnosis, group counseling, and various other methods.

In recent years there has been a movement to develop cigarettes of reduced carcinogenicity by lowering their tar and nicotine content. In addition, there has been a tendency for people to use filter-tip cigarettes.[41] However, there is evidence that people who smoke cigarettes with lowered tar and nicotine contents increase the number of cigarettes smoked, which suggests that the smoker may unknowingly compensate for the reduced tar and nicotine.[36] Nevertheless, evidence is developing that smokers of filtered cigarettes do have a lower risk of developing lung cancer.[35]

All in all, the cigarette story is without doubt the most dismal one in the history of modern public health.

Alcohol. Heavy drinkers of alcohol experience a risk of cancer of the mouth and pharynx two to six times higher than abstainers, with the difference being due mainly to the amount of tobacco smoked. It has been demonstrated that the effects of smoking and drinking alcohol have a synergistic relationship, and the combined effect of heavy smoking and drinking is to increase the incidence of cancer of the mouth and pharynx 15 times over the rate seen in persons who neither drink nor smoke.[42] Cancer of the esophagus is also increased by excessive alcohol consumption. The incidence of esophageal cancer among heavy drinkers who are moderate smokers is 25 times greater than among nondrinkers. It has been suggested that the carcinogenic agent is not the alcohol itself, but rather some as yet unidentified contaminant. (As noted earlier, an association has been found between beer consumption and rectal cancer. For some unknown reason, this does not hold for distilled liquors.)

It has been estimated that alcohol could be responsible for 7 percent of all cancer deaths among males and perhaps 2 percent of the deaths occurring among females.[42] The well-known association between tobacco smoking and consumption of alcohol is a factor that tends to obscure the role of alcohol as a cause of cancer.

Naturally Occurring Substances in Foods. The human diet is probably a major source of carcinogenic chemicals that either occur naturally or are produced from naturally occurring precursors during the preparation or storage of the food. The common fern known as "bracken" (*Pteridium aquilinium*) has been known to produce tumors and carinomas of the bladder and intestines of several species of experimental animals. This fern is used in salads in New Zealand, the United States, and Japan, where it may be a subtle contributor to the incidence of intestinal cancer.[43] During World War II it was found that liver injury sometimes developed among patients treated with tannic acids for burns. Subsequent investigations confirmed that tannic acid is toxic to the rat liver and that some forms of tannic acid were carcinogenic.[43] Of course, coffee, tea, and some wines contain tannic acids in small amounts. In addition, many spices, oils, and even some of the more common sugars are suspected of being carinogenic on the basis of laboratory studies.

Aflotoxins, a group of naturally occurring substances produced by certain fungi (*Aspergillus flavus*) found in peanuts, grains, and other foods, have been shown to be particularly potent carcinogens.[43,44] They have been identified as causing liver cancer in fish under natural conditions, and are highly carcinogenic for laboratory rats. Aflotoxins are believed to be the cause of the high incidence of liver cancer in parts of Africa and the Far East, but have not been associated with cancer in the United States. This may be due to differences in storage conditions which, in Africa and the Far East, may encourage the growth of the fungi.

In recent years there has been considerable discussion of the role of a class of compounds known as nitrosamines, some of which are known to be strong carcinogens in animals. These can be formed easily under a variety of conditions in soil and during food storage or cooking. Nitrosamines in food could be formed from nitrates and nitrites which are used as preservatives to prevent botulism.[43,45] Nitrates are of course present naturally in all foods as well, and may serve as a source of nitrosamines. These compounds have been detected in smoked foods and in fried bacon. There is no epidemiological evidence that nitrosamines cause cancer in man, but, because of their high carcinogenicity in experimental animals, the nitrosamines should receive more attention. Also, the addition of nitrates and nitrites to foods should only be permitted if they clearly can benefit health.

The practice of cooking food can result in the production of a family of carcinogenic compounds known as polycyclic aromatic hydrocarbons, of which the most common is benzopyrene. This is the carcinogen that was probably responsible for the scrotal cancer observed in English chimney sweeps 200 years ago and is known to be present in the contaminants discharged to the atmosphere when fossil fuels are burned. Although there is no evidence that the polycyclic hydrocarbons produced in food processing have caused cancer in humans, they too, like nitrosamines, must be regarded with suspicion.

Air Pollution. There is epidemiological evidence that urban residents are subject to a higher risk of lung cancer than people living in rural areas. However, there is no general agreement as to the cause of this "urban factor," and it is not certain that the increased risk is due to air pollution, although many investigators have suggested that this is so.

Three major epidemiological techniques have been used to study this question. Incidences of lung cancer in urban and rural areas have been compared. Studies have also been conducted among comparable cities in which the incidence of lung cancer is studied as a function of pollution levels, and useful information has been obtained by studying the incidence of lung cancer among migrants.

Comparisons of lung cancer mortality among rural and urban populations have been undertaken by many investigators during the past 20 years.[52] Most, but not all, investigators have found that the urban male population is at substantially higher risk than their counterparts in rural areas. Smoking habits were taken into consideration in only one study in which it was shown that the percentage of smokers in the rural areas was only slightly less than that in the urban areas (42.5 percent compared to 48.5 percent).[55] However, the number of cigarettes smoked, and the age when smoking began, may be equally important, and this information was not obtained.

A curious finding of almost all investigators has been that the urban-rural gradient has much less effect on women than on men. This can be interpreted in several ways. It is possible that men in urban areas are at higher risk because they are exposed to carcinogens in the course of their occupations. It is also possible that men may be exposed to more "second-hand" smoke in the course of their business and social activities.[35] In one Czechoslovakian study, it was found that the air of a beer hall was contaminated by BP to levels ten to 100 times higher than in city air.[52] More data of this kind are needed.

One of the most thorough of the epidemiological studies, in which the populations were adjusted for age and smoking, showed a positive correlation between the lung-cancer death rate and community size. This is shown in Table 8-3, in which it is seen that the lung cancer incidence gradually increased from 39 per 100,000 in rural areas to 52 per 100,000 in cities having populations greater than 50,000.[56]

There have been two studies in Great Britain in which it was found that the prevalence of lung cancer increases with moderate changes in population density even in rural areas. When the population density is expressed as persons per acre, mortality increases from 40 per 100,000 at a density of about .1 person per acre to 70 per 100,000 at a density of .4 person per acre.[57]

Although most studies of urban-rural differences tend to support a conclusion that air pollution by organic contaminants produced by combustion may be causally related to the risk of lung cancer, there

TABLE 8-3[56]

Lung-Cancer Cases per 100,000
(Adjusted for Age and Smoking)

Population	Cases
> 50,000	52
10,000–50,000	44
Suburban	43
Rural	39

are some notable exceptions. For example, the risk of lung cancer is high in Finland and New Orleans, but the air pollution levels are low. Studies in Nashville and Buffalo failed to show any relationship between lung cancer and pollution levels, but did suggest a strong inverse correlation with socioeconomic status. This, of course, is always a confounding factor in studies of this kind. Death rates from many diseases are higher among those who are economically less fortunate, and their economic misfortunes also tend to result in their living in more polluted neighborhoods.

Studies of the differences among migrants are made possible by the fact that there are major differences in both pollution levels and the incidence of lung cancer among many countries of the world. Among elderly residents of the United Kingdom, the lung cancer incidence is extraordinarily high (219 per 100,000). Mortality among native white South Africans is about half that rate (112 per 100,000). Among British migrants to South Africa, the mortality is intermediate, at 172 per 100,000. The tendency of migrants to assume the lung cancer risk of the population in which they assimilate has been demonstrated in studies of Italian, British, and Norwegian migrants to the United States and similar trends have been observed. These studies tend to suggest that early exposure to environmental factors has a lasting effect on lung cancer risk.[50]

In one review, published in 1959, it was concluded that "there was no firm evidence in support of the hypothesis that general urban air pollution increased the risk of lung cancer to an important degree, if at all."[46] Other investigators reexamined the question in 1972 and stated, in reference to the earlier statement that "data from our study support that conclusion: and we are unaware of any evidence which convincingly leads to a contrary conclusion."[47]

The case against the argument that the higher incidence of lung cancer in urban areas is caused by air pollution has been forcefully stated by Goldsmith and Friberg:

1. The urban factor should be largest in those states and countries where there is the heaviest urban pollution. It is not.
2. If exposure to urban pollution causes an augmentation in lung cancer, then the rates should be higher in lifetime urban residents than in migrants to urban areas. They are not.
3. Correlations of lung cancer rates with measured pollution should be found by studies in the United Kingdom where both lung cancer rates are high and pollution has been great. A positive correlation is found with population density, but not with pollution.

4. If the urban factor were community air pollution, it should affect women at least as much as men. It does not appear to.
5. If urban pollution by benzopyrene makes an important contribution to the urban excess, lung cancer in the locations most polluted by this material should be highest, and when the agent decreases, lung cancer should do so as well. This has not been shown to occur.[48]

One of the main obstacles to an understanding of the relationship between air pollution and lung cancer is the overriding influence of cigarette smoking, as was well summarized in the United Kingdom by the Royal College of Physicians in 1970:

> The study of time trends on the death rate due to lung cancer in urban areas demonstrated the overwhelming effect of cigarette smoking on the distribution of the disease. Indeed, only the detailed surveys that have taken individual smoking histories into account have succeeded in separating the relatively small influence of the "urban factors" on the overriding effect of cigarette smoking on the development of cancer of the lung.[49]

Most investigators have emphasized the relatively small contribution of the urban factor, and its tenuous association to community air pollution.[49a-e] This is in sharp contrast to the popular literature, which tends to emphasize the role of air pollution in the induction of lung cancer.[49f,g]

Among nonsmokers, the death rate from lung cancer, when adjusted for age, is three to five per 100,000 of population and among men has been found to be independent of residential history.[50] That is, no significant difference has been reported between non-smoking city or rural dwellers.[50a] However, the urban factor does increase the risk due to smoking among men who live in the city. There could be many reasons for this. We will see that there are traces of carcinogens in urban air and their effect may be synergistic with cigarette smoking. The effect may be due to the fact that people who live in the city smoke more than those who live in rural areas, or start at an earlier age.[39] It has also been noted that cigarette smoking became popular in the cities before it was popularized in rural areas.[51]

It has been demonstrated repeatedly that the particulate matter in urban air contains substances that can produce cancer in laboratory animals.[52] Other studies of various chemical fractions of the urban particulates suggest the likelihood that the most significant carinogenic action is due to polycyclic aromatic hydrocarbons or, more simply, polycyclic organic matter (POM). These compounds are characterized by complex molecular structures, are present in exceedingly low concentrations, and are difficult to analyze chemically. It has been shown that one of these compounds, benzopyrene (BP), can be detected in the air of cities throughout the United States and Europe, and that the amount present generally correlates with the concentration of smoke. BP is easier to analyze than other constituents of POM, and its presence is used as an indicator of the potential carcinogenicity of particulate contamination.

The fossil fuels are major sources of BP, and they are consumed for space heating in many cities. As a result, there is from ten to 20 times more BP present in the winter months. Coal burning produces more BP than does the burning of other fuels, but because the use of coal for space heating has been reduced during the past three decades, the average concentration of BP has diminished by about two-thirds since 1959.[52] Motor vehicles are only a minor producer of BP, but the burning of refuse can be an important source, and may be of greater significance than the production of heat and power.[52] The city-to-city differences in BP levels can be considerable. Between 1958 and 1959, the concentration of the air of Birmingham, Alabama was six times that in Los Angeles. This difference is presumably due to the greater amount of coal consumed in Birmingham. (It is worth noting that the excess risk of developing lung cancer is only about 50 percent among coal-gas workers, despite the fact that the concentrations of BP to which they are exposed are as much as 10,000 times greater than for members of the general population.)[53,54]

There is evidence that the incidence of cancer is higher in areas in which chemical industries are concentrated. This may be due to air pollution but could also be due to the employment of some members of the population by the local chemical industries. Studies of this kind have only recently been undertaken, and more research is needed before any conclusions can be drawn.[58,59]

Cancers Produced as a Consequence of Medical Practices. A variety of cancer types have been caused by medical procedures in which patients were exposed to carcinogenic drugs or radiations. Such cancers are known as "iatrogenic" (caused by physicians). Among the best-known examples in this category are cancers resulting from use of X-rays and radium, which were discovered towards the end of the nineteenth century and were soon found useful for diagnosis and treatment. Physicians administered the radiations without knowledge of their harmful effects and cancers developed from the misuse, not only among the patients, but among physicians as well. Such cases are not entirely a matter of the past. A series of bone cancers have been reported recently from West Germany, where a short-lived species of radium (Ra-224) was injected intravenously into about 2,000 patients suffering from tuberculosis and other diseases between 1944 and 1951. Fifty cases of bone cancer have developed among these patients.[60]

Thorium oxide, a mildly radioactive substance, is relatively opaque to X-rays and therefore was used in "thorotrast," which was injected into the veins of patients to assist the radiologist by increasing the contrast of X-ray images. Unfortunately, thorium tends to remain in the body and eventually caused cancers to develop, notably in the liver and sinuses.[61,62]

Another well-studied group of radiation-induced cancers are the thyroid carcinomas that are continuing to develop among children who received X-radiation for reduction of enlarged thymus glands. In the course of this procedure, the thyroid gland, an organ that has proven to be relatively sensitive to radiation, and is situated near the thymus, was irradiated sufficiently to cause some of the children to develop thyroid cancers later in life.[63] There are other examples of cancers that developed from the medical uses of ionizing radiation. The incidence of leukemia has been reported to be moderately elevated among children who were irradiated as fetuses when their mothers were subjected to pelvic X-ray examination in the course of pregnancy.[64] There is some dispute about this finding, however, since no excess of leukemias has been observed among persons irradiated *in utero* at much higher doses by the atomic bombs dropped on Japan.[65] Leukemia has also developed among patients irradiated by X-rays in the course of the arthritic disease known as ankylosing spondylitis.[66]

X-rays have been used to treat inflammation of the breasts, acne, and excessive hairiness, and a higher incidence of breast cancer has been observed later in life among women who have had these types of X-ray treatments.[67-69] Breast cancer is also more frequent among women who have been subjected to fluoroscopy because of a chronic lung disease such as tuberculosis.[67]

A major question relates to the effect of the relatively low doses of radiation involved in routine procedures such as diagnostic X-rays in cancer induction. National surveys conducted in the mid-1960s showed that many physicians were not using their X-ray equipment in a manner designed to minimize patient exposure and that the radiation dose received by the patients were often higher than would be the case if the equipment were properly used. If it is assumed that, even at low doses, the risk of cancer from ionizing radiation is directly proportional to dose and that there is not a threshold, then one must assume that *any* radiation increases the probability of the occurrence of cancer. The dose received from routine use of X-rays for diagnostic purposes is very much lower than the dose from the therapeutic procedures just discussed, and the probability of developing cancer is correspondingly lower. However, the exposed population is very much larger, since it is a rare person nowadays who does not receive a chest, dental, or other X-ray examination every year or two. It has been estimated that the per capita dose from diagnostic radiology in the United States was .020 rads in 1970.[70] For the reasons discussed in Chapter 7, there are seemingly insurmountable obstacles to determine if risk coefficients developed on the basis of high doses can be extrapolated for purposes of estimating risks at low doses. However, if modern equipment and film are used, the dose to a patient can be reduced very substantially, and in view of the uncertainty in this regard, prudence dictates that unnecessary uses of X-rays should be avoided completely. The findings that X-rays used therapeutically increases the incidence of breast cancer would certainly indicate the need for care in the application of mammographic techniques for early breast-cancer detection.

One of the saddest of the iatrogenic tragedies has been the occurrence of vaginal cancer among young women whose mothers during pregnancy were administered the synthetic estrogenlike sex hormone, diethyl stilbestrol (DES), to prevent miscarriages and to treat a number of conditions related to estrogen function. More than 100 such cases were known to have occurred by 1975, but it is estimated that from 600,000 to 2,000,000 young women may be at risk, and it is likely that even more cases will occur.[71]

There are probably other examples of iatrogenic cancers, but the extent to which they occur is not known. Because of the difficult epidemiological problems involved in such studies, adequate human investigations have been conducted only for two drugs: isoniazid (administered for treatment of tuberculosis) and the female steroidal sex hormones.[72] With respect to the latter, there is evidence that women who take estrogen to relieve the tensions of the menopause have a higher risk of developing cancer of the lining of the uterus.[71]

Suspicion of carcinogenicity has even been raised with respect to the use of drugs containing aspirin, phenacetin, and caffeine. Overuse of these drugs is not uncommon, and in extreme cases it has been associated with cancer of the kidney and bladder.[71]

It should be obvious from the foregoing that utmost caution must be employed before any new drug or therapeutic procedure is approved for use. There is a need to balance risk and benefit, a problem with which a physician is often faced in the course of his or her medical practice. There may be ample justification for administering a known carcinogen to control a disease that is placing the patient's life at immediate risk, despite the fact that treatment may involve a high probability that the patient will develop a cancer in ten or 20 years. However, if there is the slightest probability that a drug or a procedure can produce cancer, it should not be prescribed for commonly occurring minor disabilities. Between the extremes lies a large gray area in which the balancing of risks and benefits is not a simple matter and is likely to be the subject of continuing disagreement among experts.

Occupational Cancer. The observation by Potts, about 200 years ago, that cancer of the scrotum was associated with chimney sweeping was the first evidence of an association between occupation and cancer. Further knowledge was developed slowly, thereafter, and it was not until 1932 that the chemical benzopyrene was isolated from coal tar and found to be highly carcinogenic in experimental animals.[73] It was observed by 1925 that small amounts of radium deposited over a period of years in the skeletons of workers painting luminous dials could produce bone cancer, and that inhalation of radioactive gases in the course of underground mining could produce bronchiogenic lung cancer.[74,75]

The history of occupational cancer is not a happy one, and certainly it is not a story in which either government or industry should take pride. For example, the fact that coal-tar derivatives could produce bladder cancer among workers in dye-manufacturing plants was well accepted as early as 1933.[76] Some of the more dangerous compounds in this family of chemicals (benzidine and beta-naphthylamine) were banned in England and other countries in the mid-1960s.[77] Yet industry in the United States was relatively slow to respond, and as late as 1973 it was reported that 50 percent of the former employees in one benzidine plant in the United States had developed bladder cancer.[78]

Information about occupational cancer has gradually been accumulated from studies of the painful experiences of workers in many industries. Some of the occupational cancers are of types that occur so rarely in the general population that they can be readily identified even though only a few cases have been reported. Others occur ordinarily in such high incidence in the general population that sophisticated statistical techniques must be used, requiring the study of health records of large numbers of employees to identify an excess cancer risk due to occupation. As an example of the first instance, bone cancer was found among luminous-dial painters beginning in the

1920s because it is such a rare disease in the general population that even the first few cases (there were five in the original series) attracted attention and because the patients lived in a single community and had had a common history of employment. Other examples are angiosarcoma, a very rare liver cancer that has occurred among workers exposed to vinyl chloride in the plastics industry, and mesothelioma, a rare cancer of the linings within the chest and abdominal cavities, seen in workers exposed to asbestos dust. Fortunately, occupational cancers have thus far involved relatively few employees and have attack rates that usually lie between 1 and 10 percent, but which may be higher or lower.

Angiosarcoma, caused by exposure to vinyl chloride is one of the occupational cancers that has attracted considerable attention in recent years, although, by 1975, a total of only 35 cases had been recorded throughout the world.[79] The exact number of workers exposed to vinyl chloride is not known, but it must surely be greater than 10,000. The cancer developed in less than 1 percent of the exposed employees. Among the radium-dial painters, a total of about 40 cases of bone cancer developed among about 2,000 persons employed in dial-painting shops in the United States during, and immediately following, World War I (an incidence of about 2 percent).

In addition to the examples of unusual cancers that can occur among relatively small groups of employees, there is evidence that the incidence of the more common cancers such as those in the lung, stomach, and urinary tract is moderately increased among relatively large groups of workers in the textile, woodworking, and steel industries.[16,73,80,81] Also, it has been shown that higher-than-average death rates due to cancer of the bladder, lung, liver, and other sites exist in those counties in the United States that have a large concentration of chemical industries.[82] It is a reasonable assumption that this is due to occupational exposure to carcinogenic chemicals. The epidemiological studies needed to assess the overall impact of occupational factors on the statistics of cancer mortality are only being started, but there is sufficient evidence to warrant informed judgments that from 1 to 3 percent of all cancer deaths are due to occupational factors.[5]

The cancers caused by exposure to asbestos have received much attention in recent years. The history of asbestos usage, which goes back to the last third of the nineteenth century, serves to illustrate the painstakingly slow progress by which the hazards of many substances have come to be understood up to the present time.

The word "asbestos" refers to a group of fibrous natural minerals that can be spun and woven like textile fibers or pressed into forms, using cementing materials to bind the fibers. The principal uses of asbestos arise out of their remarkable heat-insulating quality and the fact that they are completely nonflammable. The asbestos minerals are widely used for fireproof roofing and house siding, interior walls, and as insulating fireproof materials around the structural steel of skyscrapers. Another major use of asbestos is for automobile brake linings. The use of asbestos increased from only 200 tons in 1868 to 3.5 million tons in 1968.[76,83]

By 1899, the first case of nonmalignant lung disease due to inhalation of asbestos dust was reported. The disease was soon given the name "asbestosis," and was observed to develop after exposures of from three to 15 years. Asbestosis, like silicosis and other diseases caused by industrial dust, is known as a pneumonoconiosis, a "dust disease of the lung." It is a disabling disease that is fatal in a high percentage of cases, but it is not a cancer.

As late as 1932, there were no indications that exposure to asbestos could cause lung cancer.[76] Lung cancer, however, began to be reported among asbestos workers in increasing incidence. In some instances, the cancer did not become manifest until ten to 12 years after cessation of work with asbestos. The period between the first exposure to asbestos and death from lung cancer has varied from 15 to 40 years.[84,85]

In the 1950s, evidence was accumulated that, in addition to causing asbestosis and lung cancer, asbestos also had the ability to produce a rare cancer of the pleura and the peritoneum, which are the linings within the walls of the chest and abdominal cavities. This disease is so rare that, when cases are reported, backtracking usually finds a history of either exposure to asbestos in the course of employment, or exposure due to living in the proximity of an asbestos plant or in the same household with an asbestos worker who contaminated the household air by dust from his work clothes.[86,87]

Two hundred years after Potts' observation of its high incidence among chimney sweeps, we still know much too little about occupational causes of cancer. Most of our information has become available because of observations made among relatively small groups of workers among whom the cancers occurred with such high frequency as to facilitate their recognition. Lung cancers among uranium miners, and workers exposed to chromium, nickel, and asbestos are such examples. In some cases, the cancers were observed because they were of such rare types, as are those found in workers in the vinyl chloride and radium industries.

A recent compilation of suspected carcinogens listed 1,500 that are used in industry.[88] In the years immediately ahead, there should be an intensive drive to complete comprehensive epidemiological studies, industry by industry, to identify the workers at risk, the agents that cause cancers, and the methods by which the carcinogenic substances can be used safely.

Chemical Contamination of Food. The question of how to judge the safety of the hundreds of artificially produced chemical substances that become incorporated into our foods is becoming increasingly difficult to answer due to the interaction of scientific, social, and even moral uncertainties.

The use of chemicals in the agricultural, food-processing, and food-distribution industries has been of enormous benefit. Per capita production of food has increased significantly in many parts of the world because chemicals used in agriculture have made it possible to increase the yield of crops on the farm and protect food from spoilage until it is consumed.

On the farm, fertilizers increase the yields of crops, and antibiotics and hormones are added to the feed of livestock to stimulate their growth. The addition of diethyl stilbestrol to the feed of steers can increase protein production, per animal, by 18 percent. When small doses of antibiotic substances are added to feed, growth of the steers is promoted to such an extent as to benefit consumers of agricultural products by more than $400 million per year in 1970.[38] Other chemicals are fed to farm animals to prevent disease, and still others to treat disease. Modern farm management has also been drastically changed by the availability of pesticides of immense potency, such as DDT, which was introduced in 1939, and the herbicide 2,4-D, which was introduced in 1941. Nearly 1,000 such compounds have been synthesized and used in agriculture throughout the world. Large quantities of pesticides have also been used to control mosquitos, lice, rats, and other vectors of disease.

When chemicals are sprayed on farm lands for purposes of pest control, both crops and animal products become contaminated, and the opportunity also exists for exposing humans who eat the contaminated products to the same chemicals. In addition, some of the chemicals cause undesirable ecological effects, as in the case of DDT, which has been much in demand as an insecticide because of its potency and persistency in the environment. (The ecological effects of DDT will be presented in the next chapter.)

Chemicals are also added to our food deliberately, for many reasons. Some are added to prevent caking or lumping, to retard deterioration due to oxidation, to impart a desired color, or to stabilize the natural color. Other compounds are added to increase shelf life and to serve as drying agents, preservatives, non-nutritive sweeteners, or for any one of a dozen other functions.[89] Still other chemicals are used in the manufacture of packaging materials, and some of these can contaminate food by contact.

Chemicals are used so widely in foods that almost all persons in the world are exposed to them. Because the latent period for development of cancer can be as long as several decades, an unsuspected carcinogenic chemical, once distributed to the population, can, if sufficiently potent, commit that population to an epidemic of cancer in the years ahead.

Experimental animals are the principal means by which chemicals are tested for carcinogenicity. The experimental doses given in these tests are likely to be as much as a million times higher than the doses humans would be expected to receive. Animal experiments are valuable for studying the mechanisms by which chemicals produce cancer, and seem to be useful for identifying the highly carcinogenic chemicals; but it is not yet possible to use data from animal experiments to define the maximum dose of a particular carcinogen to which humans can safely be exposed. Among the major problems are the differences in the susceptibility of species: A substance that produces no effect in the rat or hamster may cause one type of tumor in the mouse and another in the cat. Some of the tumors are benign, but some scientists believe that benign tumors will in time become malignant. The Environmental Protection Agency has taken the position that if a chemical is capable of producing benign tumors in experimental animals, it should be assumed that the chemical is also capable of causing cancer.[90] This philosophy has been carried further: It is possible to test a chemical for mutagenicity in bacterial cultures. This type of test is known as the "Ames test," and it is advocated by many as a test of the carcinogenicity of a chemical on the theory that the underlying biochemical processes that cause genetic changes are similar to those that cause cancer. A substance that is mutagenic in bacteria should therefore be assumed to be carcinogenic to humans.[91,92]

In one outstanding example of the use of predictive methods, it was suspected from theoretical considerations that the industrial chemical bis-chloromethyl-ether (BCME) should be a potent carcinogen.[93,94] This was confirmed initially by tests on single-celled organisms and mouse skin.[95,96] Since it was known that BCME existed in the form of a vapor that could be inhaled by industrial workers, experiments were conducted which revealed that cancer of the lung could be caused by this compound in laboratory rats.[96] This suggested the urgent need for epidemiological studies, which disclosed that lung cancer was in fact occurring in excessive incidence among industrial employees exposed to this compound.[97] This is another example in which human carcinogenicity was demonstrated after the fact.

We happen to be in a period where we will be repeatedly "locking the barn doors after the horses have been stolen." The sinister role of many chemical and physical agents in industry was formerly insufficiently understood, but we now have the opportunity to obtain valuable epidemiological information on the basis of past mistakes. In the future, it is likely that potential carcinogens will be identified *before* there is the opportunity for significant human exposure. The occurrence of occupational cancer will be eliminated, or at least minimized, by the application of more stringent industrial hygiene controls than were practiced in the past.

How Safe is "Safe"? It is not possible to state with certainty that there is any such thing as a "safe" dose of a substance capable of producing cancer. However, as the dose decreases, the risk (the probability that a cancer will develop) diminishes also, and the time it would take to produce a tumor increases.[98] The latent period, at very low doses of a carcinogen, may actually exceed the lifespan of the individual. If this is so, there is a "practical threshold," despite the fact that no threshold exists in theory.[99] It is also possible that at very low doses there is a threshold below which no cancers will be produced. Unfortunately, as was described in Chapter 7, the laboratory problems associated with the extension of dose-response curves to near-zero doses are so formidable as to have defied solution thus far. It can be shown, by statistical methods, that it would be necessary to test the carcinogenicity of a chemical on millions of animals in order to adequately define the shape of the dose-response curve. The initial cost of experimental animals would not prohibit the design of rat or mouse experiments on such a scale, but the cost of maintaining the animals for their lifetimes (a rat lives for about two years), together with the cost of pathological studies of the animal tissues, would be staggering. Even if the money could be obtained, there would not be enough pathologists to provide the skilled attention needed for the tens of millions of slides for which microscopic study would be required.

In the absence of any proof that a "safe" dose of a carcinogen exists, it is now conceded that the basic question is not, what is the "safe" dose, but what dose is "safe enough."[100–102] The scientist can give his best estimate of the risk involved in exposing people to a minute dose of a carcinogen, but the scientist cannot preempt the right to answer the question "How 'safe' is 'safe enough'?" This is a question for which the scientist is no better prepared to provide an answer than other persons of good judgment. For this reason, it is desirable that the answers to such questions reflect not only the opinions of the scientists but responsible opinions from others, as well. No such mechanism for deciding the acceptability of risk now exists.

In 1958, the Congress became impatient with the inability of scientists to provide unequivocal information about the safety of chemicals used by the food industry, and a clause was added to the amendments to the Food, Drug and Cosmetic Act [Section .409(c)(3)(A)] which provided that "no [food] additive shall be deemed to be safe if it is found to induce cancer when ingested by man or animal, or if it is found, after tests which are appropriate for the evaluation of the safety of food additives, to induce in man or animal. . . ."

This clause, which has come to be called the "Delaney Amendment," also provided that no such additive can be used in animal feeds if a residue of the additive can be found in food products obtained from the animal after slaughter.[89] Of the food additives ruled to be carcinogenic, two are of special importance because of the economic impact of the ruling. The female hormone diethyl stilbestrol (DES) had been extensively used to increase the growth rate of chickens and other farm animals. DES is the synthetic estrogenlike compound that had been found to be capable of producing cancer in humans. Residues of DES were found in animal carcasses sold at market, using ultra-sensitive radioactive-tagging techniques. The FDA took the position that human health was not in jeopardy, but the inflexibility of the Delaney clause left the agency no choice but to ban the use of DES as a growth stimulant.[101]

In the late 1960s, an important class of artificial sweeteners known as "cyclamates" was banned from the market-place because of evidence that it was carcinogenic to laboratory animals. In 1976, saccharin, the only remaining artificial sweetener of importance, was also banned. These decisions have been the subject of much debate, widely publicized in the media.

It would at first seem as though the Delaney Amendment should be totally uncontroversial: What possible justification can there be for deliberately adding substances to our food supplies that are known to produce cancer in experimental animals? However, the Amendment has, in fact, been hotly debated with powerful arguments by both its proponents and opponents. The arguments of the proponents are rather clear-cut. With possibly one exception (metallic arsenic), all chemicals that are known to have produced cancer in humans have also produced cancer in experimental animals. Prudence therefore dictates the assumption that the reverse is true, that is, that, with rare exceptions, most chemicals that produce cancer in experimental animals will also produce cancer in humans. The long incubation period, from the time of exposure to the time for cancer development, and the high probability of a fatal outcome once cancer has been induced, are additional reasons for great caution. Cancer is such a dreadful disease that it should be prevented at all costs.

Despite the seeming logic of the arguments favoring the Delaney Amendment, the opponents have much to say in their behalf. The arguments they use are influenced by the unquestionable fact that chemicals have been of enormous benefit to human health and that many lives have been extended by the chemical products used in agriculture and medicine. The opponents of the Delaney clause recognize that careful screening is necessary before human exposure to new chemicals is permitted, but they emphasize the great benefits that have come from the use of synthetic chemicals, and they urge the need for "perspective."

A major argument against the amendment is that food additives are not permissible in *any* concentration if cancer can be produced in *any* animal at *any* dose over *any* period of time.[89] Again, it would seem that this is a desirable policy. Who would want to consume a food that contains a substance that is known to have caused cancer in experimental animals? The problem is that many substances will produce cancer in *some* organ of *some* experimental animal if administered in sufficiently high doses. To illustrate, consider the decision to ban cyclamates as artificial sweeteners. From .25 gram to 1 gram of the sweetener was used in a 12-ounce bottle of "low-calorie" soft drink. Based on the animal experiments, it would be necessary for an adult to consume no less than 138 12-ounce bottles every day in order to consume the quantity of cyclamate known to have produced tumors in the rat (making allowance for the difference in weight between rat and man).

Another argument opposes the requirement that the food additive must be withdrawn if it is detected as a residue in food. Whether or not the substance can be detected depends on the sensitivity of the available chemical methods of analysis. From decade to decade, these methods have become more sensitive: The class of chemicals known as dioxins, which could not be measured quantitatively by any method in 1950, can now be measured to a level of a few parts per billion. The lower limit of detectability for DES has been reduced by a factor of 100 in the past 20 years, and the limit for DDT by about 1,000. Thus, chemicals which could not be detected in foods in 1950 can now be detected readily by present methods. A study by the President's Scientific Advisory Council in 1973 concluded, in this regard, that a "no detectable amount" clause is a refuge in the face of ignorance: "the rigid stipulations of the Delaney clause springing from presently inadequate biological knowledge places the administrator in a very difficult interpretive position. He is not allowed, for example, to weigh any known benefits to human health, no matter how large, against the possible risks of cancer production, no matter how small."[38]

The arguments over the Delaney clause will not be easily resolved. The long period of time required for cancers to develop, the difficulties of both laboratory experimentation and epidemiological studies, and the inherent statistical difficulties due to the large numbers of variables that must be taken into consideration are likely to defeat any attempts of scientists to be significantly more quantitative in their descriptions of the risk from any new chemical added to the environment. Moreover, should it prove possible to predict the number of cancers that would be produced by a given substance for any given manner of use, it is doubtful that society has, as yet, the basis for judging the acceptability of an innovation based on benefit-risk analysis. There are many who clamor for absolute safety, although we know this is unattainable. Others find it repugnant to make tradeoffs between benefits and costs when the costs are measured in the number of years that human lives might be shortened, this, despite the fact that the benefits may be translated into a greater number of years of life extension.

NOTES

1. Cairns, John. "The Cancer Problem," *Scientific American* 233:64–78 (1975).
2. World Health Organization. "Prevention of Cancer: Report of a WHO Expert Committee," Technical Report Series No. 276, Geneva (1964).
3. Boyland, E. "The Correlation of Experimental Carcinogenesis and Cancer in Man," *Progr. Exp. Tumor Res.* 11:222–234 (Karger, Basel/New York) (1969).
4. Epstein, Samuel. "Environmental Determinants of Human Cancer," *Cancer Research* 34:2425–2435 (1974).
5. Higginson, J. and C. S. Muir. "The Role of Epidemiology in Elucidating the Importance of Environmental Factors in Human Cancer," *Cancer Detection and Prevention* 1(1):79–105 (1976).
6. Doll, Richard. "Epidemiology of Cancer: Current Perspectives," *American Journal of Epidemiology* 104:396–407 (1976).
7. Natural Resources Defense Council, Inc. "Cancer: The Price of Technological Advancement?" *NRDC Newsletter* 5(2) (Summer, 1976).

8. Ember, Lois. "The Specter of Cancer," *Environmental Science and Technology* 9:1116–1121 (1975).
9. Ember, Lois. "Environmental Cancers: Humans as the Experimental Model?" *Environmental Science and Technology* 10:1190–1195 (1976).
10. Council on Environmental Quality. "Sixth Annual Report," U.S. Government Printing Office, Washington, D.C. (1975), p. 111.
11. Ashford, N. A. "Crisis in the Workplace: Occupational Disease and Injury," MIT Press, Cambridge, Mass. (1976).
12. Gillette, Robert. "Cancer and the Environment (II): Groping for New Remedies," *Science* 186:242–245 (1974).
13. American Cancer Society Inc. "Cancer Statistics, 1975—25-Year Cancer Survey," *Ca-A Cancer Journal for Clinicians* 25:2–21 (1975).
14. Eisenbud, M. "Environmental Radioactivity," 2nd ed., Academic Press, New York (1973).
15. Jacobson, A. P., P. A. Plato and N. A. Frigerio. "The Role of Natural Radiations in Human Leukemogenesis," *AJPH* 66(1):31–37 (1976).
16. N.Y. Academy of Sciences. "Occupational Carcinogenesis," Proceeding of a Conference held March 24–27, 1975 (U. Saffiotti and J. K. Wagoner, eds.), *Annals of the New York Academy of Sciences* 271 (1976).
17. Levin, D. L., S. S. Devesa, J. D. Godwin and D. T. Silverman, "Cancer Rates and Risks," 2nd ed., *DHEW Publication No. (NIH) 75-691*. U.S. Government Printing Office, Washington, D.C. (1974).
18. Lilienfield, A. M. "Cancer in the United States," Harvard Univ. Press, Cambridge, Mass. (1972).
19. U.S. Dept. of Health, Education and Welfare. "Facts of Life and Death," *DHEW Pub. No. (HRA) 74-1222*, U.S. Government Printing Office, Washington, D.C. (1974), p. 31.
20. Peto, R., F. J. C. Roe, P. N. Lee, L. Levy and J. Clack. "Cancer and Ageing in Mice and Men," *British Journal of Cancer* 32:411–426 (1975).
21. Young, J. L. and R. W. Miller. "Incidence of Malignant Tumors in U.S. Children," *Journal of Pediatrics* 86:254–258 (1975).
22. National Center for Health Statistics.
23. Miller, R. W. "Fifty-two Forms of Childhood Cancer: United States Mortality Experience, 1960–1966," *Journal of Pediatrics* 75:685–689 (1969).
24. Myers, M. H., H. W. Heise, F. P. Li and R. W. Miller, "Trends in Cancer Survival among U.S. White Children, 1955–1971," *Journal of Pediatrics* 87:815–181 (1975).
25. Hammond, E. C. "Tobacco," in: "Persons at High Risk of Cancer" (J. F. Fraumeni, ed.), Academic Press, New York (1975), p. 131.
26. Kmet, J. and E. Mahboubi. "Esophageal Cancer in the Caspain Littoral of Iran: Initial Studies," *Science* 175:846–853 (1972).
27 Kennaway, E. L. "Cancer of the Liver in the Negro in Africa and in America," *Cancer Research* 4:571–577 (1944).
28. Daniels, F. "Sunlight," in: "Cancer Epidemiology and Prevention" (D. Schottenfeld, ed.), Charles C. Thomas, Springfield, Ill. (1975), p. 126.
29. Kessler, I. I. and L. Aurelian. "Uterine Cervix," in: "Cancer

Epidemiology and Prevention'' (D. Schottenfeld, ed.), Charles C. Thomas, Springfield, Ill. (1975), p. 263.
30. Phillips, R. L. "The Role of Life-style and Dietary Habits in Risk of Cancer Among Seventh-Day Adventists," *Cancer Research* 35:3513–3522 (1975).
31. Breslow, N. E. and J. E. Enstrom. "Geographic Correlations between Cancer Mortality Rates and Alcohol-Tobacco Consumption in the United States," *Journal of the National Cancer Institute* 53:631–639 (1974).
32. Higginson, J. "Present Trends in Cancer Epidemiology," in: *Proceedings of the Eighth Canadian Cancer Research Conference* (J. F. Morgan, ed.), Pergamon Press, Oxford (1969), pp. 40–75.
33. Müller, F. H. "Tabakmissbrauch und lungencarcinom," *Z. Krebsforsch* 49:57 (1939).
34. Ochsner, A. and M. DeBakey. "Symposium on Cancer: Primary Pulmonary Malignancy," *Surg. Gynecol. Obstet.* 68:435–451 (1939).
35. U.S. Department of Health, Education and Welfare "The Health Consequences of Smoking, 1975," Center for Disease Control, Atlanta, Ga., U.S. Government Printing Office, Washington, D.C. (1975).
36. Wynder, E. L., K. Mabuchi and D. Hoffman. "Tobacco," in: "Cancer Epidemiology and Prevention" (D. Schottenfeld, ed.), Charles C. Thomas, Springfield, Ill. (1975), p. 102.
37. Hammond, E. C. "Smoking in Relation to the Death Rates of One Million Men and Women," *National Cancer Institute Monograph* 19:127–204 (1966).
38. President's Science Advisory Committee. "Chemicals & Health," Report of the Panel on Chemicals and Health, Science and Technology Policy Office. National Science Foundation, U.S. Government Printing Office, Washington, D.C. (September, 1973).
39. U.S. Surgeon General. "Smoking and Health: Report of the Advisory Committee to the Surgon General of the Public Health Service," U.S. Government Printing Office, Washington, D.C. (1964).
40. Horn, D. "Determinants of Change," in: "The Second World Conference on Smoking and Health" (R. G. Richardson, ed.), Pitman Medical and Scientific Publishing Co., Ltd., London (1972).
41. Gori, G. B. "Low-Risk Cigarettes: A Prescription," *Science* 194:1243–1246 (1976).
42. Rothman, K. J. "Alcohol," in: "Persons at High Risk of Cancer" (J. F. Fraumeni, ed.), Academic Press, New York (1975), p. 139.
43. National Academy of Sciences. "Toxicants Occurring Naturally in Foods," 2nd ed., Committee on Food Protection, National Research Council, Washington, D. D. C. (1973).
44. Berg, J. W. "Diet," in: "Persons at High Risk of Cancer" (J. F. Fraumeni, ed.), Academic Press, New York (1975), p. 201.
45. Environmental Protection Agency. "Scientific and Technical Assessment Report on Nitrosamines," *EPA-600/6-77-001*, Office of Research and Development, Washington, D.C. 20460 (1976).
46. Cornfield, J., W. Haenszel, E. C. Hammond, A. M. Lilienfeld, M. B.

Shimkin and E. L. Wynder. "Smoking and Lung Cancer: Recent Evidence and a Discussion of Some Questions," *Journal of the National Cancer Institute* 22:173–203 (1959).

47. Hammond, E. C. "Smoking Habits and Air Pollution in Relation to Lung Cancer," in: "Environmental Factors in Respiratory Disease" (D. H. K. Lee, ed.), Academic Press. New York (1972), pp. 177–198.

48. Goldsmith, J. R. and L. T. Friberg. "Effects on Human Health" in: "Air Pollution," 3rd ed. (A. Stern, ed.), Academic Press, New York (1977), Vol. II.

49. Royal College of Physicians of London. "Air Pollution and Health," Summary and Report on Air Pollution and its Effect on Health by the Committee of the Royal College of Physicians of London on Smoking and Atmospheric Pollution. Pitman Medical and Scientific Publishing Co., Ltd., London (1970).

49a. Higgins, I. T. T. "Epidemiology of Lung Cancer in the United States," in : "Air Pollution and Cancer in Man" (U. Mohr, D. Schmahl and L. Tomatis, eds.), *IARC Scientific Publication No, 16*, International Agency for Research on Cancer, World Health Organization, Lyon (1977), p. 191.

49b. Saracci, R. "Epidemiology of Lung Cancer in Italy," in: "Air Pollution and Cancer in Man" (U. Mohr, D. Schmahl and L. Tomatis, eds.), *IARC Scientific Publication, No. 16*, International Agency for Research on Cancer, World Health Organization, Lyon (1977), p. 205.

49c. Saxén, E., L. Teppo and T. Hakulinen. "Epidemiology of Lung Cancer in Scandinavia," in: "Air Pollution and Cancer in Man" (U. Mohr, D. Schmahl and L. Tomatis, eds.), *IARC Scientific Publication No. 16*, International Agency for Research on Cancer, World Health Organization, Lyon (1977), p. 217.

49d. Waterhouse, J. A. H. "Epidemiology of Lung Cancer in England and Wales," in: "Air Pollution and Cancer in Man" (U. Mohr, D. Schmahl and L. Tomatis, eds.), *IARC Scientific Publication No. 16*, International Agency for Research on Cancer, World Health Organization, Lyon (1977), p. 229.

49e. Bogovski, P., M. Purde and M. Rahu. "Some Epidemiological Data on Lung Cancer in the USSR," in: "Air Pollution and Man" (U. Mohr, D. Schmahl and L. Tomatis, eds.), *IARC Scientific Publication No. 16*, International Agency for Research on Cancer, World Health Organization, Lyon (1977), p. 241.

49f. *New York Times* "21 Cigarettes," Editorial (February 19, 1975).

49g. Esposito, John C. "Vanishing Air," The Ralph Nader Study Group Report on Air Pollution, Grossman Publishing Co., New York (1970).

50. Shy, C. M. "Lung Cancer and the Urban Environment: a Review," in: "Clinical Implications of Air Pollution Research," Publishing Sciences Group, Inc., Acton, Mass. (1976). p. 3.

50a. Haenszel, W., D. B. Loveland and M. G. Sirken. "Lung-Cancer Mortality as Related to Residence and Smoking Histories," I. White Males. *Journal of the American Cancer Institute* 28(4):947–1001 (1962).

51. Higginson, J. and O. M. Jensen. "Epidemiological Review of Lung Cancer in Man," in: *Proceedings of the Symposium Air Pollution and Cancer in Man, Publication No. 16*, International Agency for Research on Cancer, Lyon, France (1977), p. 169.

52. National Academy of Sciences. "Particulate Polycyclic Organic Matter," Committee on Biologic Effects of Atmospheric Pollutants, NAS Printing Office, 2101 Constitution Ave., Washington, D.C. (1972).

53. Waller, R. E. "The Combined Effects of Smoking and Occupational or Urban Factors in Relation to Lung Cancer," *Ann. Occ. Hyg.* 15:67–71 (1972).

54. Hoffman, D. and E. L. Wynder. "Environmental Respiratory Carcinogenesis," in: "Chemical Carcinogens" (C. E. Searle, ed.), American Chemical Society, Washington, DC (1976), pp. 324–365.

55. Prindle, R. A. "Some Considerations in the Interpretation of Air Pollution Health Effects Data," *Journal of the Air Pollution Control Association* 9:12–19 (1959).

56. Hammond, E. C. and D. Horn. "Smoking and Death Rates—Report on 44 Months of Follow-up of 187,783 Men," *Journal of the American Medical Association* 166:1159–1172, 1294–1308 (1958).

57. Curwin, M. P., E. L. Kennaway and N. M. Kennaway. "The Incidence of Cancer of the Lung and Larynx in Urban and Rural Districts," *British Journal of Cancer* 8: 181–198 (1954).

58. Blot, W. J., L. A. Brinton, J. F. Fraumeni and B. J. Stone. "Cancer Mortality in U.S. Counties with Petroleum Industries," *Science* 198:51–53 (1977).

59. Mason, T. J., F. W. McKay, R. Hoover, W. J. Blot and J. F. Fraumeni. "Atlas of Cancer Mortality for U.S. Counties: 1950–1969," *DHEW Publication No. (NIH)75-780*, U.S. Department of Health, Education and Welfare, Washington, D.C.

60. Spiess, H. and C. W. Mays. "Bone Cancers Induced by ^{224}Ra (Th X) in Children and Adults," *Health Physics* 19:713–729 (1970).

61. Faber, M. "A Follow-up of 1000 Thorotrast Cases in Denmark," *Proceedings of the First International Congress of Radiation Protection*, Pergamon Press, New York (1968), pp. 1521–1524.

62. Battifora, H. A. "Thorotrast and Tumors of the Liver," in: "Heptocellular Carcinoma" (K. Okuda, ed.), John Wiley and Sons, New York (1976), pp. 83–93.

63. Hempelmann, L. H., W. J. Hall, M. Phillips, R. A. Cooper and W. R. Ames. "Neoplasms in Persons Treated with X-Rays in Infancy: Fourth Survey in 20 Years," *Journal of the National Cancer Institute* 55:519–530 (1975).

64. Stewart, A. and G. W. Kneale. "Radiation Dose Effects in Relation to Obstetric X-rays and Childhood Cancers," *The Lancet* (1970), p. 1185.

65. Jablon, S. and H. Kato. "Childhood Cancer in Relation to Prenatal Exposure to Atomic-Bomb Radiation," *The Lancet* (1970), pp. 1000–1003.

66. United Nations Scientific Committee on the Effects of Atomic Radia-

tion. "Ionizing Radiation: Levels and Effects," 27th Session, Supplement No. 25 (A/8725). United Nations, New York (1972).

67. Simon, N. and S. M. Silverstone. "Radiation as a Cause of Breast Cancer," *Bulletin of the New York Academy of Sciences* 52(7):741–751 (1976).

68. Simon, N. "Breast Cancer Induced by Radiation," *Journal of the American Medical Association* 237(8):789–790 (1977).

69. Shore, R. E., L. H. Hempelmann, E. Kowaluk, P. S. Mansur, B. S. Pasternack, R. E. Albert and G. E. Haughie. "Breast Neoplasms in Women Treated with X-Ray for Acute Postpartum Mastitis," *Journal of the National Institute* 59(3):813–822 (1977).

70. U.S. Department of Health, Education and Welfare, "Gonad Doses and Genetically Significant Dose from Diagnostic Radiology: U.S., 1964 and 1970," *HEW Publication (FDA), 76-8034*, U.S. Government Printing Office, Washington, D.C. (1976).

71. Clayson, D. B. and P. Shubik. "The Carcinogenic Action of Drugs," *Cancer Detection and Prevention* 1(1):43–77 (1976).

72. Hoover, R. and J. F. Fraumeni. "Drugs," in: "Persons at High Risk of Cancer" (J. F. Fraumeni, ed.), Academic Press, New York (1975), p. 185.

73. Eckhardt, R. E. "Occupational Cancer Now," in: "Environment and Cancer," *Proceedings of the 24th Annual Symposium on Fundamental Cancer Research*, Williams and Wilkins Co., Baltimore, Md. (1972), p. 93.

74. Evans, R. D. "The Radium Standard for Boneseekers—Evaluation of the Data on Radium Patients and Dial Painters," *Health Physics* 13:267–278 (1967).

75. Archer, V. E., J. D. Gillman, and J. K. Wagoner. "Respiratory Disease Mortality among Uranium Miners," *Annals of the New York Academy of Sciences* 271:280–293 (1976).

76. Hueper, W. C. "Occupational Tumors and Allied Diseases," Charles C. Thomas, Springfield, Ill. (1942).

77. Clayson, D. B. "Case Study 2: Benzidine and 2-Naphthylamine—Voluntary Substitution or Technological Alternatives," *Annals of the New York Academy of Sciences* 271:170–175 (1976).

78. Wagoner, J. K. "Occupational Carcinogenesis: The Two Hundred Years Since Percivall Pott," *Annals of the New York Academy of Sciences* 271:1–4 (1976).

79. International Agency for Research on Cancer (IARC). "Report of a Working Group on Epidemiological Studies on Vinyl Chloride Exposed People," IARC Internal Technical Report No. 75/001, Lyon (January 8–9, 1975).

80. Dinman, B. C. "The Nature of Occupational Cancer," Charles C. Thomas, Springfield, Ill. (1974).

81. Cole, P. and M. B. Goldman. "Occupation," in: "Persons at High Risk of Cancer" (J. F. Fraumeni, ed.), Academic Press, New York (1975), p. 167.

82. Hoover, R. and J. F. Fraumeni. "Cancer Mortality in U.S. Counties with Chemical Industries," *Environmental Research* 9:196–207 (1975).
83. U.S. Department of the Interior, Bureau of Mines. "Mineral Facts and Problems," Bureau of Mines *Bulletin* #650, U.S. Government Printing Office, Washington, D.C. (1970).
84. Buchanan, W. M. "Asbestosis and Primary Intrathoracic Neoplasms," in: "Biological Effects of Asbestos," *Annals of the New York Academy of Sciences* 132 (Art. 1):1–766 (1965), p. 507.
85. Hammond, E. C., I. J. Selikoff and J. Churg. "Neoplasia among Insulation Workers in the United States with Special Reference to Intra-Abdominal Neoplasia," *Annals of the New York Academy of Sciences* 132 (Art. 1):1–766 (1965), p. 519.
86. New York Academy of Sciences. "Biological Effects of Asbestos," Proceedings of a Conference held on October 19–21, 1964 (H. E. Whipple and P. E. van Reyen, eds.), *Annals of the New York Academy of Sciences* 132 (*Art. 1*):1–766 (1965).
87. Selikoff, I. J. "Environmental Cancer Associated with Inorganic Microparticulate Air Pollution," in: "Clinical Implications of Air Pollution Research," Publishing Sciences Group, Inc., Acton, Mass. (1976), p. 49.
88. Christensen, H. E., T. T. Luginbyhl and B. S. Carroll. "Suspected Carcinogens: A Subfile of the NIOSH Toxic Substances List," National Institute for Occupational Safety and Health, U.S. Department of Health, Education and Welfare. U.S. Government Printing Office, Washington, D.C. (June 1975).
89. Food and Drug Administration. "Study of the Delaney Clause and Other Anti-Cancer Clauses," Part 8 in: "Agriculture-Environmental, and Consumer Protection Appropriations for 1975," Hearings before a Subcommittee of the Committee on Appropriations, House of Representatives, 93rd Congress, 2nd Session, U.S. Government Printing Office, Washington, D.C. (1974), p. 163–591.
90. Environmental Protection Agency. "Interim Procedures & Guidelines for Health Risk and Economic Impact Assessments of Suspected Carcinogens," Washington, D.C. (1976).
91. McCann, J., E. Choe, E. Yamasaki and B. N. Ames. "Detection of Carcinogens as Mutagens in the *Salmonella*/Microsome Test: Assay of 300 Chemicals," Part I, *Proceedings of the National Academy of Sciences* 72(12):5135–3139 (1975).
92. McCann, J. and B. N. Ames. "Detection of Carcinogens as Mutagens in the *Salmonella*/Microsome Test: Assay of 300 Chemicals: Discussion," Part II, *Proceedings of the National Academy of Sciences* 73(3):950–954 (1976).
93. Van Duuren, B. L. "Carcinogenic Epoxides, Lactones, and Halo-ethers-and-their Mode of Action," *Annals of the New York Academy of Sciences* 163(2):633–651 (1969).
94. Nelson, N. "The Chloroethers—Occupational Carcinogens: A Summary of Laboratory and Epidemiology Studies," *Annals of the New York Academy of Sciences* 271:81–90 (1976).

95. Mukai, F. and W. Troll. "The Mutagenicity and Initiating Activity of Some Aromatic Amine Metabolites," *Annals of the New York Academy of Sciences* 163:828 (1969).
96. Laskin, S., M. Kuschner, R. T. Drew, V. P. Cappiello and N. Nelson. "Tumors of the Respiratory Tract Induced by Inhalation of Bis(chloromethyl) Ether," *Arch. Env. Health* 23:135–136 (1971).
97. Albert, R. E., B. S. Pasternack, R. E. Shore, M. Lippmann, N. Nelson and B. Ferris. "Mortality Patterns among Workers Exposed to Chloromethyl Ethers: A Preliminary Report," *Environmental Health Perspectives* 11:209–214 (1975).
98. Albert, R. E. and B. Altshuler. "Assessment of Environmental Carcinogen Risks in Terms of Life Shortening," *Environmental Health Perspectives* 13:91–94 (1976).
99. Evans, R. D. "Radium in Man," *Health Physics* 27:497–510 (1974).
100. National Council on Radiation Protection and Measurements. "Review of the Current State of Radiation Protection Philosophy," *NCRP Report No. 43,* Washington, D.C. (1975).
101. Burger, E. J. "Protecting the Nation's Health," D. C. Heath and Co., Lexington, Mass. (1976).
102. Lowrance, W. W. "Of Acceptable Risk," Wm. Kaufmann, Inc., Los Altos, Cal. (1976).

CHAPTER 9

The Organic Chemicals

The organic chemical industry has played an important role in producing the drugs that protect us from many diseases, the pesticides and herbicides that have increased agricultural yields, and the many forms of plastics that provide us with inexpensive durable fabrics, furniture, floor coverings, and a myriad of other products we encounter in our daily lives. Hundreds of thousands of new chemicals have been produced, and many of these have found their way to the market place. They include highly specialized pharmaceuticals manufactured by the ounce, and products such as synthetic rubber, and some products derived from petroleum that are produced in hundred-ton quantities. The synthetic organic chemicals have been produced at an exponentially increasing rate since World War II, from a little more than 2.5 million tons in 1940 to about 90 million tons in 1970. (See Figure 9–1.)

The organic chemical industry, from its earliest days, has been beset with occupational health problems, ranging from relatively mild skin rashes to fatal diseases. In recent years there has also been increased awareness of the effects of many organic compounds on wildlife, as well as the effects that may be produced in the general population by long-term exposures to contaminants in air, water, or food. The public is exposed to many fewer compounds than industrial workers, and at far lower levels of concentration, but there are many reasons for being more cautious when the general public is involved.

A review of even a representative sample of all of the organic compounds used by modern industry would make this book unmanageable in size. However, two compounds, DDT and the PCBs, are of special interest, and are selected here for special attention. Together, they illustrate many of the problems common to synthetic organic compounds generally.

227

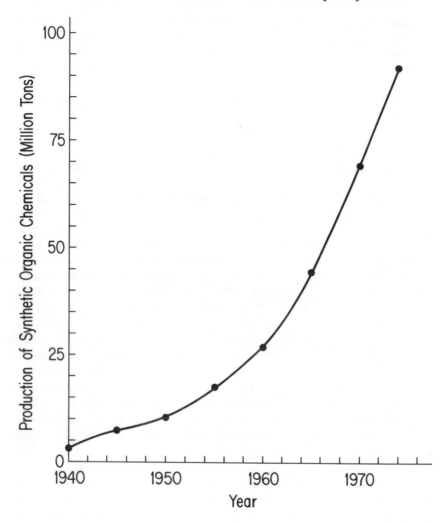

Figure 9-1. Production of synthetic organic chemicals in the United States between 1940 and 1974.

The PCBs and DDT are both members of a class of chemicals known as halogenated hydrocarbons. These are organic compounds in which hydrogen atoms are replaced by one or more of the halogens—chlorine, bromine, or fluorine. An example of a chlorinated hydrocarbon that was widely used at one time for household and industrial purposes is carbon tetrachloride, a compound in which the four hydrogen atoms in methane, CH_4, are replaced by chlorine to form CCl_4. Carbon tetrachloride was once popular as a solvent, degreaser, and spot remover both in industry and in the home. A few decades ago, it was realized that CCl_4 was capable of producing serious liver injury among persons exposed to its vapors, and its use for these purposes was substantially reduced in favor of less toxic substances by midcentury.

Other halogenated hydrocarbons can be prepared from methane by replacing the hydrogen atoms by fluorine and chlorine. These compounds include the Freons, CCl_3F (Freon 11), and CCl_2F_2 (Freon 12), which are widely used as refrigerants in air-conditioners and refrigerators. The Freons have also been used as the propellant gas in home atomizers such as those used for antiperspirants, shaving-cream dispensers, and insect repellants. Although carbon tetrachloride is highly toxic, the Freons are not, but they have attracted attention in recent years because they are chemically inert, persist in the atmosphere, and diffuse into the stratosphere where, by dissociation under the influence of sunlight, they produce free chlorine. We will see in Chapter 14 that free chlorine has the ability to reduce the ozone content of the stratosphere which, in turn, may result in increased transmission of harmful ultraviolet radiation to the earth's surface. The use of Freon as the propellant for many atomizer applications was banned by the Environmental Protection Agency (EPA) in 1977.

DDT

Of all the environmental issues that involve the use of chemicals, none persisted for so long and excited such extreme points of view as the controversy over DDT.

Paul Müller of Switzerland discovered in 1939 that DDT (an abbreviation for dichlorodiphenyl-trichloroethane), was a remarkably effective insecticide. An attractive property was its persistence: Its insecticidal potency after application did not diminish as rapidly as that of other insecticides. It was widely used during World War II for mosquito eradication in the malaria-ridden theaters of war and was credited with the quick control of a typhus epidemic in Italy in 1943 and 1944. Considering that typhus claimed millions of lives in Eastern Europe after World War I, the fact that this louse-borne disease could be controlled by dusting the wartime refugees with DDT was of major public health significance.

Other chlorinated hydrocarbons such as aldrin, dieldrin, endrin, heptachlor, and chlordane were soon developed for use as pesticides, but none could compete with DDT to control mosquitos, which in many areas of the world are the most obnoxious of all insects. Mosquitos are not only the carriers of malaria, but also carry yellow fever, dengue fever, encephalitis, and other diseases. In the United

States, DDT was widely used for control of cotton pests.

Early studies of the toxicity of DDT indicated that it was relatively nontoxic to mammals, but it became apparent in the mid-1950s that it was capable of causing serious ecological damage. Useful insects were being eliminated along with the pests, and birds and fish were being poisoned by consuming DDT-contaminated food organisms such as insects, seeds, and aquatic life. It was also learned that DDT was a contaminant of human foods and was present in body fat, where most of the chlorinated hydrocarbons tend to collect. Although no known cases of DDT injury were reported among the general public, concern was suddenly aroused because of DDT's possible carcinogenicity. Most of the studies of mammalian cancer were negative, but there were a few of reports that tumors, both malignant and benign, had developed in some experimental animals. Widespread publicity was given to the prospect that this ubiquitous contaminant would, in time, cause cancer in the general population. This was stressed in Rachel Carson's *Silent Spring,*[1] as well as in the publications of other popular writers, and some scientists.

Although *Silent Spring* was concerned mainly with the effects of pesticides on wildlife, the book included a chapter on cancer that has served well to alert the world to the dangers of environmental cancer generally, and in the long run it may be judged to have performed a valuable service for that reason. However, the chapter is replete with statements that have made the book so controversial. For example, the statement was made that *"today more school children die of cancer than of any other disease."* This happens to be a true statement, but is not relevant to the question of chemical carcinogenicity and is very misleading. Today cancer *is* a major cause of death among children, second only to accidents. It was less in importance on the list of childhood diseases in past generations because of the prevalance of deaths due to the diarrheas, respiratory infections, diphtheria, polio, and other diseases that have since been eliminated. It was noted in the last chapter that there are certain cancers—fortunately, comparatively rare—that occur mainly in children, but there is no evidence that these cancers are due to chemicals introduced recently into the environment. In fact, we learned that cancer mortality among children in the United States in recent decades has been reduced, due to advances in treatment.

The question of the carcinogenicity of DDT divided able scientists into two camps: those who believed DDT was a human carcinogen and those who did not. We will see that this disagreement was to arise again and again in connection with other pesticides, food additives, and industrial chemicals. As we saw in the last chapter, there are severe statistical obstacles inherent in proving "no effect," and many scientists are reluctant to accept evidence of tumors developed by laboratory animals exposed to high doses as evidence that a substance can be carcinogenic to humans when it is administered in low doses.[2] In the case of DDT, there was also disagreement as to the significance of the observed tumors. The fact that high doses of DDT produced liver tumors in rats was reported as early as 1947, and many subsequent studies confirmed this finding, but there was persistent disagreement as to whether the tumors were potentially malignant. A number of committees appointed to study the matter concluded that a prudent course of action required the assumption that some of the benign tumors would in time become malignant.[3]

Other pesticides were becoming available, but they were usually more expensive, sometimes less effective, and often far more toxic to humans.[4] In time, many of them were also found to produce tumors in animals.

It is significant that 30 years after DDT was first introduced, there is no evidence that it or its related organochlorine compounds have contributed to the production of human liver cancer of the kind seen in experimental animals. As a matter of fact, deaths due to liver cancer in the United States have *decreased* about 30 percent from 8.4 per 100,000 in 1944, when DDT was first introduced, to 5.6 in 1972. This is all the more significant in view of the increased age of the population since 1944.[5]

The epidemiological evidence that DDT is not a direct threat to human safety, insofar as acute effects are concerned, is impressive. Eighty-seven deaths were reported from pesticides of all kinds in the United States in 1969. The one case involving DDT was a two-year-old child who drank milk to which her five-year-old brother had added an exterminating fluid containing this pesticide. The death was caused by the insecticide solvent rather than by the DDT.[6] By 1971, after DDT had been widely used for nearly 30 years, the World Health Organization (WHO) was able to find only one death and three cases of acute poisoning caused by DDT. All four episodes were due either to a suicide attempt or to an accident.[7] This is an excellent record, considering that the less persistent but more toxic pesticides are estimated by WHO to cause about 500,000 cases of poisoning annually, with no less than 5,000 deaths.[4] When DDT substitutes were introduced in one WHO program, all of the sprayers showed neurological symptoms of insecticide poisoning within two weeks. Twenty out of 5,000 villagers also showed such symptoms.

The persistence of DDT in the environment is an asset to its function as a pesticide, but it is disadvantageous ecologically because DDT is readily absorbed by living organisms. When DDT is sprayed onto a field or is washed into a lake or estuary, the pesticide passes stepwise from the lower life forms to the higher forms. At each step, a process of biological magnification takes place.[8,9] In one example, DDT was found in plankton at a concentration of .04 parts per million. As it passed gradually up the food chain, it was concentrated, by birds, to more than 100 times the concentration originally contained in the plankton. This process was illustrated in Figure 7–3. The accumulation of DDT in birds became so high as to affect their ability to reproduce. In some way not fully understood, DDT was found to be associated with thinning of egg shells, causing premature breakage and loss of chicks.[10,11] DDT has also been shown to cause impairment of reproductive and liver functions in birds and, at higher doses, to produce neurological damage.[12] As a result, bird populations in the United States and England in the 1950s and 1960s decreased. This was widely attributed to the effects of DDT and other chlorinated organic pesticides.[12] Not all scientists agreed that this was the case, and some have attributed the declines to the natural fluctuations in weather and to human predation.[13,14] However, the evidence for severe reductions in the populations of many birds including the hawks, eagles, robins, and others is impressive, although it is possible that in a few instances, declines due to other factors were erroneously attributed to DDT.

Concern also developed because DDT was detectable in the oceans, and laboratory studies had shown that the photosynthetic activity of plankton was reduced by DDT.[15] Because of the importance of the marine plankton as a source of oxygen, it was suggested that DDT could reduce the world's oxygen supply.[16,17] However, even in badly contaminated coastal waters, DDT contamination was far below the level which would have been required to reduce oxygen production by plankton.[18,19]

By the mid-1960s it was apparent that DDT was in additional trouble. The insect pests were gradually developing resistance to the new pesticides—DDT included—as a result of genetic mutation. In some areas of the world, the malaria-eradication programs met with diminished success after having produced spectacular results for the two decades following the introduction of DDT after World War II. More than 200 species of insects, including the common housefly, have developed such resistance to DDT.[20]

Despite all the drawbacks, WHO has stood firmly by its position that this pesticide is important in disease control, stating that the concept of malaria eradication rests completely on its (DDT) continued use. The record of safety of DDT to man has been outstanding during the past 20 years, and its low cost makes it irreplaceable in public health at the present time. Limitations on its use would give rise to greater problems in the majority of the developing nations.[12]

Following the publication of *Silent Spring,* opposition to the use of DDT in the United States mounted rapidly and led to bitter debates between the opponents and proponents of the continued use of DDT. In 1966, a small group of scientists and a lawyer on Long Island, New York banded together to prevent the Suffolk County Mosquito Control Commission from spraying DDT onto the local marshes. In what may have been the first successful attempt by scientists to use the law to achieve an environmental goal, the State Supreme Court supported the petition. The group, "by coupling science with the law, had accomplished in a few weeks, what conservationists on Long Island had previously been unable to achieve during the preceding ten years."[21,22] This modest effort led to formation of the Environmental Defense Fund (EDF) in the fall of 1967. This coalition of scientists and lawyers moved into Michigan and then Wisconsin in successful forays that led to the banning of DDT in those states. Finally, EDF took the issue to the federal government and succeeded in obtaining an EPA ban. Although the examiner issued an opinion that all "essential" uses of DDT should be retained, his recommendation was reversed by the EPA Administrator at the time, William D. Ruckelshaus.[23] The use of DDT was banned by EPA order on December 31, 1972 except for very specialized uses. An important opinion that influenced the decision of the hearing examiner was that the use of DDT did not involve a risk of human cancer. The EPA administrator did not agree with this finding. The question of the carcinogenicity of DDT was basic to the final decision.

Eighty million pounds of DDT were being used in the United States in 1959. Actually, its use declined thereafter to less than 25 million pounds per year when the ban took effect. Consumption has been less than 500,000 pounds per year since 1972, mainly for use against the tussock moth in the northwestern part of the United States.[24]

Contamination of both wildlife and human food with DDT has diminished rapidly since the curtailment of its use. The maximum DDT concentrations in the fatty tissues of migratory songbirds varied between 18 and more than 30 parts per million during the period between 1964 and 1972. However, in 1973 the maximum concentration had dropped to about 3 parts per million.[25] DDT residues in starlings decreased 36 percent from 1972 to 1974.[26] Lake trout in Lake Michigan contained almost 20 parts per million in 1970, but less than ten parts per million in 1974, and Coho salmon from Lake Michigan dropped from 15 parts per million in 1970 to five parts per million in 1973. Lake Michigan chubs, which contained ten parts per million in 1970, dropped to less than two parts per million in 1974.

The brown pelican population off the coast of southern California had been greatly reduced in numbers by 1969, apparently due to the effect of DDT on the egg shells. The pelicans absorbed the DDT from anchovies contaminated by the effluence of a large sewage-treatment plant near Los Angeles. This source was eliminated in the spring of 1970, and by 1974 the DDT content of the anchovies was reduced to less than 4 percent of the 1969 values, and the hatchability of the pelican eggs had greatly improved.[27]

The rapid decline in DDT contamination was reassuring in view of ominous forecasts, made on the basis of early data, that it would take ten years or longer for DDT to be reduced by 50 percent if all discharges to the environment were stopped.[28] The data obtained in recent years suggest that the half-time (or the time it takes to reduce the quantity of a substance by 50 percent) for disappearance of DDT in fish and birds ranges from one to two years. However, there is evidence that the DDT content of soil may be declining more slowly. In Arizona, where a moratorium on the use of DDT was declared in 1969, the residues of DDT in the local soil were essentially unchanged four years later.[29] The Arizona findings may have been affected by the extraordinary dryness of the area. Elsewhere in the United States, DDT residues in soil dropped about 36 percent during the four-year period 1969 through 1973.

For many years the Food and Drug Administration has been monitoring "market basket" samples of foods taken from market shelves. The data on the amount of DDT these samples contained in the period 1965 through 1974 indicate that the DDT residues declined rapidly during that period. The daily intake was estimated at about 70 milligrams per day in 1966, and had diminished to about five milligrams in 1974. This corresponded to a half-time of about two years.

The DDT content of human fat in recent years is shown in Figure 9–2, in which it is seen that the DDT burden in subjects above 15 years of age continued to rise until 1971, when it reached a peak of almost eight parts per million. Two years later, the DDT content had diminished only about 25 percent in this age group. However, the amount of DDT in the fat of children younger than 14 years of age has

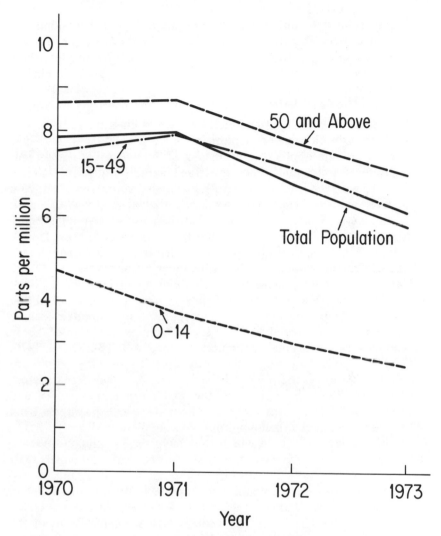

Figure 9-2. DDT in human fatty tissues by age between 1970 and 1973.[24]

been lowered more substantially, and by 1973 was less than three parts per million. This can be interpreted to mean that DDT persists for a long time in human fat and that, once deposited, its rate of removal does not reflect the rate at which the DDT intake is diminishing. However, in the younger members of the population, whose fat was formed when the dietary DDT content had already diminished, the levels of DDT in fat did not achieve the high levels of the older persons who had been exposed in earlier years. There is no evidence that the presence of DDT in human fat is harmful at the levels reported in the general population.

THE POLYCHLORINATED BIPHENYLS (PCBs)

The PCBs are a family of chemicals known as chlorinated biphenyls and which with chlorinated naphthalenes were used early in World War II to provide nonflammable insulation for electrical equipment. The products were sold under the trade name "Halowax," and a number of workers who were exposed to their vapors and dusts died from liver damage. The chlorinated naphthalenes proved to be so toxic to humans working with them that their use as insulating materials was voluntarily discontinued, but the chlorinated biphenyls now known as PCBs (polychlorinated biphenyls) continued to be used in electrical equipment, and for numerous other purposes.

The PCBs comprise more than 200 chemical species, each distinguished by differences in the number of chlorine atoms and the position in which they have been incorporated into ringlike organic molecules known as "phenyls." Large-scale production of these compounds began around 1930 and reached a maximum in 1970, afterwhich production was voluntarily curtailed by the Monsanto Chemical Company, the principal manufacturer, because of developing concern about the effects of the PCBs on human health and wildlife. Sale of PCBs was thereafter limited to sealed systems from which environmental contamination could be minimized.

During the many years in which they have been manufactured, the PCBs had been found to have remarkable properties that favored their use for many purposes. Because the compounds are excellent electrical insulators and are not flammable, they have been used as dielectrics in transformers and large electrical condensers. These are sealed systems in which the PCBs can remain in place for the life of the equipment and can then be purified for reuse when the equipment is scrapped. However, some leakage can be expected to occur during the lifetime of the equipment, and PCBs may eventually enter the environment via sewers or surface runoff to streams.

PCBs are also used in small electrical capacitators, such as those in fluorescent-lamp boosters, that have a relatively short life and ultimately find their way to the scrap heap for incineration or land disposal. The compounds are also used in lubricating and cutting oils, as plasticizers in paints, in pesticides, in the manufacture of paper, adhesives, and plastics, as well as in heat transfer and hydraulic fluids. The many uses of the PCBs, plus the fact that they are highly stable chemicals, assured their role as ubiquitous environmental contaminants.

The danger of PCBs to humans came to worldwide attention as a result of an accident in Japan in 1968, in which rice oil was contaminated by PCBs leaking from a heating system.[30] The first cases had unusual skin complaints, and upon investigation it was found that all patients had consumed rice oil that was contaminated by PCBs. The disease was soon found to involve more than skin complaints and was given the name "yusho" (oil disease). The patients suffered from fatigue, nausea, swelling of the arms and legs, and liver injury. Babies born to mothers with yusho were smaller than normal. In addition, their subsequent growth and development were diminished, all of which indicated that the PCBs had caused fetal injury after being transported through the placenta. By 1973, about 1,200 cases of yusho had been reported as a result of this mishap.[31]

Shortly after the yusho incident occurred, it became apparent that the analytical procedures used to detect DDT were being affected by the presence of another class of chlorinated organic compounds whose behavior in the analytical procedures was very similar to that of DDT. The results of tests for DDT were being biased by the presence of substances which were soon found to be PCBs. Thus, coincidentally with worldwide awareness of the human toxicity of PCBs, it was found that they, like DDT, were persistent in the environment and were present as contaminants of fresh water and marine organisms.

Like DDT, the PCBs tend to accumulate in the fatty tissues of plants and animals. They are also capable of biomagnification: PCBs, at each step in the food chain, can be concentrated from ten to 100 times the concentration in the previous step. At the top of the food chain, in the sea birds and large carnivorous fish and sea mammals, the concentration can be 10 million times that in the surrounding water.[32] As a result, although the concentration of PCBs in the ambient water might be only one part per billion, fish from the waters and bays of Europe, North America, and Japan contain several parts per million. Concentrations as high as 800 parts per million have been found in the fish from some inland waterways.[33] The higher levels tend to be found near centers of industrial activity, but worldwide contamination exists at levels on the order of one part per million.

The toxicity of PCBs is variable, depending on the biological species and the particular molecular form of PCB to which they are exposed.[34-36] Some organisms are able to tolerate concentrations of PCBs that kill others. One of the most susceptible is the pink shrimp, among which complete mortality can be produced in water containing

100 parts per billion. The stage of development is also important: Many fish that can tolerate several parts per million as adults are seriously affected in earlier stages of growth. Among the mammals, the mink has proved to be markedly sensitive, a fact that was discovered in 1967 when commercial mink ranchers fed their stock food prepared from Coho salmon contaminated by PCBs in a Lake Michigan tributary. Subsequent laboratory experiments demonstrated that PCBs were extraordinarily toxic to mink, and even more so to the embryos.[37]

The acute toxicity of PCBs to humans was demonstrated in the yusho incident, but in that instance the levels of exposures were high and therefore do not shed light on the toxicity of PCBs to humans at the levels "normally" found in the environment. The main source of PCBs for humans is the diet, and FDA market-basket studies since 1969 have shown that fish are the major source of PCBs.[38,39] The daily intake of PCBs by the average teenage American male has dropped from 15 micrograms per day in 1971 to about 8.7 in 1974 and 1975. This is due to reduced contamination in food other than fish and poultry, which have not changed significantly.

Studies in Canada[40] and the United States[41] have shown that PCBs tend to concentrate in body fat and, because of the high fat content of human milk, can be passed in this way to breast-fed babies. The actual PCB content of the fat in human milk is approximately the same as in other fatty tissues of the body and has averaged about 1.7 parts per million in North America during the period between 1969 and 1974. Samples containing as much as ten parts per million have been reported. People who eat fresh fish, particularly from lakes known to be contaminated with PCBs, tend to have the highest levels.

There is reason to be concerned about these reports. Experiments with rhesus monkeys have shown that severe PCB intoxication occurs in the rhesus infants born to mothers fed PCB-contaminated food. The levels of PCBs in the rhesus milk are comparable to the highest values found in human milk in the United States and Canada. Nevertheless, the consensus among informed individuals is that the presence of PCBs in human milk does not warrant changes in nursing practice.[42] However, an advisory committee to the New York State Health Department recommended that if mothers have been consuming fish from waters known to be contaminated with PCBs on a regular basis, nursing should be discontinued after three months.[43]

It has also been found that 2.5 to five parts per million of PCBs in the diet of rhesus monkeys caused toxic symptoms to develop within one to two months.[44] The rhesus monkey is a relatively sensitive species, and is affected at concentrations as low as 1 percent of those that affect rats.[45]

On the assumption that a teenager consumes 2,000 grams of food per day, his daily intake of PCBs is of the order of 1 percent of that known to have caused toxic symptoms in monkeys. This may seem to be a safe level, but it should be noted that the market-basket estimates are applicable to "average" people, and not to people who consume large quantities of fresh-water fish. Moreover, since PCBs are passed to breast-fed infants, infants may be relatively heavily exposed.

The infants of rhesus monkey mothers fed about 2.5 parts per million of PCB in their diet were undersized at birth, generally debilitated, and subject to frequent infections. We cannot be certain that somewhere in the United States there are not nursing mothers with a preference for fish contaminated with PCB, whose nursing babies are showing the effects of PCB toxicity.

The basic concerns about PCBs are over the relative susceptibility of humans compared to rhesus monkeys, and to what extent there are deviations from the national average. There have been no reports of PCB poisoning in nursing infants, but there had been no epidemiological studies as of the end of 1977. Only by careful and sufficiently extensive studies can assurances be gained that occasional cases of PCB poisoning are not occurring in the infant population. The fact that these studies have not been done is an example of the ineffectiveness of the means for epidemiological research for environmental health in the United States.

The PCBs are a particularly difficult class of compounds to evaluate from the toxicological point of view. As noted earlier, their chemical complexity is such that they exist in more than 200 forms (isomers), and the toxicity of a particular isomer, as well as the rate at which it will degrade in the environment, depends on the chemical species involved. A further complication arises because PCBs can be transformed into a class of highly toxic impurities, polychlorinated dibenzofurans (PCDFs). These PCDFs can be formed in the process of manufacture or by subsequent degradation processes. It is known that the contaminated rice oil that caused yusho in Japan contained high levels of dibenzofurans, and it has been suggested that because the PCBs were used as a heat-transfer medium, the high temperatures may have degraded the PCBs to produce PCDFs. The toxicity of PCBs may thus depend, not only on their particular chemical identity, but also on the extent to which they are contaminated with dibenzofurans.[46] The role of PCDFs in the epidemiology of PCB poisoning is not well understood. Unfortunately, the PCDFs are extremely difficult to measure in trace amounts and very few studies on the effects of PCDFs have been reported.

In 1973, the FDA issued regulations that banned the industrial uses of PCBs in establishments manufacturing, handling, or storing human food, animal feeds, and paper food-packaging materials.[47] The regulations also proposed temporary limits on PCBs in food. The most important limitation was that pertaining to fish: a maximum of five parts per million in the edible portions. The Canadian authorities subsequently reduced the permissible PCB concentration in fish to two parts per million, following which the United States FDA proposed a similar change.[48]

The PCB content of a fish depends to a large extent on its feeding habits. For example, the American shad spends most of its life in the open ocean but comes to the estuaries to spawn. Since the shad are usually caught for market on their way to their spawning grounds, they have relatively low levels of PCBs. Catfish and eels, which tend to feed among the sediments of the estuaries, are among the fish that contain the highest concentrations.

It is ironic that, at the very time when benefits were being expected from the enormous expenditures for the clean waters programs of many states, the PCB and other similar problems have developed and, for some time to come, will deny the use of important waterways for some of the very purposes for which billions were spent for pollution control. As an example, the total expenditures by local, state and federal governments for water-pollution control in the Hudson River estuary were in excess of 3 billion dollars by 1976. However, the new sewage-treatment plants were not designed to remove toxic substances that accordingly are allowed to enter the lakes, rivers, or bays along with the treated sewage.

Because of the high concentrations of PCBs in Hudson River fish, it was decided to ban commercial fishing for an indefinite period starting late in 1975. The principal species affected was the striped bass, which is highly prized by both sports and commercial fishermen. Sports fishing was allowed, but the fishermen were cautioned not to eat more than one fish meal per week and were forbidden to sell the catch. Eels were not allowed to be taken by either sportsmen or commercial fishermen. An exception to the ban on commercial fishing was made in the case of shad which, for the reasons noted, proved not to be excessively contaminated.

The main sources of PCB contamination of the Hudson were believed to be two General Electric Company plants at Fort Edward and Hudson Falls, located in the upper reaches of the river. A prolonged hearing resulted from charges by New York State that the company polluted the waters of the river to such an extent as to be injurious to wildlife, and that the company should eliminate its discharge of PCBs and "restore the health of the Hudson River and other natural resources to the extent that PCB discharges have despoiled them. . . ."[49] The hearing that followed illuminated the history of PCB discharges by the company in a manner that was neither a credit to the company nor to the state and federal governments. As stated in the opinion of the hearing examiner, "these unlawful consequences are the product of both corporate abuse and regulatory failure: corporate abuse in that G.E. caused the PCBs to be discharged without exercising sufficient precaution and concern; regulatory failure in that G.E. informed the responsible Federal and state agencies of its activities, and they too exercised insufficient caution and concern. . . ."

The General Electric Company had purchased more than 41,000 tons of PCBs of various types from the Monsanto Chemical Company since 1966. The material was used in the manufacture of electrical condensers, and during much of the time it used them, the company had been discharging an average of about 30 pounds per day of PCBs to the river.

As a result of the 1972 amendments to the Federal Water Pollution Control Act, the states were required to adopt control programs based on effluent limitations. As a result, New York State required that a certificate be obtained that authorized the discharge of individual pollutants. A federal discharge permit was required following certification by the state. General Electric applied for such a permit in late 1971. Two years later, the state certified that no effluent standards were applicable to the proposed discharges, and the U.S. Environmental Protection Agency regional office issued a draft permit that proposed to authorize a daily average discharge of 30 pounds from the two plants. The draft permit also proposed that the chlorinated hydrocarbon discharges be reduced to zero within 21 months of the permit's issuance.[49]

A requirement for zero discharge is unrealistic. Large industrial plants can reduce their discharges to any desired level short of zero. The discharge of 30 pounds per day could have been reduced to three pounds per day, .03, .003 pounds, or lower, but some finite discharge would remain that could be detected. In practice, a waste-discharge limit should be based on the risks to human health and wildlife, technical feasibility, the characteristics of the receiving body of water, and cost. If the plant is to operate, the permissible discharge must be expressed in finite terms. It can be reduced to whatever level is required, but not to zero.

When the permit was finally issued in December of 1974, it ordered that the discharge of PCBs be reduced to 3.5 ounces by mid-1977. The position that General Electric took at the hearing rested substantially on the premise that since the Department of Environmental Conservation and the Environmental Protection Agency were fully aware of the PCB discharges, the discharges could not be illegal. The hearing examiner commented that "G.E.'s argument has more than superficial appeal."

By 1972 it was well known by the State Department of Environmental Conservation that PCBs were present in striped bass in concentrations ten times or more the FDA's limit of five parts per million. The high levels persisted through 1977, despite the fact that G.E. quickly reduced its releases. The PCBs were deposited in the sediments, where they have served as a reservoir for pollution of the biota, and it is not known how long it will take for the river to cleanse itself.

The state had the power all along to correct the problem, the existence of which was known six years before the matter received public attention. Since the matter of PCB pollution was a national problem, the state looked to the federal government for policy, but received only a guideline for the maximum permissible PCB content of fish, and no detailed instructions as to how the guideline should be interpreted. The federal government had the power to stop the sale of fish in interstate commerce, but failed to do so. The company presumably thought the problem could not be a serious one because they applied for and received permission to discharge PCBs in the stated amount. However, the large companies that manufactured and used the PCBs had the means to evaluate the impact of PCB discharges to the estuary. They did not do so on the assumption that generic questions of this kind have been dealt with traditionally by government.

Problems similar to those created by PCBs and DDT have been caused by other chlorinated organic compounds. These include kepone, and mirex. Most of the offending organic chemicals have been chlorinated hydrocarbons that are both toxic and persistent in the environment.

The episodes of pollution by chlorinated hydrocarbons served to emphasize the need for a National Toxic Substances Control Act that would prevent potentially catastrophic pollution with chemical wastes. A law finally was signed by President Ford in 1976. It requires that before they can be approved for use, all new chemicals must be tested for their human toxicity and ecological effects. This is a law that was long overdue and, if wisely administered, can greatly reduce, if not eliminate entirely, the danger that a new chemical will damage human health or wildlife. However, if it is administered too zealously, the world may be denied the use of new chemicals that could be of great benefit to mankind. Great administrative wisdom will be needed to provide the kind of balance that will give the protection needed, while at the same time assuring that the incentives to produce beneficial new chemicals will not be destroyed.

NOTES

1. Carson, Rachel. "Silent Spring," Houghton Mifflin Co., Boston, Mass. (1962).
2. Kilgore, W. W. and M. Li. "The Carcinogenicity of Pesticides," in: "Residue Reviews: Residues of Pesticides and Other Contaminants in the Total Environment" (F. A. Gunther, ed.), Springer-Verlag, New York/Heidelberg/Berlin (1973), Vol. 40, p. 141.
3. U.S. Environmental Protection Agency. "Report of the DDT Advisory Committee," Washington, D.C. (1971).
4. National Academy of Sciences. "Pest Control: An Assessment of Present and Alternative Technologies," Vol. V: "Pest Control and Public Health." National Research Council, Environmental Studies Board, Washington, D.C. (1976).
5. Deichmann, W.B. "Liver Cancer Deaths in the Continental USA from 1930 to 1972," *Am. Indus. Hyg. Assn. J.* (September 1976), p. 495–498.
6. Hayes, W. L. "Mortality in 1969 from Pesticides, Including Aerosols," *Arch. Env. Health* (March–April, 1976), pp. 61–72.
7. World Health Organization. "Health Hazards of the Human Environment," B. Carcinogens, Geneva (1972), pp. 219–228.
8. Bevenue, A. "The 'Bioconcentration' Aspects of DDT in the Environment," in: "Residue Reviews: Residues of Pesticides and Other Contaminants in the Total Environment" (F. A. Gunther, ed.), Springer-Verlag, New York/Heidelberg/Berlin (1976), Vol. 61, p. 37.
9. Woodwell, G. M. "Toxic Substances and Ecological Cycles," *Scientific American* 216(3):24–31 (1967).
10. Ware, G. W. "Effects of DDT on Reproduction in Higher Animals," in: "Residue Reviews: Residues of Pesticides and Other Contaminants in the Total Environment" (F. A. Gunther, ed.), Springer-Verlag, New York/Heidelberg/Berlin (1975), Vol. 59, p. 119.
11. Cooke, A. S. "Shell Thinning in Avian Eggs by Environmental Pollutants," *Environmental Pollution* 4:85–152 (1973).
12. "U.S. Department of Health, Education and Welfare Report of the Secretary's Commission on Pesticides" (December, 1969), p. 50.
13. Robinson, J. "Organochlorine Insecticides and Bird Populations in Britain," in: "Chemical Fallout: Current Research on Persistent Pesticides" (M. Miller and G. Berg, eds.), Charles C. Thomas, Springfield, Ill. (1969), p. 113.

14. Devlin, R. M. "DDT: A Renaissance?" *Environmental Science and Technology* 8:322–325 (1974).
15. Wurster, Charles F. "DDT Reduces Photosynthesis by Marine Phytoplankton," *Science* 159:1474 (1968).
16. Wagner, R. H. "Environment and Man," W. W. Norton, New York (1971).
17. Commoner, Barry. "The Closing Circle," Knopf, New York (1971).
18. Massachusetts Institute of Technology. "Man's Impact on the Global Environment: Report of the Study of Critical Environmental Problems" (SCEP) (1970).
19. "DDT Residues in Marine Phytoplankton," in: "Residue Reviews: Residues of Pesticides and Other Contaminants in the Total Environment" (F. A. Gunther, ed.), Springer-Verlag, New York/Heidelberg/Berlin (1972), Vol. 44, p. 23.
20. National Academy of Sciences. "Pest Control: An Assessment of Present and Alternative Technologies." Vol. 1. "Contemporary Pest Control Practices and Prospects: The Report of the Executive Committee." National Research Council, Environmental Studies Board, Washington, D.C. (1975).
21. Wurster, Charles F. "DDT Goes to Trial in Madison," *BioScience* 19:809–813 (1969).
22. Wurster, Charles F. "American Scientists Take Legal Action to Improve Pesticide Regulation," presented at the International Conference on The Responsibility of Science in Modern Society, Florence, Italy (October 4, 1976).
23. Federal Register. "Environmental Protection Agency, Consolidated DDT Hearings," *Federal Register* 37 (131):13369–13375 (July 7, 1972).
24. Council on Environmental Quality. "Sixth Annual Report," U.S. Government Printing Office, Washington, D.C. (1975), p. 368.
25. Johnston, D. W. "Decline of DDT Residues in Migratory Songbirds," *Science* 186:841–842 (1974).
26. Nickerson, P. R. and K. R. Barbehenn. "DDT Residues in Starlings, 1974," *Pest. Monit. J.* 9(1):1 (1975).
27. Anderson, D. W., J. R. Jehl, R. W. Risebrough, L. A. Woods, L. R. Dewees and W. G. Edgecomb. "Brown Pelicans: Improved Reproduction off the Southern California Coast," *Science* 190:806–808 (1975).
28. Woodwell, G. M., P. P. Craig and H. A. Johnson. "DDT in the Biosphere: Where Does It Go?" *Science* 1974:1101–1107 (1971).
29. Ware, G. W., B. J. Estesen and W. P. Cahill. "DDT Moratorium in Arizona—Agricultural Residues after 4 Years," *Pest. Monit. J.* 8(2):98 (1974).
30. Higuchi, Kentaro (ed.). "PCB Poisoning and Pollution," Kodansha Ltd., Tokyo and Academic Press, New York (1976).
31. Kuratsune, Masanori. "Epidemiologic Studies on Yusho," in: "PCB Poisoning and Pollution" (K. Higuchi, ed.), Kodansha Ltd., Tokyo and Academic Press, New York (1976).

32. Nisbet, I. C. T. and A. F. Sarofim. "Rates and Routes of Transport of PCBs in the Environment," *Environmental Health Perspective,* Experimental Issue No. 1 (April, 1972).
33. Panel on Hazardous Trace Substances. Report of Panel on Hazardous Trace Substances to Ad Hoc Committee on Environmental Health Research (David P. Rall, Chairman), *Environ. Res.* 5:249–362 (1972).
34. Peakall, D. B. "Polychlorinated Biphenyls: Occurrence and Biological Effects," in: "Residue Reviews: Residues of Pesticides and Other Contaminants in the Total Environment" (F. A. Gunther, ed.), Springer-Verlag, New York/Heidelberg/Berlin (1972), Vol. 44, p. 1.
35. *Environmental Health Perspectives,* Experimental Issue No. 1 (April, 1972).
36. Environmental Protection Agency. *Proceedings of the National Conference on Polychlorinated Biphenyls* (November 19–21, 1975), Chicago, Ill., *EPA-560/6-75-004,* Environmental Protection Agency, Office of Toxic Substances, Washington, D.C. (March 1976).
37. Kornreich, M., et al. "Environmental Impact of Polychlorinated Biphenyls" *MITRE Technical Report 7006.* The MITRE Corporation, McLean, Va. (May, 1976).
38. Kolbye, A. C. "Food Exposures to Polychlorinated Biphenyls," *Environmental Health Perspectives,* Experimental Issue No. 1 (April, 1972).
39. Jelinek, C. and P. E. Corneliussen. "Levels of PCB's in the U.S. Food Supply," in: *Proceedings of the National Conference on Polychlorinated Biphenyls* (November 19 25, 1975), Chicago, Ill., *EPA-560/6-75-004,* U.S. Environmental Protection Agency, Office of Toxic Substances Washington, D.C. (March, 1976).
40. Grant, D., J. Mes and R. Frank. "PCB Residues in Human Adipose Tissue and Milk," in: *Proceedings of the National Conference on Polychlorinated Biphenyls* (November 19–21, 1975), Chicago, Ill. *EPA-560/6-75-004,* U.S. Environmental Protection Agency, Office of Toxic Substances, Washington, D.C. (1976), p. 144.
41. Kutz, F. W. and S. C. Strassman. "Residues of Polychlorinated Biphenyls in the General Population of the United States," in: *Proceedings of the National Conference on Polychlorinated Biphenyls* (November 19–21, 1975), Chicago, Ill., *EPA-560/6-75-004,* U. S. Environmental Protection Agency, Office of Toxic Substances, Washington, D.C. (1976), p. 139.
42. Miller, R. W. "Pollutants in Breast Milk," *Journal of Pediatrics* 90(3):510–512 (1977).
43. N.Y. State Health Planning Commission. "Report of the Ad Hoc Committee on the Health Implications of PCBs in Mothers' Milk," Albany, N.Y. (January, 1977).
44. Allen, J. R. "Response of the Nonhuman Primate to Polychlorinated Biphenyl Exposure," Research Activities at Regional Primate Centers, *Federation Proceedings,* Vol. 34, No. 8, pp. 1675–1679 (1975).
45. McNulty, Wilbur P. *Proceedings of the National Conference on Poly-*

chlorinated Biphenyls (November 19–21, 1975), Chicago, Ill., EPA-560/6-75-004, U.S. Environmental Protection Agency, Office of Toxic Substances, Washington, D.C. (1976), p. 347.

46. Kuratsune, M., Y. Masuda, and J. Nagayama. "Some of the Recent Findings Concerning Yusho," in: *Proceedings of the National Conference on Polychlorinated Biphenyls* (November 19–25, 1975), Chicago, Ill., *EPA-560/6-75-004*, U.S. Environmental Protection Agency, Office of Toxic Substances, Washington, D.C. (1976).

47. Environmental Protection Agency. *Proceedings of the National Conference on Polychlorinated Biphenyls*, (November 19–25, 1975), Chicago, Ill., *EPA-560/6-75-004*, U.S. Environmental Protection Agency, Office of Toxic Substances, Washington, D.C. (1976).

48. Department of Health, Education and Welfare, Food and Drug Administration, "Polychlorinated Biphenyls (PCB's)," *Federal Register* 42(63):17487–17494 (April 1, 1977).

49. State of New York, Department of Environmental Conservation, Interim opinion and order, File No. 2833 (February 9, 1976).

CHAPTER 10

The Metals: Lead and Mercury

The metals are among the most important of the natural resources on which modern technology is based. The uses of iron, copper, zinc, nickel, chromium, lead, arsenic, mercury, and uranium are generally well known to the general public. Other metals, like beryllium, zirconium, and cadmium are less well known but also serve important industrial uses. In Table 5–4 were listed the remarkable assortment of metals that are used in so seemingly simple a device as the telephone hand-set.

It has been known at least since the time of the ancient Greeks that two metals, lead and mercury, have toxic properties that affected the health of workmen exposed to their dusts or fumes. Arsenic, another metallike element that has long been known to be highly toxic, has been widely used for many purposes. Until the advent of the organic chemicals, arsenic was a constituent of the important pesticide, lead arsenate.[1,2a]

Much has been learned about the toxic effects of the metals in this century, mainly as a result of the studies of occupational diseases among metal workers exposed to their dusts and fumes. Certain forms of nickel, chromium, and arsenic are associated with high incidences of cancer. Beryllium produces two forms of lung disease, and cadmium, vanadium, and zinc can produce toxic reactions, particularly when inhaled as a freshly formed fume. Low-level chronic exposure to cadmium can also produce toxic reactions of a delayed type.

Because both lead and mercury are of contemporary interest and have been the subject of controversy in recent years, they have been selected for special attention in this chapter.

247

LEAD

Lead is a metal that has served many useful purposes for thousands of years. It occurs in nature mainly in the form of the mineral galena, a lead sulfide that the ancients learned to convert into metallic lead and into the white or red oxides that have been used ever since for pigments in paints and ceramic glazes. Lead is a major constituent of many alloys and, until comparatively recently, it was used in making water pipes in metallic form. (The word "plumbing" is derived from *plumbum*, which is the Latin word for lead.) Lead is also used in soldering compounds, in storage batteries, and for type metal. In recent decades, one of the most important uses of lead has been as an "anti-knock" compound in gasoline. In 1968, this accounted for nearly 20 percent of the 1,300,000 tons utilized in the United States. However, the use of lead as a gasoline additive has recently been curtailed.[3]

Natural sources of lead exist but apparently do not contribute significantly to the lead present in air, water, or human tissues. However, smelting, the combustion of leaded gasoline, the aging of leaded paints, and the incineration of wastes containing lead add to the lead content of the atmosphere, and ultimately to an increased burden of lead in the biosphere and in the human body. An additional source of atmospheric contamination has been combustion of coal, which contains traces of lead that are volatilized when the coal is burned.

The toxic effects of lead were reported by Hippocrates as early as 370 B.C.[4] There were occasional references made to occupational lead poisoning afterwards, but the subject was not studied extensively until 1839, when Tanquerel des Planches published a report of 1,200 lead-poisoning cases in Paris. It was not until the last quarter of the nineteenth century that the subject of occupational lead poisoning began to attract serious attention in England, and the first studies in the United States were not undertaken until about 1910 by Alice Hamilton.[5]

Severe cases of lead poisoning were certainly relatively common in the United States prior to World War II, but the absence of a national reporting system precludes any understanding of the incidence of the disease. However, lead poisoning has been a reportable disease in the United Kingdom since 1896, and the vast improvement in factory hygiene in that country is readily discerned from the trend in the reported cases. In 1900, there were about 1,000 cases of lead poisoning in the United Kingdom, including 40 deaths. There were only five cases in 1960, with no fatalities.[6]

Workers in lead smelters, storage-battery plants, paint-manufacturing plants, and other lead-using industries are at risk primarily due to dusts that contain lead which they inhale in the course of their work. In former times, it was thought that the dust was ingested due to contamination of hands, food, and tobacco, but studies undertaken early in this century showed that if the dust content of the air is maintained below certain levels, lead poisoning can be avoided. Lead poisoning has also been reported in communities in which the public has been exposed to lead in food or water.

The effects of lead poisoning can be acute or chronic. A prominent symptom of acute poisoning is colic, in which severe abdominal pain is the cardinal symptom. Lead poisoning can interfere with human fertility and result in neurological injury, producing a complicated set of signs and symptoms among which may be convulsions. This can lead to death or permanent paralysis and brain damage, and mental retardation in children. A form of chronic lead poisoning, now also seen far less frequently than in the past, is a palsy, commonly known as "wrist drop," caused by paralysis of muscles of the forearms.

Industrialization has resulted in worldwide fallout of traces of lead that can be demonstrated in a number of ways.[6a] In recent years it has become possible to drill deep cores from the Greenland icecap, and analyze ancient strata for various trace metals. The lead content of the ice, formed of frozen precipitation, can be associated with the age of a given strata. Analysis of ice cores representing 2,700 years of precipitation shows that the lead content of the ice increased only slightly until the eighteenth century, but has since risen twentyfold, presumably due to the increase in industrial activity. A major part of this increase has occurred since about 1940 and is presumably due to the relatively recent use of lead additives in gasolines.

Although levels of low-level lead contamination have increased globally in recent decades, it is known that exposure to lead in many societies was much higher in the past, and that lead poisoning of entire communities sometimes occurred.

It has been suggested that lead poisoning was a factor in the decline of the Roman Empire because this metal was used widely among upper-class Romans in wine-making equipment. The proponents of this line of reasoning cite the fact that spontaneous abortions are known to have occurred with higher-than-normal frequency among women exposed to lead in industry, and conclude that the reduced fertility observed among the Roman aristocracy was also the result of exposure to lead.[7] The defect in this reasoning is that in the modern literature a higher than normal incidence of spontaneous abortions among female lead workers was associated with exposures that were so high as to also result in colic, convulsions, wrist drop, and other consequences of severe lead absorption. In the ample recorded history of ancient Rome, there are no reports that these severe symptoms of lead poisoning were prevalent. However, this subject may continue to attract attention in view of a recent study in which it was shown that reproductive ability is reduced among men occupationally exposed to lead compounds.[8]

Outbreaks of colic due to the exposure of the general community to lead were reported in France in 1617, and in Devonshire, England in 1703, but it was not until 1766 that these cases were recognized as lead poisoning. George Baker, in a classic essay published in 1767, reported that during the period between 1762 and 1767, 285 cases had been admitted to Exeter Hospital, and that during a single year (1766 to 1767) 80 cases were admitted to the hospital in Bath. Although the outstanding complaint among the population of Devonshire was that of colic, the only cases that were admitted to the hospital were those that had progressed to the point of paralytic weakness of the arms. The total number of cases must have been very much larger, since only the most heavily exposed would develop paralysis, but many more would demonstrate lesser symptoms such as colic. Baker's studies led him to the conclusion that the disease was lead poisoning caused by metallic lead used in fabricating the grindstones for the cider mills.[9,10] An interesting sidelight is the possibility that Benjamin Franklin, who developed a surprisingly high degree of interest in lead poisoning, may have been instrumental in calling to the attention of the English investigators the relationship between lead and what was called "dry bellyache" in colonial America, a disease similar to the Devonshire colic. In a 1786 letter to an English correspondent, Franklin reported his observation that in America the bellyaches occurred only among drinkers of New England rum and that the disability was lead poisoning, due to contamination of the rum because the stills were made with lead tubes fabricated into condenser coils.[11] Lead has also been shown to be a contaminant of old wines,[12] and poisoning from contamination of spirits continues to occur, mainly from lead in automobile radiators used by "moonshiners" for their illicit stills.[13]

In the early nineteenth century, lead-lined water tanks were considered to be a hazard by the Board of Health of the City of Paris. In its report for the decade of 1829 to 1839, this group called attention to the fact that the Parisian bakers lined their water tanks with lead, which was observed to oxidize to such an extent that a deposit of lead oxide accumulated on the bottoms of the tanks. The Board, no doubt erroneously, inferred that the lead-oxide deposit was totally insoluble and that for this reason there would be no hazard if the water drawn from the tank were free of sediment. With this in mind, they ordered that the cock from which the water was drawn should be at least three inches from the bottom of the tanks so as to always be above the level of the sedimentary deposit. As in the case of most reports of this kind, no quantitative data are given as to the lead content of the Parisian bread.[14]

People nowadays are often surprised to learn that prior to the advent of organic pesticides in the 1940s, the most frequently used pesticidal spray was lead arsenate. This compound of two highly toxic substances was used widely, and in the early 1940s about 50 million pounds of lead were employed in its manufacture. Not only would the farm employees be exposed to the sprayed lead arsenate, but residues on the fruits and vegetables were a source of exposure to the consumers. Both the farm employees and consumers were found to have elevated levels of lead and arsenic in their blood and urine, but the amounts were apparently insufficient to result in measurable physical damage.

It is known that some old houses may still use lead plumbing and that tap water can become contaminated from this source; cases of poisoning from use of lead plumbing have been reported recently from Scotland.[15] The extent to which such plumbing is still in use is not known, nor are adequate data available to describe the extent to which the water passing through these pipes becomes contaminated with lead. This is a matter that should be investigated thoroughly in the United States where lead pipes are known to exist in many old homes.

Another source of lead exposure is the use of lead pigments for ceramic glazes. Occasional fatal cases of lead poisoning have been attributed to this practice, particularly when the glazes were imperfectly fired.[16,17] This is another potential source of contemporary lead exposure that has not been adequately evaluated.

The most stubborn of the contemporary lead-poisoning problems is that the walls and ceilings of many old houses and apartment buildings have been coated with lead paint that peels because of dilapidation and is then nibbled by children. The most important white pigment for many years was "white lead," the basic lead carbonate, which was first produced in Holland during the sixteenth century and was introduced to the United States in the late 1800s. White lead was for a long time the principal pigment used for both interior and exterior paints, and prior to about 1940, the pigment contained from 40 to 60 percent lead. By 1941, a far superior white pigment, titanium oxide, was developed, and quickly replaced lead for this purpose.

However, the titanium-oxide paint in most cases was simply applied over the lead paints, so the potential for future exposure remained.

In 1914 it was observed in Baltimore that lead poisoning was occurring among young children. Most of those affected were between one and two years old, and fewer were older than five years.[18] It was soon found that lead poisoning was developing among children who had the habit of eating flaking paint and chewing cribs, toys, furniture, and woodwork, all of which at that time were painted with lead-bearing coatings. During the 20-year period between 1931 and 1951, 293 cases of juvenile lead poisoning were discovered in Baltimore, of whom 85 died.[19] In more recent years, cases of severe lead poisoning seem to have been less frequent. In New York City, for example, the Health Department has reported that fatal cases have been reduced to about one or two per year, compared to about 20 per year two decades ago. Although juvenile lead poisoning is generally thought to be a problem of big-city ghettos, this is by no means true. Studies in Portland, Maine, showed that the risk is not related to the size of the city, but to the age of the housing, and the standard of maintainance.

The importance of the lead-paint problem is illustrated by estimates that as many as 600,000 children are at risk in the United States.[20] At first glance, the solution would seem to be obvious: Simply remove the lead paint or cover it. But the very size of the population at risk immediately hints at the number of residences that require rehabilitation. It has been estimated that it costs about $2,000 per housing unit to remove the leaded paint. If we assume there are 400,000 units requiring rehabilitation, the cost would be nearly $1 billion. The houses in question are the older houses, most often in the inner city, in dilapidated neighborhoods that nowadays are inhabited by impoverished blacks and Puerto Ricans. The rates of abandonment of unprofitable rental buildings by owners is already very high and to require that every apartment-house owner should remove all leaded paint would only increase this rate of abandonment. The situation is one that could clearly be ameliorated by federal subsidy, but we have here one of the many examples of *bona fide* environmental problems which clearly affect the health of people, but to which such a low priority is assigned that adequate funds are not made available for solving the problem. The cost of reducing the sulfur content of fossil fuels, reducing the emissions from automobile engines, and providing high-degree secondary treatment for sewage being discharged into estuarine waters are only a few examples of programs that are costing many tens of billions of dollars but for which only relatively imperceptible benefits to health can be foreseen. If environmental priorities could be established in an orderly way, the problem of lead poisoning among juveniles could be swiftly eliminated.

Changing technology caused the lead content of many types of paints to be reduced from 50 percent to about 1 percent in the early 1940s. This was not because of the concern over the health hazard at the time, but because the titanium-oxide pigment proved to be superior to lead oxide. Unfortunately, there has been insufficient epidemiological research to determine whether the 1 percent lead-bearing paint is hazardous. One would like to know what is the lowest concentration of lead in paint that has been associated with lead intoxication among children. In 1957, one study concluded that lead intoxication did not occur when the dried paint contained less than 1 percent lead.[21] This is an important conclusion which has neither been confirmed nor refuted by subsequent studies. One thing, however, is certain: The number of *severe* cases of juvenile lead poisoning is diminishing.

There have been only a few investigations of the more subtle effects of lead. Because lead is a neurological poison, it is possible that impaired mental development may occur in children exposed to lead in quantities insufficient to produce the classical clinical symptoms of lead poisoning. Such evidence has, in fact, recently come to light.[22,23] Unfortunately, studies which should have been undertaken decades ago have begun only recently.

The use of lead as an additive to gasoline has been the principal source of lead as an air pollutant in urban areas and near arterial highways in recent years. In 1968, more than 98 percent of the total emission of lead to the atmosphere (362 million pounds) was attributed to this source, but it has since been suggested that the true figure is somewhat lower due to the fact that important sources other than the automobile were overlooked when the estimate was made.[24] In 1972 the National Academy of Sciences estimated that the air over the largest American cities had a concentration of lead 20 times greater than over sparsely populated areas of the country, and 2,000 times greater than the air over the mid-Pacific Ocean.[25]

The concentration of lead to which the general population has been exposed from leaded gasoline, even in densely trafficked urban areas, is very much less than the exposures that are known to produce signs of lead poisoning. It is rare to find average daily exposures greater than about 2 micrograms of lead per cubic meter of air, a fraction of the maximum permissible level for occupational exposure, which is 50 micrograms per cubic meter of air for a 40-hour week. While the amount of lead to which humans have been exposed, due to leaded gasoline, has been very much less than that associated with lead poisoning, the question arises as to whether the lower concentrations may produce some subtle effect on behavior or health that is not yet recognized. Among other things, it is possible that a metal such as lead can act synergistically with some other substance in the environment.

During the early 1970s there was intense public debate over the question of whether lead should be banned from gasoline. The issue was resolved when the automobile industry found that in order to meet the automobile emissions standards it would be necessary to pass the exhaust gases across catalytic surfaces that would be poisoned by lead. In anticipation of this, and to some extent because of apprehensiveness over the possible health effects of low levels of lead exposure, some cities including New York passed regulations which reduced the permissible lead content of gasoline. This resulted in a reduction in airborne lead that was reflected promptly by a lowering of the lead burden in New York residents.[26]

The atmosphere is not the major source from which humans absorb lead. They take in between 100 and 500 micrograms each day in their food, an amount that evidently has not changed for at least 30 years.[25] The exact sources of lead in food are not known. Because people usually are not exposed to lead from the atmosphere in concentrations greater than 1 microgram per cubic meter the dietary sources of lead are somewhat more important than the atmospheric, even in heavily trafficked regions. This conclusion is supported by analyses of children's teeth. The teeth of children exposed to leaded paint contained more lead than the teeth of children exposed to air pollution from high-density traffic. There was no measurable difference in the lead content of the teeth of children from the sparsely trafficked areas or heavily trafficked areas in which leaded paint was known to be absent.[27]

Hair can be analyzed to determine an individual's exposure to some metals, and samples of human hair collected at various times from 1871 to 1971 have been analyzed for lead. It has been found that the lead content of the samples of hair dating from the nineteenth century is more than ten times that of the more recent samples.[28] Bones also can be analyzed to indicate if the individual has been exposed to lead, for lead is one of several metals that tend to deposit in the skeleton. The lead content of samples of preserved bones of thirteenth- and nineteenth-century Poles has been reported higher by a factor of as much as 100 than bone collected recently from the same part of Poland.[29] Investigators have published similar results from analysis of bones in England, the United States, and Germany.

These reports simply suggest that there have been times in the past when lead was used in ways that caused people to be more heavily exposed than inhabitants of modern industrialized communities. However, there are also examples of ancient and modern primitive societies in which lead exposure has been comparatively lower. Measurements of lead in the teeth of contemporary Mexican Indians living far from urbanized or industrialized areas are comparable with the levels observed in twelfth-century Peruvian Indians and ancient Egyptian mummies, both of which are low in comparison with tooth measurements made in modern urban areas.[30]

MERCURY

Mercury is unique because it is the only metal that exists in liquid form at normal temperatures. The popular name "quicksilver" is said to have been given to it in the time of Aristotle because it was believed to be "living silver" in the purest of forms. It occurs in nature primarily in the form of cinnabar, the red sulfide of mercury (HgS) that has been mined in Almaden, Spain for nearly 2,400 years. Cinnabar is a chemically inactive compound that may not be toxic itself. However, in many mines, cinnabar coexists with small globules of metallic mercury, the vapors of which contaminate the mine air. It was found at Almaden, centuries ago, that mercury poisoning could be controlled by limiting the time the workers spent within the mines to 32 hours per month.[4,5] This was the only possible method of control in the absence of mechanical ventilation, which nowadays makes it possible to dilute the mercury vapor to acceptable levels.

Mercury was sought in ancient times, in its metallic form, because of its curious and attractive properties. In the form of cinnabar it may have been useful as a pigment, possibly for cosmetic purposes, and its properties are said to have encouraged its use for occult purposes.[31]

The first important application for the metal was probably made in 1643, when Torricelli found that mercury in a sealed, inverted tube always maintained a level equivalent to about 760 millimeters at sea level, an observation that soon led to development of the mercury barometer and other devices for measuring small fluctuations in pressure, such as the manometer. The mercury thermometer was perfected by Fahrenheit in Holland, in 1714. Mercury barometers, manometers, and thermometers are used to this day and are a source of potential danger to laboratory workers.

The toxic action of mercury depends on the form and quantity in which it is absorbed. Until comparatively recently, the principal mode of exposure was to the vapors emanating from the metallic form of mercury. When a small amount of mercury is spilled accidentally on the floor or workbench, it tends to divide into small particles which collectively present substantial surface area from which mercury vapor can emanate. It was soon observed that ingestion of the liquid mercury was of itself not hazardous because it is poorly absorbed from the gastrointestinal tract. However, when the vapor of mercury is inhaled, it readily passes from the lungs to the bloodstream. The toxic effects of exposure to mercury vapor include inflammation of the gums, psychic irritability, kidney injury, tremors, and a variety of other symptoms of neurological damage.

Some of the soluble inorganic salts of mercury have been known to be highly toxic for hundreds of years, and mercuric chloride was a popular but painful means for committing suicide until recently.[31] In spite of this, mercuric chloride has been used to treat syphilis for hundreds of years, in the form of pills and salves. The salves were frequently applied barehanded by physicians who, like their patients, absorbed the mercury into their bodies. Because the toxic properties of mercury were then not well understood, the patients often developed severe ulcers of the mouth and throat. "The remedy is more formidable than the disease," reported an early nineteenth-century text on the use of mercury in the treatment of venereal disease.[32]

However, some compounds of mercury are not at all toxic. Mercurous chloride (HgCl), for example, commonly known as "calomel," has been widely used as a cathartic with little or no toxic effects. Many other mercury compounds have been used for medical purposes, some up to the present time. The fact that mercury would form amalgams with gold and silver was discovered hundreds of years ago and led to the important use of mercury in dental practice. The silver fillings with which most people are familiar are prepared by mixing mercury and silver to form an amalgam, which hardens into a durable filling. Mercurochrome, an organic compound of mercury with mild properties, was introduced in the first quarter of this century as a substitute for the tincture of iodine, and to this day is often self-applied by children. Systemic absorption of mercury is known to occur in minor amounts after use of both mercurochrome and dental amalgams, but mercurialism is not known to have resulted from this practice.

In the seventeenth century, the French Huguenots first used mercury nitrate, $Hg(NO_3)_2$, in the manufacture of felt. This compound has a carrot-yellow color which gave the name "carroting" to the process. The felt was used mainly in the hat industry and it is widely, though perhaps incorrectly, believed that Lewis Carroll's description of the Mad Hatter referred to the characteristic irritability and tremors that were known to be endemic among hatmakers due to their exposure to mercury.[31] The U.S. Public Health Service studied 544 hatters from five factories in the United States and reported, in 1941, that 59 had signs of chronic mercurialism from contamination of the workroom atmospheres with mercury vapor.[2] A nontoxic substitute for the mercury nitrate was found that made it possible to eliminate the use of this hazardous material in hat making. There are other examples of processes in which substitute materials or procedures have eliminated the hazard from mercury. Mercury was used for centuries in mirror making until it was replaced by silver. Gilders formerly used gold amalgam (gold and mercury) for plating, but this process has been replaced by electroplating gold directly, without the need for the amalgam.

While some uses of mercury have been eliminated, others have been developed. The production of chlorine, an important industrial chemical, has grown steadily during the past quarter century and has resulted in an increased demand for mercury, which is used in the electrolytic process by which chlorine is made.[33] High-purity alkali products required for rayon manufacture and other purposes are also produced in the same electrolytic process. However, this demand for mercury has been reversed recently because processes that have more recently been developed do not require mercury.

There are also comparatively new uses for mercury in the electrical industry. High-intensity mercury lamps are used for industrial and highway illumination, and mercury is used in compact throw-away batteries that are installed in such devices as photographic light meters, hearing aids, portable radios, and minicalculators. The common household silent light switch uses a globule of mercury to make and break electrical contact.

The mercury-vapor turbine was an interesting application of mercury that was abandoned partly because of the inability of the designers to confine the toxic vapors. An experimental power plant was built about 1925, in Hartford, Connecticut, that was more efficient than steam turbines. Mercury is also used in the catalytic production of important plastics such as urethane and polyvinyl chloride. Another major development is the use of organic mercury compounds in agricultural pesticides, an application to which we will refer again.

Mercury is probably unique among the toxic metals in the number and severity of incidents of mass poisoning that have been reported from time to time. In 1810, containers of mercury broke loose in the hold of the British sloop *Triumph*, resulting in the death of all the cattle that were being carried aboard the ship, and sickness of 200 sailors, three of whom died.[4] In 1903, a fire in a mercury mine at

Idrija, Yugoslavia, caused acute mecurialism in 900 people and many cattle.

Despite the foregoing, the subject of mercury poisoning received comparatively little public attention until very recently. Modern industrial hygienic procedures and the substitution of nontoxic materials for mercury for some purposes has eliminated the complaints that were common among industrial workers exposed to mercury in former times. Some of the newer uses of mercury presented industrial health problems, but these were quickly solved, and by midcentury, mercury had ceased to be an important industrial hygiene problem. Moreover, there was then no reason to believe that the public ever was at risk.

However, beginning in 1953, a number of severe outbreaks of mercury poisoning were reported. Two of these were in Japan, in the cities of Minamata and Niigata. A third was in Alamogordo, New Mexico, and others occurred in Iraq, Pakistan, and Guatemala. These outbreaks involved symptoms that were different from those of classical mercury poisoning in several respects. As noted earlier, the classic symptoms of mercury poisoning involved inflammation of the mouth, tremors, psychic irritability, and kidney injury, as well as cases of severe brain damage. In the Japanese episodes, progressive blindness, deafness, lack of coordination, and intellectual deterioration also were observed.[34] One-hundred thirty one cases, with 46 deaths, were reported among the residents of Minamata, a fishing town on the southwest coast of the Japanese island of Kyushu.[35] In 1965, 47 similar cases were reported from the city of Niigata with six deaths.

The Minamata outbreak was the first to be studied. All of the cases were among fishermen's families who consumed substantial quantities of fish or shellfish gathered in Minamata Bay. It was not until 1958 that it was learned that a factory on the bay was using mercury as a catalyst for the manufacture of two widely used chemicals, vinyl chloride and acetaldehyde. Traces of mercuric chloride were found in the effluents to the bay, but this compound, although moderately toxic, could not of itself be considered to be the agent that was causing the observed symptoms. Upon further investigation, it was found that the mercuric chloride was deposited in the sediments, where it was converted to methylmercury, a highly toxic organic compound, by microbial action. Unlike other mercury compounds, methylmercury can pass across the brain barrier, which accounted for the severe neurological symptoms seen in the affected population. Methylmercury also can pass the placental barrier and affect the fetal brain. In Minamata there were 23 cases with symptoms resembling cerebral palsy in babies who had not consumed mercury-contaminated fish, but who had been poisoned during their fetal life.

The outbreak in Niigata occurred in 1965, more than a decade later. The circumstances were nearly identical: An inorganic mercury compound was used as a catalyst in plastic manufacture and was present in liquid wastes. Upon entering the bay, the inorganic mercury compound was deposited in the sediments where the mercury was converted to the methyl form by microbial action.

The main lesson learned as a result of the investigations in Japan was that relatively innocuous forms of mercury, discharged from industrial sources, could be converted to highly toxic organic forms by microbial action. The methylmercury thus formed could then be taken up by aquatic organisms and passed along the food chain to fish or shellfish that are consumed by humans.

At the same time that the cases of mercury poisoning were being detected in Japan, a number of cases were discovered in Iraq where several hundred farm families were poisoned when they ate seed treated with a fungicide containing an organic mercury compound instead of using the seed for planting, as was intended. Similar outbreaks were also reported in Pakistan, Guatemala, and New Mexico.

The incident in the United States (New Mexico) involved only a single family, but the circumstances were particularly pathetic and were widely publicized in the press. A poor farmer had somehow accumulated floor sweepings from a granary that had stored seed treated with a fungicide containing organic mercury. Following customary practice, the seed had been dyed red to indicate that it had been so treated and should be used only for planting, and not for consumption by humans or farm animals. The farmer nevertheless fed the seed to hogs which were later butchered and eaten by his family. The children of the family were promptly affected to the extent that they were badly crippled and blinded by irreversible neurological damage.

This episode occurred in 1969, at a time when interest in the environment was developing throughout the world. Both the scientific and lay media were providing extensive coverage of environmental issues and these issues were receiving increasing attention at every level of government. The widespread publicity associated with the New Mexico case called attention to the considerable scientific literature concerning mercury poisoning that had already been published about the Japanese incidents. A worldwide search then was begun to detect the presence of methylmercury, first in the effluents from industrial plants using mercury and then elsewhere. By 1970, it had been demonstrated that methylmercury was unexpectedly present in fish from many bodies of water. In 1969 Scandinavian investigators reported that mercury was present in marine fish off the west coast of Norway and in freshwater fish from lakes far removed from industry. High mercury levels were also demonstrated to exist in seabirds feeding off the Swedish coast, and in aquatic mammals such as the sea otter.[36] It was then discovered that tuna and swordfish, harvested commercially and canned for domestic consumption throughout the world, contained significant quantities of mercury.

Certain basic questions had to be answered. What were the industrial sources of environmental mercury pollution? Were the episodes of mercury poisoning in Minamata-Niigata, New Mexico, Iraq, and elsewhere the result of localized pollution, or was the pollution more general? What should be the acceptable level of mercury in food, and how could mercury contamination be kept at this level? Does mercury occur in fish naturally, and if so, at what levels? Were the reported outbreaks of mercury poisoning the only ones that had occurred? Were more subtle, and as yet unrecognized, effects of mercury poisoning occurring in the general population? A wave of apprehension swept the world.

Action was immediately taken to minimize the discharge of mercury from industrial sources. The regulatory authorities were readily able to identify the major users of mercury and they undertook the necessary environmental studies to determine if significant quantities of mercury were being discharged from the industrial plants. Measures were adopted to minimize the discharges. Orders were also given to cease the use of organic mercurials for seed treatment. These were straightforward professional decisions demanded by logic. The U.S. Food and Drug Administration (FDA) recommended that the maximum permissible concentration of mercury in fish be limited to .5 parts per million. Consumption of fish was restricted in many parts of the world where the mercury levels exceeded these limits. The regulatory authorities in the United States placed a ban on the sale of swordfish, because the mercury content of this popular seafood consistently exceeded .5 parts per million.

The exact process by which swordfish was removed from the food market is not clear. The guideline for mercury and fish, which was adopted by FDA in May, 1969, was evidently promulgated without going through the usual procedures of publishing the regulation in the *Federal Register* and holding public hearings. The circumstances have suggested to legal analysts that the swordfish industry may have been sacrificed "in the continuing battle between Federal agencies and their consumer and environmental critics."[52,53] Similarly, it is not clear how swordfish returned to the marketplace—quietly, gradually, and without public notice.

When fish from the oceans and freshwater lakes, far removed from industrial sources of contamination, were found to contain mercury, it appeared to some that the world was already committed to an irreversible menace of recent industrial origin. This possibility was strengthened when it was realized that coal contains about one part per million of mercury that is volatilized in the course of combustion. To many, the world was faced with a tragic fait accompli. A University of Illinois chemist, who appeared on a national television program "Is Mercury a Menace?" is reported to have expressed the view that it was not sufficient to merely stop putting mercury into the environment. "It's too late. There is so much material that's been deposited. What we have to do is come up with good methods of removing it from the environment."[37] This would be a formidable task indeed! Senator Philip Hart (Democrat from Michigan) is quoted as having said, after two days of Congressional testimony by various witnesses, that the matter of mercury contamination may well constitute "the greatest environmental crisis in our history." "A national disaster," said Ralph Nader, the consumer advocate.[38]

While the doomsayers received media attention, the professionals were hard at work on a more orderly course of action. In 1971, the Swedes published an informative monograph based on their extensive analysis of the problem.[36] The United States government then appointed an expert committee to confer with Scandinavians and report independently.[34] A well-respected student of mercury toxicity who reviewed the problem early in 1971 noted that "A calm view of the present state of affairs regarding mercury in the environment suggests that the best way to deal with the problem is to apply the techniques of epidemiology, preventive medicine, public health, and industrial hygiene that have been effective in meeting hazards in the past."[39]

The presence of methylmercury in lakes, bays, and oceans throughout the world was certainly cause for concern. In fact, the Swedish studies had shown that the mercury content of fish-eating Swedes was as high as some of the lower values reported among the affected Japanese populations. Yet no evidence of mercurialism was found in the Swedish population. Tunafish is a popular staple in the menus of dieting Americans, and a large fraction of the nation's tuna consumption was known to be eaten by a relatively small group in the population who should have been systematically studied for possible symptoms of mercury poisoning. Although no epidemiological studies were undertaken in the United States, studies in other countries of populations which subsist mainly on seafood high in methylmercury failed to demonstrate evidence of mercury poisoning.[40]

It began to be evident that the tragic episodes of severe mercurialism in Japan, the United States, and other parts of the world were localized incidents. The more generalized presence of mercury in the food chain and in human tissues was found to be due largely to natural sources, of which coal combustion was only a minor part. Scientists turned to the museums and medical schools for specimens that could be used to measure the amounts of mercury which had been present in the past. It was found that the mercury levels of museum specimens of tuna caught 62 to 90 years previously, and in a swordfish caught 25 years previously, contained mercury in the same amounts as those that had been caught recently.[41] Nor was there any significant difference noted in the mercury content of specimens of fish that fed on the deep ocean bottom which were caught 90 years ago, compared with those specimens of the same species that had just been caught.[42] Pathological specimens of human tissues taken at autopsies between 1913 and 1970 not only failed to show an increase in the mercury content of recent specimens but, surprisingly, showed a rather substantial decrease.[43] As in the case of lead, people had been exposed to mercury in ways which had been discontinued. The estimated releases of mercury to the environment had increased about fourfold since 1913,[44] but the mercury content of the human tissues had apparently decreased.

In what is undoubtedly the most ingenious of the attempts to reconstruct the levels of mercury contamination in the past, an 1,100-year record of mercury in the food chain of an insect-eating species of bat was documented by measurements of dated strata of guano (bat manure) cores recovered from a cave in south central Arizona.[45] It was found that the mercury followed a consistent downward trend over the period of time. The amount of mercury was found to have increased during the past 116 years, but the current levels are well below the highs recorded 1,100 years ago. Some natural factor, such as volcanic activity, has been proposed as having been the cause of the higher levels in early times.

As in the case of the retrospective lead studies, chemists turned also to the Greenland icecap where the snows of prehistory are perpetuated in the frozen strata.[46] The same strata that had shown a systematic increase in lead content, beginning in about 1720, failed to show any change in the mercury content from about 800 B.C. to 1950. One would expect that mercury in the Greenland icecap would originate from atmospheric contamination, the principal anthropogenic source being combustion of coal.[47-49] However, natural sources, including volcanic activity and emanation of mercury vapor from the earth's crust, are probably more important.[50,51]

A few short years after the intense excitement of 1969 and 1970, the "mercury crisis" again was receiving only scant attention in the popular press. Swordfish slowly found its way back to the American restaurants and marketplaces, although the FDA did not rescind its order. It gradually became evident that, like many other elements, mercury is ubiquitous in nature. It is present in the earth's crust at an average concentration of about .5 parts per million, although there are wide deviations from this value. Mercury of natural origin finds its way into the lakes from the bottom rocks and sediments and into the oceans by way of the suspended solids transported by the rivers of the world. It has been noted that man's activities are unlikely to affect the mercury concentrations of the oceans, since there are about 70 million tons of mercury already present from natural sources. This amount is at least 1,000 times greater than the 20,000 tons of mercury used annually by man.[44]

The subject of mercury poisoning should not be taken lightly. Hundreds of people died unnecessarily, and many more were maimed. Fortunately, the problem was evidently both localized and controllable, but the publicity given the episodes was overly alarming, and the subject was out of perspective. In a world in which there is already much about which to despair, a somewhat unnecessary wave of fear affected the sense of well-being of millions of people.

NOTES

1. Whorton, James. "Before Silent Spring," Princeton Univ. Press, Princeton, N.J. (1974).
2. Neal, P. A. et al. "Mercurialism and its Control in the Felt-Hat Industry," *Public Health Bulletin No. 263*, U.S. Public Health Service, Federal Security Agency (1941), pp. 1–132.
2a. Nelson, W. C. et al. "Mortality Among Orchard Workers Exposed to Lead Arsenate Spray," *Journal of Chronic Disease* 26:105 (1973).
3. U.S. Bureau of Mines. "Minerals Yearbook: Metals, Minerals, and Fuels," Vols. I, II U.S. Government Printing Office, Washington, D.C. (1969).
4. Hunter, Donald. "Diseases of Occupations," English Univ. Press, Ltd. (1975).
5. Hamilton, Alice. "Exploring the Dangerous Trades," Little, Brown & Co., Boston (1943).
6. Lane, Ronald E. "Health Control In Inorganic Lead Industries," *Arch. Env. Health* 8:243–250 (1964).
6a. Patterson. Claire C. "Contaminated and Natural Lead Environments of Man," *Arch. Environ. Health* 11:344 (1965).
7. Gilfillan, S. C. "Lead Poisoning and the Fall of Rome," *J. of Occup. Med.* 7(2):53–60 (1965).
8. Lancranjan, Ioana, Horia I. Popescu, Olimpia Gavanescu, Iulia Klepsch

and Maria Serbanescu. "Reproductive Ability of Workmen Occupationally Exposed to Lead," *Arch. Environ. Health* 30:396 (1975).

9. McCord, Carey P. "Lead and Lead Poisoning in Early America," *Industrial Medicine and Surgery* 22 (9,11,12) (1953) and 23:(1,2,3,4) (1954).

10. Baker, George. "An Essay Concerning the Cause of the Endemial Colic of Devonshire" (1967) (reprinted by the Delta Omega Society, 1958).

11. Franklin, Benjamin. "Lead Poisoning," letter written to Benjamin Vaughan from Philadelphia in 1786 in: "The Ingenious Dr. Franklin" (N.G. Goodman, ed.), Univ. of Pennsylvania Press, Philadelphia, Pa. (1956).

12. Ball, G. V. "Two Epidemics of Gout," *Bulletin of the History of Medicine* 45:401–408 (1971).

13. Ball, G. V. and L. B. Sorensen. "Pathogenesis of Hyperuricemia in Saturnine Gout," *New England Journal of Medicine* 280 (22):1199–1201 (1969).

14. Chadwick, Edwin. "Report on the Sanitary Condition of the Labouring Population of Gt. Britain" (1842) (edited with introduction by M. W. Flinn), Edinburgh Univ. Press, Edinburgh, (1965).

15. Goldberg, A. "Drinking Water as a Source of Lead Pollution," *Environmental Health Perspectives* 7:103 (1974).

16. Hughes, J. T., J. J. Horan, and C. P. Powles. "Lead Poisoning Caused by Glazed Pottery: Case Report," *New Zealand Medical Journal* 84 (573):266–268 (1976)

17. Klein, Michael, Rosalie Namer, Eleanor Harpur and Richard Corbin. "Earthenware Containers as a Source of Fatal Lead Poisoning," *New England Journal of Medicine* 283:669-672 (1970).

18. Williams, H., E. Kaplan, C. Couchman and R. R. Sayers. "Lead Poisoning in Young Children," *Pub. Health Rep.* 67:230–236 (1952).

19. National Academy of Sciences. "Report of the Ad Hoc Committee to Evaluate the Hazard of Lead in Paint," prepared for the Consumer Product Safety Commission (November 1973).

20. Gilsinn, J. F. "Estimates of the Nature and Extent of Lead Paint Poisoning in the United States," *NBS Technical Note 746*, National Bureau of Standards (1972).

21. Chisolm, J. J. and H. E. Harrison. "The Exposure of Children to Lead," *Pediatrics* 18:943–958 (1956).

22. Albert, R. E., R. E. Shore, A. J. Sayers, C. Strehlow, T. J. Kneip, B. S. Pasternack, A. J. Friedhoff, F. Covan and J. A. Cimino. "Follow-up of Children Overexposed to Lead," *Environmental Health Perspectives* (May 1974), pp. 33–39.

23. Beattie, A. D., M. R. Moore, A. Goldberg, M. J. W. Finlayson, E. M. Mackie, J. F. Grahma, J. C. Main, D. A. McLaren, R. M. Murdoch and G. T. Stewart. "Role of Chronic Low-Level Lead Exposure in the Aetiology of Mental Retardation," *The Lancet* (March 15, 1975), pp. 589–592.

24. Goldwater, Leonard. "An Assessment of the Scientific Justification for Establishing $2\mu g/m^3$ as the Maximum Safe Level for Airborne Lead," *Industrial Medicine and Surgery* 41 (7):13 (1972).

25. National Academy of Sciences. "Lead: Airborne Lead in Perspective," Committee on Biologic Effects of Atmospheric Pollutants, NAS Printing Office, Washington, D.C. (1972).
26. Bernstein, D. M. "The Influence of Trace Metals in Disperse Aerosols on the Human Body Burden of Trace Metals," Ph.D. Dissertation, New York Univ., New York (1977).
27. Lockeretz, William. "Lead Content of Deciduous Teeth of Children in Different Environments," *Arch. Environ. Health* 30:583 (1975).
28. Weiss, D., B. Whitten and D. Leddy. "Lead Content of Human Hair (1871–1971)," *Science* 178:69 (1972).
29. Jaworowski, Zbigniew. "Stable Lead in Fossil Ice and Bones," *Nature* 217:152 (1968).
30. Shapiro, Irving M., G. Mitchell, I. Davidson and S. H. Katz. "The Lead Content of Teeth," *Arch. Environ. Health* 30:483 (1975).
31. Goldwater, Leonard J. "Mercury: A History of Quicksilver," York Press, Baltimore, Md. (1972).
32. Mathias, Andrew. "The Mercurial Disease," Philadelphia, Pa. (1811).
33. U.S. Department of the Interior Bureau of Mines. "Mineral Facts and Problems," *Bureau of Mines Bulletin #650*, U.S. Government Printing Office, Washington, D.C. (1970).
34. Nelson, Norton, T. C. Byerly, A. C. Kolbye, L. T. Kurland, R. E. Shapiro, S. I. Shibko, W. H. Stickel, J. E. Thompson, L. A. Van Den Berg and A. Weissler. "Hazards of Mercury: Special Report to the Secretary's Pesticide Advisory Committee, Dept of Health, Education and Welfare, Nov. 1970," *Environmental Research* 4:69 (1971).
35. Tsubaki, Tadao and Katsuro Irukayama. "Minamata Disease: Methylmercury Poisoning in Minamata and Niigata, Japan," Kodansha Ltd., Tokyo (1977) and Elsevier Scientific Publishing Co., Amsterdam/ Oxford/New York.
36. Friberg, L., G. Lindstedt and G. Nordberg. "Mercury in the Environment" (L. Friberg and J. Vostal, eds.), The Karolinska Institute, Department of Environmental Hygiene, Stockholm, Sweden (November 1971).
37. Zwick, David and Marcy Benstock. "Water Wasteland," Bantam, New York (1972).
38. Montague, Katherine and Peter Montague. "Mercury," Sierra Club, San Francisco/New York (1971).
39. Goldwater, Leonard J. "Mercury in the Environment," *Scientific American* 224 (5):15 (1971).
40. Stopford, W. and L. J. Goldwater. "Methylmercury in the Environment: A Review of Current Understanding," *Env. Health Persp.* 12:115–118 (1975).
41. Miller, G. E., P. M. Grant, R. Kishore, F. J. Steinkruger, F. S. Rowland and V. P. Guinn. "Mercury Concentrations in Museum Specimens of Tuna and Swordfish," *Science* 175:1121 (1972).
42. Barber, R. T. and A. Vijayakumar. "Mercury Concentrations in Recent and Ninety-year-old Benthopelagic Fish," *Science* 178:636 (1972).

43. Kevorkian, Jack; D. P. Cento, J. R. Hyland, W. M. Bagozzi and E. van Hollebeke. "Mercury Content of Human Tissues during the Twentieth Century," *A.J.P.H.* (April, 1972), p. 504.
44. Korringa, P. and P. Hagel. "Mercury Pollution: A Local Problem," *Proceedings of the Symposium on Problems of the Contamination of Man and his Environment by Mercury and Cadmium, EUR 5075*, Commission of the European Communities, Director General of Scientific and Technical Information and Information Management, Luxemburg (1974).
45. Petit, Michael G. "A Late Holocene Chronology of Atmospheric Mercury," *Environ. Res.* 13:94–101 (1977).
46. Dickson, E. M. "Mercury and Lead in the Greenland Ice Sheet: A Reexamination of the Data," *Science* 177:536 (1972).
47. Anderson, W. L. and K. E. Smith. "Dynamics of Mercury at Coal-Fired Power Plant and Adjacent Cooling Lake," *Environmental Science and Technology* 11 (*1*):75–80 (1977).
48. Billings, C. E., A. M. Sacco, W. R. Matson, R. M. Griffin, W. R. Coniglio and R. A. Harley. "Mercury Balance on a Large Pulverized Coal-Fired Furnace," *Journal of the Air Pollution Control Association* 23:773 (1973).
49. Electric Power Research Institute. "Health Effects of Mercury," *EPRI EC-224*, Palto Alto, Cal. (July, 1976).
50. Olafsson, J. *Nature* (London) 225:138 (1975).
51. Robertson, D. E., E. A. Crecelius, J. S. Fruchter and J. D. Ludwick. "Mercury Emissions from Geothermal Power Plants," *Science* 196:1094–1097 (1977).
52. Krenkel, Peter A. "Mercury Waste Treatment Methodology and the Potential for the Decontamination of the Environment," "Critical Reviews in Environmental Control" (Conrad P. Straub, ed.), CRC Press Inc., Cleveland, (1974), p. 314.
53. Harvard Law Review. "Health Regulation of Naturally Hazardous Foods: the FDA Ban on Swordfish," *Harvard Law Review* 85:1025 (1972).

CHAPTER 11

Sulfur Oxides and Particulates
From the Combustion of Coal and Oil

When coal or oil are burned, the products of combustion include carbon dioxide, sulfur dioxide, several forms of nitrogen oxides, and a mixture of dusts, fumes, and smokes that together are called the *particulate emissions*. Sulfur dioxide (SO_2) and the particulates are by far the most important of these waste products from the point of view of human health. Huge quantities of these fuels are burned for space heating in the northeastern and central regions of the United States, and for power generation throughout the country. Sulfur dioxide is also produced copiously by smelting and by volcanos.

Particulates are formed not only when coal and oil are burned, but also originate from automobile traffic, industrial activity, incineration of wastes, and many natural sources. The ash content, which is one measure of the tendency of the coal to produce particulates, ranges from less than 3 percent to as much as 50 percent.[1] The sulfur content of coal ranges from less than 1 to more than 5 percent.

Petroleum oil is burned in several grades, of which there are two basic types, residual and distillate. The latter is low in both sulfur and ash, and is ordinarily used for domestic heating. The residual oils have varying amounts of sulfur and ash, depending on the source of the oil and the extent of pretreatment. The sulfur content can range from less than .1 percent up to 6 percent,[1] and the ash from less than .05 percent up to .10 percent.[2]

When it is released to the atmosphere, sulfur dioxide can gradually be converted to sulfuric acid (as a mist) or to other particulate forms of sulfates. The rates at which these processes occur are not known and are probably variable, depending on the chemical and physical properties of the airborne dust with which the sulfur dioxide comes in

267

contact, and on the temperature and humidity of the atmosphere. For example, the atmosphere contains traces of ammonia, zinc, iron, and vanadium, which can combine either directly with sulfur dioxide or act catalytically to convert it to sulfate forms. The various forms of sulfur oxides are frequently symbolized as SO_x.

For a fixed rate of emission into the atmosphere of fumes from a point source of pollution, such as a smokestack, the maximum concentration, at ground level on level terrain, is inversely proportional to the square of the stack height, and occurs at greater distances from the point of release. Thus, per unit of emissions, an apartment house emitting sulfur dioxide from a rooftop chimney that is 100 to 200 feet above ground will have a much greater effect on the air people breathe than an electrical generating station whose emissions are discharged into the atmosphere 400 feet above ground. (Because boiler gases from a power plant are so hot, they are convected to great altitudes, and the "effective stack height" may be as high as 1,000 to 1,500 feet. It is for this reason that power plant emissions, although they may comprise 40 to 50 percent of the total emissions in a given area, are usually less important, insofar as ground-level concentrations are concerned, than the emissions from apartment houses and public buildings.)

EFFECTS OF SULFUR DIOXIDE ON HUMAN HEALTH

Sulfur dioxide is a gas that is frequently encountered in industrial processes. When inhaled in sufficiently high concentrations, the gas impairs the ventilatory function of the lungs by constricting the bronchial passageways. In lower concentrations, it is irritating to the eyes and the upper respiratory tract. The maximum allowable concentration for occupational exposure to sulfur dioxide is five parts per million, a concentration that would be highly irritating to the eyes and upper respiratory tract of most people. However, workers seem to develop a high degree of tolerance for the unpleasant effects of sulfur dioxide, and many old timers in industry are convinced that it is good for colds!

Sulfur dioxide is very soluble and when inhaled, tends to be absorbed in the upper bronchial passageways. In 1957, it was shown that the broncho-constricting effect in experimental animals increases markedly when the sulfur dioxide is combined with an aerosol.[3] The extent of this synergism depends on many factors, including relative humidity and the chemical characteristics of the particulates associated with the sulfur dioxide. The effect may be due to the conversion of the gaseous sulfur dioxide to particulate form as either sulfate or sulfite. It is also possible that some dusts are capable of adsorbing the sulfur dioxide, allowing penetration to the lower respiratory tract where the effects on bronchoconstriction are more pronounced.

Very few systematic measurements of the extent of sulfur dioxide or particulate pollution in community atmospheres were made prior to 1952, when interest increased enormously because of the occurrence in London of a several-day period of meteorological stagnation. During this episode, air-pollution levels rose to such high levels that nearly 4,000 premature deaths occurred. These casualties were almost entirely among persons with advanced lung or heart disease. Meteorological stagnation associated with high death rates were known to have occurred elsewhere, including the Meuse Valley in Belgium in 1930 and in Donora, Pennsylvania in 1948, but these incidents involved far less severe consequences.

The episode of the London fog that occurred in 1952 caused the death rate among the elderly to increase immediately. Furthermore, it remained higher than usual for more than a week following the breakup of stagnation after four days. The number of deaths due to bronchitis increased nearly tenfold, and deaths from coronary disease more than doubled. Mortality from pneumonia and other respiratory diseases increased about fivefold. The normal number of deaths for that season of the year was approximately 887, but there were 2,484 deaths during the week following the onset of the episode. The total number of people who ultimately died because of the fog amounted to between 3,500 and 4,000. Most of the victims were advanced in age, but the number of deaths from bronchitis and pneumonia among persons under the age of 45 was also three times higher than the normal rate. The relationships between the number of daily deaths and the atmospheric concentrations of sulfur dioxide and smoke are shown strikingly in Figure 11–1.[4]

The 1952 London episode resulted in worldwide concern about air pollution. A Clean Air Act was passed in England in 1956, and air-pollution controls were imposed in the United States. Specialists in Great Britain and the United States adopted different strategies to control air pollution. In Great Britain, no standards were established to limit either the ambient air concentration of sulfur dioxide or the amount of sulfur contained in the fuels used. The British program was directed at getting rid of smoke, and this was done by specifying that only "clean fuels (those low in volatile matter) could be burned in certain areas where high levels of air pollution from coal burning were known to exist. The clean fuels had a lower sulfur content as well, but this was coincidental to the main objectives of burning fuels low in

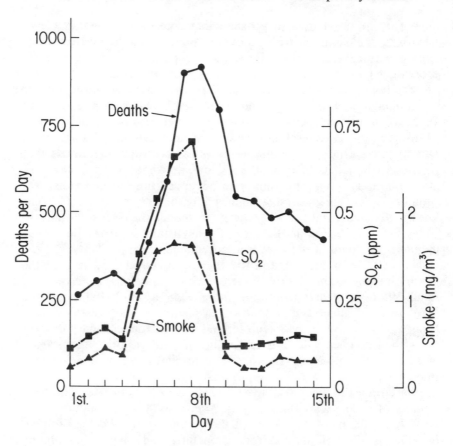

Figure 11-1. Deaths per day and pollution levels during the London fog of December, 1952.[5]

volatile matter. As a result of these measures, airborne particulates in London were reduced by two-thirds during the decade between 1958 and 1968.[4] Although control over sulfur dioxide was not a direct objective, the concentrations of this gas have also been reduced, although not so markedly as the concentrations of smoke.[5] The reduction in the amount of sulfur dioxide is attributed in part to the fact that the cleaner fuels contain less sulfur as well as less volatile matter and in part to the fact that by elimination of smoke, more sunshine is admitted to the streets of London. This hastens the breakup of temperature inversions and results in the more rapid dilution of the sulfur dioxide.

There is also evidence that London air had been gradually becoming cleaner for many years prior to the act.[6] This can be seen in

Figure 11–2, in which are plotted the wintertime smoke levels at Kew since 1922. A long-range trend is evident, which may have been due to the gradual use of cleaner fuels prior to passage of the Act. The effects of the scarcity of coal during the Great Depression and World War II should also not be overlooked. The need for corrective action was indicated strongly by the resurgent levels following the end of World War II in 1945.

A test of the corrections of the British approach may have taken place inadvertently in December 1962 when atmospheric conditions developed in London that were very similar to those of December 1952. The two fogs were of similar duration and the concentrations of sulfur dioxide rose to almost identical levels. However, the *smoke* levels during the 1962 fog were much lower. Compared to the 4,000 excess deaths that took place in 1952, excess mortality was limited to 700 during the 1962 fog. Londoners should be exposed to even less risk by now, since both the smoke levels and the sulfur levels have been reduced even further since 1962.

In addition to the two major episodes that resulted in high mortality, it had been known for years that in the cities of England a positive correlation existed between disability due to respiratory disease and the varying peaks of air pollution that occurred many times each year. By 1969 it was reported that this association could no longer be demonstrated.[5]

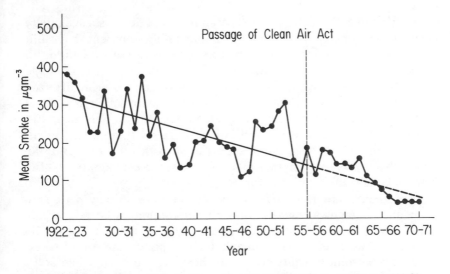

Figure 11-2. Average smoke levels at Kew between October and March 1922-23 through 1970-71.[6]

The situation in the United States is somewhat different than in the United Kingdom, where the incidence of chronic respiratory disease has always been relatively high. (Also, it will be seen that the health effects of urban air pollution have been much more difficult to document in the United States.)

The authorities in the United States, unlike the British, adopted air quality standards for both sulfur dioxide and particulates. The main emphasis however, was placed on restricting the amount of the sulfur content of fossil fuels. This was concentrated upon because particulates originate from tens of thousands of oil burners, apartment-house incinerators, automobiles, and many other sources that are far more difficult to control. It proved easier to order a few large oil refiners to reduce the sulfur content of the oil they supply than to attempt to regulate the performance of thousands of smoke-producing oil burners.

Although there are many important air pollutants, only sulfur dioxide has been monitored for any length of time in the United States. This is because relatively simple and reasonably reliable methods of monitoring this pollutant became available many years before techniques were available to measure carbon monoxide, airborne particulates, lead, or other air contaminants. Thus, when statisticians first began to examine the health effects of air pollution, only data on the presence of sulfur dioxide were available, and when they studied its day-to-day fluctuations in relation to health effects, a correlation could be demonstrated. Caution should have been indicated before drawing the conclusion that sulfur dioxide was causally related to the sickness or death. It is immediately apparent that sulfur dioxide is associated with other factors that were not measured, but which could have been the cause of the effects on health. For example, sulfur dioxide in metropolitan atmospheres originates primarily from the combustion of fossil fuels, which release many other pollutants, such as inorganic particulates, nitrogen oxides, and copious quantities of trace elements. The concentrations of these substances tend to rise and fall with the concentration of sulfur dioxide, the concentration of which is thus also an indicator of the relative amounts of other pollutants as well.

The concentration of SO_2 is apt to increase during periods of atmospheric stagnation when atmospheric ventilation is reduced and pollutants of all kinds tend to accumulate. During some seasons of the year, this atmospheric condition may include pollen and fungal spores which are harmful to some people. In heavily trafficked urban areas, automobile tailpipe emissions also accumulate, as do the pollutants from industrial plants. Thus, air pollutants of all kinds would also be expected to be higher during periods when the sulfur dioxide levels are increased. In the temperate climates, these periods of meteorological stagnation may also coincide with heat waves, and it is well known that the death rate increases during such periods because of the effects of elevated temperature.[7] It also happens that those who are most susceptible to the effects of high temperatures are individuals with preexisting heart and lung disease, and these members of the population are also susceptible to the atmospheric irritants such as the sulfur oxides.

Studies that have been made in the United States of the effects of the pollution levels encountered in urban air under so-called normal conditions are far from unequivocal. Some investigators have found no association between sulfur dioxide and either sickness or death. Others have demonstrated that such relationships do exist. It is interesting that as late as 1965 (when air pollution in most cities of the United States was very much greater than in more recent years), the Environmental Pollution Panel of the President's Scientific Advisory Committee was able to state, with respect to air pollution, that "while we all hear and many believe that long continued exposure to low levels of pollution is having unfavorable effects on human health, it is heartening to know that careful study has so far failed to produce evidence that this is so, and such effects if present must be markedly less notable than those associated with cigarette smoking. Attempts to identify possible effects of ordinary air pollution on longevity or on the incidence of serious disease have been inconclusive."[8]

It is difficult to determine how dangerous air pollutants are to health for other reasons. For example, the principal lung diseases that have been suspected to be due in part to air pollution are cancer, emphysema, and bronchitis. However, a fundamental complication in attempting to quantitate the relationships between air pollution and these diseases is due to the overwhelming effects of cigarette smoking. Smokers are far more susceptible than nonsmokers to bronchitis, emphysema, and cancer. Another difficulty in assigning responsibility to specific pollutants arises from the fact that the individual pollutants unquestionably interact with each other. The interaction of particulates and sulfur dioxide has been mentioned. The sulfur dioxide-particulates combination reduces the ventilatory ability of the lungs by its broncho-constricting action. This may add to the effect of carbon monoxide, which reduces the capacity of the red blood cells to transport oxygen to the tissues. Both carbon monoxide and sulfur dioxide place a burden on the cardio-respiratory system, as do cigarette smoking, allergy-producing substances, and respiratory infections.

There has been a 25-year trend in many cities towards the use of "cleaner" fuels, in the United States due in part to the ready availability of gas or oil, as well as to the growing desire to reduce air pollution. Fuel oil and natural gas have gradually displaced coal for power generation and space heating in many parts of the country, mainly for reasons of convenience and economy. The improvement in air quality in many cities has been dramatic wherever data are available. New York City has collected air-pollution data since the mid-1950s, as can be seen in Figure 11–3, which shows a spectacular drop in the concentration of sulfur dioxide, beginning about 1964.

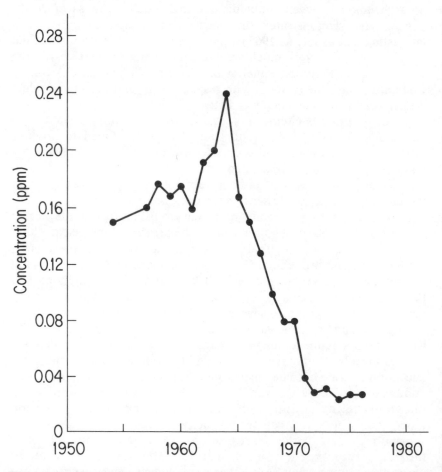

Figure 11-3 Sulfur dioxide concentrations in New York City 1954-1976. Note the rapid drop beginning in 1964, several years prior to development of federal regulations.[13]

In New York City, important studies of the need for air-pollution control and the methods by which control could be achieved, were conducted in 1965 and 1966, and two reports published during those years[9,10] laid the groundwork for a phased program of air-pollution control which, among other actions, would have limited the sulfur content of fuels burned to 1 percent. With only minor exceptions, the burning of coal was eliminated. As a result of these actions, the emissions of sulfur dioxide into the city's atmosphere were limited to such an extent that by the end of 1969 the annual average concentration of sulfur dioxide had been lowered by 56 percent. The New York City government was planning to adopt an ambient-air-quality standard of .1 parts per million sulfur dioxide, which could have been achieved by limiting the sulfur content of oil burned in New York City to 1 percent. However, this was during a period when there was a tendency for various governmental agencies to outdo one another in setting clean air goals. Within a matter of months, early in 1970, the target of .1 parts per million, which would have represented a substantial reduction over the preexisting sulfur dioxide concentrations, was reduced more than threefold to .03 parts per million by the requirements of the federal government. To meet this goal, it was necessary that the maximum concentration of sulfur in New York City fuels be limited to .3 percent.

To some, the more stringent federal restrictions might have seemed to have been a wise move. From one point of view, exposure to sulfur dioxide should be reduced to the lowest possible level, if there is any possibility at all that it could be harmful to human health. Some may believe that every city should have the cleanest air possible. However, air-pollution abatement measures are extremely costly, and the costs increase exponentially as the goals become more strict. No satisfactory estimate of the additional costs incurred by the people of New York City is available, but there have been informed judgments that it cost about $100 million per year to reduce the sulfur content of fuels from more than 2 percent to 1 percent as mandated in the original New York City Clean Air Act of 1968. It is estimated that it cost as much as $200 million per year, at the 1970 prices of oil, to reduce the sulfur content to .3 percent. Considering that this cost goes on year after year, it represents a major investment on behalf of the public health. Annual investments of this kind should not be made in the absence of clear-cut evidence that meaningful improvements in the public health will result. Reliable evidence is not at hand, although many studies have been undertaken in recent years.

Reports published in the 1960s and early 1970s purported to show an *association* between the daily concentrations of sulfur dioxide and "excess mortality" (which more properly should have been called "premature mortality") in New York City.[11,12] The term "excess mortality," which is frequently used in the air pollution literature, should more properly be replaced by "life shortening." The evidence suggests that the mortality due to the combined effects of sulfur oxide and particulates occurs among individuals with advanced cardiorespiratory disease who were already nearing the end of life. ". . . the most we can say with a high degree of probability is a certain portion of deaths would not have occurred at the time they did in the absence of pollution."[14] The air pollution is an additional stress that may make the difference between life and death for some of the chronically sick. What is *not* known is the degree to which life is shortened. It may be a matter of days, weeks, months, or years, but the data are lacking. All that is known is that elderly persons with pre-existing cardiovascular disease are the ones primarily affected. From the point of view of a policy maker seeking to establish environmental priorities, it makes a great deal of difference whether the life to be saved is that of an ailing octogenarian with a life expectancy of two months (which is likely to be the case for many of the cases reported as excess mortality during air pollution episodes) or a vigorous 40-year-old who develops lung cancer and dies in six months. At any rate, the association was interpreted by some as evidence that sulfur dioxide was the *cause* of increased death rates. Because the conversion was made from coal to oil, and because, to a greater extent, control measures were applied in 1968, the concentrations of sulfur dixoide in New York City air diminished rapidly, and by 1976 had been reduced about 90 percent from an annual average of .24 parts per million to slightly less than the ambient-air-quality standard of .03 parts per million stipulated by the federal government.[13]

If premature mortality was, in fact, due to sulfur dioxide at the high concentrations that existed in the early 1960s, it should have been possible to demonstrate in the 1970s that air pollution was having less of an effect on health than previously. Further investigation by one of several investigators who were responsible for the original reports that seemed to show a relationship between sulfur dioxide and excess mortality indicated that, despite the marked diminution there had been no change in the pattern of daily mortality.[14,14a] It was demonstrated by statistical analysis that less than 1 percent of the daily deaths seemed to be associated with pollution, and that sulfur dioxide was not the causative factor. During the same period, the particulates had been reduced somewhat, but not nearly to the same extent as sulfur dioxide.

Excessive mortality is only one index by which the effects of a pollutant can be measured. Another is the incidence and severity of respiratory disease.

A number of epidemiological studies had produced evidence of the effects of sulfur dioxide on community health during the 1970s, but in none of the studies was it possible to rule out the influence of socioeconomic factors or other air pollutants on the incidence of respiratory disease. The important relationship between socioeconomic factors and health is well known: The prevalence of many diseases is inversely correlated with economic level. One would expect that there is an inverse correlation between exposure to air pollution and income, and this has indeed been demonstrated in one area, the New York region.[15]

Respiratory disease is the most common cause of sickness in the United States and accounts for 50 to 80 percent of all childhood illnesses.[16] Although most such illnesses are minor in severity, they may contribute to the development of chronic respiratory disease in adult life. In 1970, the EPA (Environmental Protection Agency) began an important investigation known as the Community Health and Environmental Surveillance System (CHESS) in which studies of the relationship of pollution to respiratory-tract disease were undertaken in New York City, Salt Lake City, Los Angeles, Chicago, Birmingham, Charlotte, and Chattanooga.[17] In each of the areas selected for study, sample populations were identified which were exposed to different levels of air pollution. The pollutants of interest were sulfur dioxide, sulfates, particulates, nitrogen oxides, nitrates, ozone, carbon monoxide, and hydrocarbons.

Efforts were made to compare populations that were similar in socioeconomic and ethnic characteristics, age distribution, and other factors which could influence the observed health effects. For example, the health of sample populations in New York City was compared to residents of Riverhead, a seaside community of 6,000 inhabitants located about 70 miles to the northeast, near the end of Long Island. However, some of the factors involved in such studies are very subtle and cannot be quantified.

When a study looks back in time and attempts to reconstruct the incidence of disease in relation to environmental factors of interest, it is said to be "retrospective." Such a study was conducted in New York by querying the parents of a sample of elementary-school children about the frequency of treatment by a physician for pneumonia, croup, bronchitis, or other chest infections during the three-year period June, 1969 to May, 1972, when the questionnaire was distributed. It was found that in the communities having the higher levels of air pollution, the rates of "any lower respiratory disease," croup, bronchitis, and "other" chest infections were increased. However, the frequency of hospitalization for pneumonia were significantly increased among the younger children in the low-pollution community. The higher frequency of some respiratory diseases among the children living in the more-polluted areas, would be "as expected." However, one would not expect that the higher levels of pollution in New York City were the reasons for the *lower* rates of hospitalization for pneumonia.[16] This peculiar finding may have been due to differences in the diagnostic preference of local physicians in the low-pollution area, or differences in the criteria for hospitalization. This is an example of the subtle ways in which socioeconomic environmental factors can influence the results of an epidemiological study.

It is difficult to isolate each of the many factors that may contribute to the causes of respiratory disease. Among those mentioned in the reports of the CHESS studies are parental smoking habits, the presence of gas stoves (which are known to produce nitrogen oxides) for cooking or space heating, socioeconomic factors that may not have been identified, and biases that can be introduced by the diagnostic preferences of physicians.

Additionally, CHESS included a prospective study which considered the acute respiratory illness experience among families living in areas having different levels of exposure to sulfur dioxide and particulates. The study was begun in September, 1970 and ended in early May, 1971. The subjects selected lived within 1.5 miles of an air-monitoring station and resided at that address for at least one year. These families were telephoned every two weeks and were questioned about the presence of illness in the family and the dates when medical attention was received. The study concluded that there was a higher incidence of respiratory disease in the more polluted areas, where an average family of four might expect to require one extra visit to a physician each year and might experience five extra days of restricted activity due to acute respiratory disease. As before, the large number of variables other than air pollution that affect the occurrence of respiratory disease or, for that matter, the "perception" of respiratory disease, make it difficult to agree that the observed differences were, in fact, due to differences in air pollution.

The CHESS studies have been criticized by many epidemiologists. It has been said that the work was done in haste, that there were basic methodological defects, and that the data were overinterpreted:[18-24]

The need for a great deal of information in the shortest possible time has meant that the EPA has been forced to attempt too much too superficially. A more deliberate approach with outside consultation might have led to a more solid body of knowledge. As it is, the studies have a number of deficiencies which makes evaluation difficult. The samples studied, the response rates in certain categories, the methods and procedures which have been used and the analysis of the results can all be criticized. It is particularly disquieting, in view of these deficiencies, that there has been a rather marked tendency to over-interpret the data and in particular to select findings which point to an effect of pollution on health and ignore those which do not.[23]

A Congressional investigatory committee concluded that "technical errors in measurement, unresolved problems in statistical analysis, and inconsistency in data in the 1974 CHESS monograph render it useless for determining what precise levels of specific pollutants represent a health hazard."[24] While the report of this investigatory committee was highly critical of the CHESS program, it also noted that EPA was operating under an unrealistic self-imposed timetable, with severe financial restrictions.

The annual average concentration of sulfur dioxide reported by the CHESS study during 1971 in the New York high-pollution area was a little more than twice that in the area of low pollution. But in the mid-1960s, the annual average sulfur dioxide concentrations in the more polluted parts of New York City were nearly ten times the 1971 levels. If a demonstrable difference in the incidence of disease exists when the concentrations vary by only a factor of 2, the incidence of respiratory disease prior to the marked improvement that began a decade earlier should have been dramatically higher, but no data exist on which to base a conclusion.

Beginning about 1973, the U.S. Environmental Protection Agency began to express increasing concern about the possible effects of atmospheric sulfates.[25] Their concern has been reinforced by an independent investigation,[25a] but the findings have not been generally accepted.[25b] One of the most perplexing observations is that although the concentrations of sulfur dioxide have decreased markedly during the past 10 years in many cities, there has been no corresponding decrease in the concentration of sulfates. In New York, the concentrations of sulfur dioxide have been reduced by more than 90 percent, whereas the sulfates have been reduced by only about 60 percent. The sulfate levels tend to be higher (7 to 20 micrograms per cubic meter) in urban areas of the eastern United States than in rural eastern areas (6 to 9 micrograms per cubic meter) and, western rural areas are particularly low (one to two micrograms). The latter concentrations approach the natural background of sulfates.

The complicated interrelationships between sulfur dioxide, atmospheric particulates, and sulfates (alone or in combination) are not understood. There are even uncertainties about how the sulfates should be measured. The data gathered thus far have been obtained by analyzing filter papers through which air has been drawn and on the surface of which the particulates are collected. The sulfates may actually form on the filter-paper surface by the reactions that take place between the particulates already deposited and the sulfur dioxide that is present in the air drawn through the filter. Such reactions would be aided by moisture and by the presence of metals that either react directly with the sulfur dioxide or serve as catalysts to accelerate reactions with other metals. By this line of reasoning, the sulfur dioxide could be oxidized to any one of several forms of sulfate that could remain on the filter paper. Thus, the subsequent sulfate analysis might give a false measure of the sulfate actually present in the atmosphere.

In summary, limitations in epidemiological methods have thus far made it difficult to differentiate the health effects of air pollution from those due to other factors. Nevertheless, there is persistent though inconclusive evidence that, after correcting for socioeconomic factors and cigarette smoking, there does appear to be an effect on health by air pollutants. Granted this much, one wishes to know the magnitude of the effect, and which specific agents are responsible.

The magnitude of the effect cannot be stated with certainty. After reviewing the world literature on the subject, the National Academy of Sciences concluded, very tentatively, that "it seems reasonable to assign *no more* than 10 percent of chronic bronchitis in the general population to the effects of sulfur oxides" (emphasis added). This report suggests a similar estimate for acute respiratory illness, with one to two days lost per year for each working person. The report concludes that these figures should be interpreted with caution.[18] These estimates are based on data collected in communities in which substantial improvement in air quality has already taken place. There is no way of showing that the effects were more prevalent in the past, when the sulfur oxides (sulfur dioxide or sulfates) were present in much higher concentrations.

There remains the problem of apportioning the apparent detrimental effects among sulfur dioxide, sulfates, and particulate pollution. This is a matter of great practical importance because it determines the priorities that should be assigned to various control strategies. As noted earlier, the British chose to control particulates and ignored the sulfur oxides. We have discussed the evidence that they were justified in this approach. In the United States, standards were established for both sulfur dioxide and particulates (though not for sulfates, due to lack of data) but the main emphasis has been on the control of sulfur oxides.

The subject of the health effects of sulfur oxides and particulates is one of the most confusing of the many contemporary environmental issues. The fact that the matter has resulted in such intense debate in the mid-1970s is particularly astonishing because such dramatic reductions in the sulfur dioxide concentrations in many cities in the United States had already taken place, with no indication of any health benefits. This does not imply that such benefits had not occurred. After all, the CHESS studies began after much of the improvement had taken place, and similar studies had not been conducted in prior years.

The findings of the CHESS studies have been released mainly in government reports. Often, the findings are first presented in press releases or news conferences. Obviously, these procedures bypass the more traditional means of making scientific information available. It is normal practice for scientists to publish the results of their research in technical journals that require that articles submitted for publication be reviewed by experts. This system of peer review has been an important factor in assuring the quality of scientific literature. A disadvantage of this system is that it may take as long as 18 months for an article to survive the review procedure and finally appear in print. The contribution often receives a critical review after publication by readers who submit letters to the journals, and these letters are usually published along with the responses of the original authors.

Regrettably, environmental issues have been debated at such a frenetic pace in recent years that scientists now frequently bypass the tedious and time-consuming procedures required by scientific journals. This is less a criticism of the authors of the scientific reports than of the fact that interest in environmental matters has been so intense that public-interest groups and environmental organizations are unwilling to wait until a report has been published in the more conventional way. In so doing, they run the risk that studies will be conducted and reported with somewhat less than the desired scientific discipline.

METHODS OF SULFUR OXIDE CONTROL

High-sulfur fuel oils can be desulfurized at the refinery. Coal desulfurization will probably be accomplished ultimately by means of processes that liquefy or gasify coal, but these have not yet been demonstrated on a large scale.

Tall stacks have been a popular method of dispersing pollutants on the theory that, if the emissions can be diluted sufficiently, they will do no harm. In a few cases, utilities use high-sulfur fuels during periods of favorable weather and are prepared to switch to low-sulfur fuels when necessary. Computerized meteorological and air-sampling networks relay information to a central control station, so that the system operators can be alerted if the air-quality standards are likely to be exceeded. When this happens, the utilities can either reduce the rate of power generation or switch to lower-sulfur fuels. Systems of this type are feasible but are rarely employed. Alternatively, a utility can buy electricity from neighboring generating systems located in areas where the meteorological conditions are more favorable.

One strategy for maximizing the limited resources of low-sulfur fuels is to use high-sulfur fuels in places where this will cause no significant deterioration of air quality. For example, power plants or commercial buildings located in rural areas where there are few other sources of sulfur oxides could be permitted to burn fuels with higher sulfur content without causing the air quality standards to be exceeded. Conversely, the use of fuels with high-sulfur content would not be permitted in large urban or industrial areas in which there are many other sources of sulfur oxides. However, the Congress and the EPA have discouraged the use of high-sulfur fuels even in remote areas, with or without stacks, because of its belief that the sulfur dioxide would be transported for great distances and gradually convert to sulfates that could add to the contamination of distant localities. The validity of this argument is highly speculative because the sulfates would exist in very dilute form at such great distances. However, it is known that the acidity of rainfall has been increasing in many parts of the world, and this is thought to be due to combustion of sulfur-containing fuels, sometimes at great distances from the places where acid rain is occurring. Acid rain is said to be changing the characteristics of some lakes and has resulted in destruction of fish and plants.[26] This phenomenon has been most pronounced in Scandinavia, where the rain acidity has been attributed to sulfur oxides introduced into the atmosphere in Western Europe and carried to Norway and Sweden by the southerly winds.

The policy of the EPA has been to require power plants to install mechanical equipment, called scrubbers, to remove sulfur dioxide from flue gases. There are three main objections to these devices at the present time: They create huge quantities of liquid and solid waste; their mechanical reliability has not been demonstrated for large plants over a sufficiently long period of time; and they are very expensive.

The majority of flue-gas desulfurization plants now in operation or under construction depend on reactions between the flue gases and limestone for the removal of sulfur dioxide. However, other processes were being investigated, and by late 1975 there were ten different systems at or near the prototype demonstration stage.[27] In the process that has attracted the most attention, lime or limestone, as a finely divided powder suspended in water, is intimately mixed with the exhaust gas in the scrubber. The slurry thus formed is then passed to a tank in which the chemical reaction between the sulfur dioxide and lime produces a precipitate that is dewatered and removed from the system as solid wastes. All of this sounds very simple, but the process can be accomplished only with equipment of enormous size, the initial cost of which, for a 1,000-megawatt electric generator, will be in the range of $75 to $125 per kilowatt of plant capacity. The desulfurization installation for a modern 1,000-megawatt generating station will thus cost in the neighborhood of $100 million, which is a significant fraction of the total plant construction cost. The operating costs of sulfur scrubbers, in addition to the cost of amortizing the capital investment, can add up to 30 percent to the cost of generating electricity. This is roughly ten times greater than the cost of control by tall stacks.[27,28] In addition, these scrubbers use 3 to 5 percent of the total electricity generated.

The sludges produced can either be stored in ponds or used as a source from which the sulfur may be removed for sale. The enormity of the problem of the sludge disposal is illustrated by the fact that a 2,000-megawatt, coal-fired station burning high-sulfur coal will produce about 80,000 cubic feet of sludge per day. At Shippingport, Pennsylvania a new plant is being built at which scrubber sludges will be piped seven miles to a valley that is being dammed and will be filled for five miles to a depth of 400 feet during the next 25 years. The long-term environmental effects of sludge storage are not fully understood. For example, it is not certain that microbial action on stored sludges will not produce hydrogen sulfide.[29] These sludges, the product of a system intended to abate air pollution, may become sources of potentially serious water and ground pollution.

Both the economics of the process and the sludge disposal problem will be assisted if the sulfur can be recovered either as elemental sulfur, sulfuric acid, ammonium sulfate (which can be used as a fertilizer), or gypsum (calcium sulfate). The scrubbers, expected to be in operation by 1980, could produce about 50 percent of the sulfuric acid used in the United States. Alternatively, it would be possible to produce substantial quantities of fertilizer or gypsum (which is used in wallboard and other building materials). However, both the engineering feasibility and economics of these opportunities for by-product production remain to be demonstrated.

The sulfur content of fuel oil consumed in the United States is limited by state and federal laws to .3 to 1 percent by weight. Since the only major source of oil in which the sulfur content is within this range is Libya, most crude oil requires desulfurization. In 1973 it was reported that it cost 87¢ per barrel to desulfurize crude oil to a level of 1 percent sulfur and $1.18 per barrel to desulfurize it to a content of .3 percent. At that time, oil was selling at about $2.50 per barrel, but the price was raised more than fourfold to about $11.50 per barrel following the Mideast war in late 1973. The cost of desulfurization, which raised the cost of oil by 35 to 47 percent, based on the early 1973 prices, was relatively less (7.5 to 10 percent), based on the 1975 price of about $11.50 per barrel.[30] However, the cost of desulfurization is substantial in absolute terms. Not the least of the components of the cost is the energy required for desulfurization. It is estimated that the energy required to desulfurize crude oil to a level of between .3 and 1 percent of sulfur amounts to 10 percent of the energy value of the fuel oil produced.[31] Metropolitan New York consumed 140 million barrels of oil with a level of sulfur of .3 percent in 1973. The yearly cost of desulfurization, based on a charge of $1.18 per barrel, would thus have been about $165 million in that year, had desulfurization been necessary.[32]

The complexity of the petroleum industry, the rapidly changing pace of technological developments, and changing oil prices make it difficult to estimate the cost of desulfurization, which is certainly no less than $1 to 2 billion per year for the United States. Since the cost of installing flue-gas desulfurization scrubbers could easily reach $1 to $2 billion per year in the immediate future, the total annual cost of desulfurizing oil and coal can be estimated to be $2 to $4 billion. These costs will be borne mainly by the electrical-utility industry and will be passed on to the consumer as an added cost for electricity and the products produced with electrical energy. Whether a commitment of this magnitude is justified in the interest of public health remains to be demonstrated.

NOTES

1. Magee, E. M., H. J. Hall and G. M. Varga. "Potential Pollutants in Fossil Fuels," Environmental Protection Agency *Report No. EPA-R2-63-249* (June, 1973).
2. Goldstein, H. L. and C. W. Siegmund. "Influence of Heavy Fuel Oil

Composition and Boiler Combustion Conditions on Primary Particulates," *Environ. Sci. & Tech.* 10:1109 (1976).

3. Amdur, M. O. "The Influence of Aerosols upon the Respiratory Response of Guinea Pigs to Sulfur Dioxide," *Amer. Ind. Hyg. Assoc.* 18:149–155 (1957).

4. Royal College of Physicians of London. "Air Pollution and Health," Summary and Report on Air Pollution and its Effect on Health by the Committee of the Royal College of Physicians of London on Smoking and Atmospheric Pollution, Pitman Medical and Scientific Publishing Co. Ltd., London (1970).

5. Waller, R. E., P. J. Lawther and A. E. Martin. "Clean Air and Health in London," in: Preprints of Papers (Part 1), Clean Air Conference Eastbourne (October 21-24 1969), National Society for Clean Air, Field House, Breams Buildings, London EC4. Printed in Great Britain by Spottiswoode, Ballantyne & Co., Ltd., London/Colchester (1969), p. 72.

6. Auliciems, A. and I. Burton. "Trends in Smoke Concentrations before and after the Clean Air Act of 1956," *Atmospheric Environment* 7:1063–1070 (1973).

7. Marmor, M. "Heat Wave Mortality in New York City, 1949 to 1970," *Arch. Env. Hlth.* 30:130–136 (1975).

8. President's Science Advisory Committee. "Restoring the Quality of our Environment," Report of The Environmental Pollution Panel, PSAC, The White House, U.S. Government Printing Office, Washington, D.C. (1965).

9. Special Committee of the City Council of the City of New York to Investigate Air Pollution. "Air Pollution in New York City: An Interim Technical Report," *The City Record* (June 24, 1965).

10. Mayor's Task Force on Air Pollution in the City of New York. "Freedom to Breathe," City of New York, New York (1966).

11. Beuchley, R. W. et al. "SO_2 Levels and Perturbations in Mortality, a Study in the New York-New Jersey Metroplis," *Arch. Environ. Hlth.* 27:134–37 (1973).

12. Schimmel, Herbert and Leonard Greenburg. "A Study of the Relation of Pollution to Mortality: New York City, 1064–1968," *Journal of the Air Pollution Control Association* 22(8):607 (1972).

13. Eisenbud, Merril. "Levels of Exposure to Sulfur Oxides and Particulates in New York City, and their Sources," *Proceedings of the New York Academy of Medicine Symposium on Environmental Effects of Sulfur Oxides and Related Particulates* (in press).

14. Schimmel, Herbert and T. J. Murawski. "The Relation of Air Pollution to Mortality," *J. of Occup. Med.* 18:316–333 (1976).

14a. Schimmel, Herbert and Peter Bloomfield. "Evidence for the Influence of Sulfur Oxides and Related Particulates on Mortality," *Proceedings of the New York Academy of Medicine Symposium on Environmental Effects of Sulfur Oxides and Related Particulates* (in press).

15. Zupan, J. M. "The Distribution of Air Quality in the New York Region," Johns Hopkins Univ. Press, Baltimore, Md. (1973).

16. Hammer, D. I., F. J. Miller, A. G. Stead and C. G. Haynes. "Air Pollution and Childhood Lower Respiratory Disease: Exposure to Sulfur Oxides and Particulate Matter in New York, 1972," presented by Hammer at the American Medical Association Air Pollution Medical Research Conference (December 5–6, 1974), San Francisco, Cal. (1974).

17. Community Health and Environmental Surveillance Studies (CHESS). "Health Consequences of the Sulphur Oxides: A Report from the CHESS Program 1970–1971." Draft Monograph (1972).

18. National Academy of Sciences. "Air Quality and Stationary Sources Emission Control," Report by NAS Commission on Natural Resources prepared for Committee on Public Works, U.S. Senate (94th Cong.) pursuant to S. Res. 135. Committee Print *Serial No. 94-4*, U.S. Government Printing Office, Washington, D.C. (1975), p. 40.

19. Health Research Council. Study on "Health Impact of Stationary Source Fossil Fuel Air Pollution Control Measures in New York City," HRC's Working Group on Environmental Pollution. (Meeting was held March 6, 1975 at which time this working group presented its conclusions to NYC officials. This paper is a report by a special panel offering critique of conclusions presented at meeting, including CHESS studies.) (1975).

20. Committee on Science and Technology, U.S. House of Representatives, 94th Congress. "Research and Development Related to Sulfates in the Atmosphere." Hearings before the Subcommittee on the Environment and the Atmosphere. *Serial No. 94-36* (July 8, 9, 11, 14, 1975).

21. Greenfield, Attaway and Tyler, Inc. "A Detailed Critique of the Sulfur Emission/Sulfate Health Effects Issue." Prepared for the Edison Electric Institute and the Electric Utility Industry Clean Air Coordinating Committee (April 22, 1975).

22. Tabershaw/Cooper Associates, Inc. "A Critical Evaluation of Current Research Regarding Health Criteria for Sulfur Oxides." Prepared for the Federal Energy Administration (April 11, 1975).

23. Higgins, I. T. T. and B. G. Ferris. "Epidemiology of Sulphur Oxides and Particles," in: *Proceedings of the Conference on Health Effects of Air Pollutants*, Serial No. 93-15, U.S. Government Printing Office, Washington, D.C. (November, 1973), p. 247.

24. Committee on Science and Technology U.S. House of Representatives, 94th Congress. "The Environmental Protection Agency's Research Program with Primary Emphasis on the Community Health and Environmental Surveillance System (CHESS): an Investigative Report." Report prepared for the Subcommittee on Speical Studies, Investigations and Oversight and the Subcommittee on the Environment and the Atmosphere of the Committee on Science and Technology, U.S. Government Printing Office, Washington, D.C. (November, 1976).

25. Love, G. J. "Epidemiologic Studies of Adverse Health Effects Associated with Exposure to Air Pollution," in: *Proceedings of Recent Advances in the Assessment of the Health Effects of Environmental Pollution*," *EUR 5300*, Commission of the European Communities,

Directorate General Scientific and Techn. Inform. and Management, Luxembourg (1975), vol. 1, p. 301.

25a. Lave, L. B. and E. P. Seskin. "Air Pollution and Human Health," Johns Hopkins Univ. Press, Baltimore, Md. (1977).

25b. New York Academy of Medicine. *Proceedings of the New York Academy of Medicine Symposium on Environmental Effects of Sulfur Oxides and Related Particulates* (in press).

26. Likens, Gene E. "Acid Precipitation," *C&EN Special Report*, American Chemical Society, Washington, D.C. (November 22, 1976).

27. Devitt, T. W., L. V. Yerino, T. C. Ponder and C. J. Chatlynne. "Estimating Costs of Flue Gas Desulfurization Systems for Utility Boilers," *J. Air Poll. Control Assn.* 26:204–215 (1976).

28. Yeager, K. E. "Stacks vs. Scrubbers," in: *EPRI Research Progress Report FF-3*, Electric Power Research Inst., Palo Alto, Calif. (July, 1975), p. 2.

29. Hammond, Allen L. "Coal Research (II): Gasification Faces an Uncertain Future," *Science* 193:750–753 (1976).

30. American Petroleum Institute. "Basic Petroleum Data Book," American Petroleum Institute, Washington, D.C. (December, 1975).

31. Scott, R. W. "Estimated Cost of Desulfurizing Fuel Oil," personal communication from Coordinator for Conservation Technology, Exxon Research and Engineering Co., Florham Park, N.J. (August 6, 1976).

32. U. S. Department of the Interior, Bureau of Mines. "Sales of Fuel Oil and Kerosene in 1973."

CHAPTER 12

Air Pollution From Automobiles

Most automobiles and trucks are powered by internal-combustion engines which are the source of three major air pollutants: unburnt gasoline vapors, carbon monoxide, and nitrogen oxides.

The unburnt or partially burnt gasoline vapors that evaporate from the gas tank, the carburetor, and from drippings at the gas pump, are characterized generally as "hydrocarbons." Gasoline vapor also passes between the piston and the cylinder to enter the crankcase, from which, in older cars, it was vented to the atmosphere. The hydrocarbon vapors are of themselves innocuous at the concentrations found in the general atmosphere, but they react with nitrogen oxides and sunlight in photochemical reactions that produce irritating compounds classed as "oxidants." It is these compounds that cause the smogs of some cities to irritate the eyes.

Carbon monoxide is produced by the incomplete combustion of gasoline, and we will see that this is one of the most important of the urban air pollutants.

When combustion occurs at a high enough temperature, as in the internal-combustion engine, the nitrogen of the atmosphere is partially oxidized and discharged as nitrous oxide (NO), which further oxidizes to nitrogen dioxide (NO_2) and other compounds of nitrogen when released to the atmosphere. The resulting mixture is referred to as *nitrogen oxides*.

It is widely stated that the automobile is the principal source of air pollution in the United States, and this would seem to be supported by the data in Table 12–1.[1] The table summarizes the estimated weights of air pollutants discharged to the atmosphere in the United States during 1970, when the federal clean-air programs were just getting underway. Transportation accounted for 143.9 million tons of pollution, or 54 percent of the total emissions of 263.9 million tons. How-

TABLE 12.1
Estimated National Air Pollution Emissions, 1969
Millions of Tons Per Year*

Source	Carbon monoxide	Sulfur oxides	Hydro-carbons	Nitrogen oxides	Particles
Transportation	111.5	1.1	19.8	11.2	0.8
Fuel combustion from stationary sources	1.8	24.4	0.9	10.0	7.2
Industrial processes	12.0	7.5	4.8	0.2	14.4
Solid waste disposal	7.9	0.2	2.0	0.4	1.4
Miscellaneous	18.2	0.2	9.2	2.0	11.4
Total	151.4	33.4	36.7	23.8	35.2

*National Academy of Sciences, 1974.

ever, these figures are of themselves misleading because the several pollutants listed are not equally hazardous. The federal standards of permissible air quality permit concentrations of carbon monoxide about 100 times higher than the permissible concentration of sulfur dioxide. The annual emission of 111 million tons of carbon monoxide is thus no more hazardous than about 1 million tons of sulfur dioxide. The quantity of sulfur dioxide discharged annually from all sources is more than 30 times this amount. Thus, if one uses the standards of ambient air quality as a basis for comparing the health hazards of the various pollutants, carbon monoxide is less significant than the sulfur oxides when weighted for its relatively low toxicity. Although the total pollution from automobiles is higher than that from any other single manmade source, it is not the most important source when adjusted for the relative toxicity of the various component pollutants.

The figures given in Table 12-1 are for tailpipe emissions only, which may not be the principal contribution of the automobile to air pollution. Moving automobiles resuspend settled dust, abrade road surfaces, abrade the rubber from tires, and spread asbestos from brake linings. Studies in New York City suggest that the automobile is responsible, in one way or another, for 30 percent of the airborne dust.[2]

The automobile has also been a source of atmospheric lead pollution by reason of the organic lead compounds which have been added to gasoline to improve its antiknock characteristics. However, lead is gradually being eliminated from gasoline, primarily because it tends to poison the catalysts used to control the tailpipe emissions (Chapter 10).

CARBON MONOXIDE

Until comparatively recently, carbon monoxide was thought to be a relatively inert compound that was remarkable only for its extraordinary affinity for hemoglobin, the oxygen-carrying pigment in the red blood cells that transport oxygen from the capillaries of the lungs to the various organs of the body. When carbon monoxide is absorbed into the bloodstream, it combines with hemoglobin to such an extent as to cause interference with oxygen transport to the tissues.

Carbon monoxide has unquestionably been a ubiquitous pollutant of the human environment since man first learned to build fires. This gas is produced copiously by the cooking and heating fires of primitive people, and studies in villages in the highlands of New Guinea revealed that people were being exposed to concentrations of carbon monoxide higher than are found in the streets of American cities.[3] The fact that carbon monoxide is an important product of coal combustion apparently accounts for the relatively high concentrations of the gas reported on the streets of Paris and New York more than 50 years ago.[6] Coal was then widely used for home heating, and the smoke discharged from domestic chimneys resulted in generalized carbon-monoxide pollution to such an extent that city dwellers were exposed to higher levels of contamination that at present.

At high concentrations, carbon monoxide can be rapidly lethal. Until the early 1950s, the gas supplied to kitchen stoves in many eastern cities was manufactured from coal, and contained 6 to 30 percent carbon monoxide. Charcoal braziers and coal or oil stoves used in badly ventilated places have been throughout history a common cause of asphyxiation in many parts of the world. Deaths caused by inadvertent exposure to manufactured gas were also frequent and, in New York City alone, accounted for the death of 350 persons per year prior to the introduction of natural gas, which is primarily methane and is relatively nontoxic.

Many types of modern heating devices are known to be particularly dangerous.[4] In 1971, a death was caused by carbon monoxide in a family camping trailer in which charcoal was used for heating purposes during the night. The conditions were subsequently reproduced, and carbon monoxide tests indicated that dangerous concentrations can be reached in a matter of minutes after a heater is ignited.[5]

Carbon-monoxide exposure is a frequent occupational hazard for garage attendants (because the automobile exhaust can be lethal in badly ventilated spaces) and those working in vehicular tunnels, as well as workers in a wide variety of industrial processes.

The combination of carbon monoxide and hemoglobin produces carboxyhemoglobin (COHb), with equilibrium being reached in a few hours. The equilibrium, expressed as the percent of carboxyhemoglobin, is a function of the concentration of carbon monoxide to which the person is exposed, and the respiration rate. Industrial workers are permitted to inhale up to 50 parts per million of carbon monoxide for eight hours, which results in a concentration of 7.36 percent. The earliest effects of carbon-monoxide poisoning on a healthy adult are encountered at two or three times this level in the form of moderate shortness of breath, accompanied by slight headache. Unconsciousness and death can result from exposure to a concentration of a few thousand parts per million.

Cigarette smoking is one of the most common sources of carbon-monoxide exposure. Mainstream cigarette smoke contains carbon monoxide at a concentration of about 40,000 parts per million, and cigarette smokers commonly have carboxyhemoglobin levels up to 5 or 6 percent.[7] There are relatively few data on the levels of carbon dioxide in indoor air due to cigarette smoking, but in one report, it was demonstrated that concentrations in public places can exceed the outdoor concentrations by several parts per million.[8] In one experiment in which 80 cigarettes and two cigars were smoked in a badly ventilated room, the concentrations of carbon monoxide rose to 38 ppm, and the carboxyhemoglobin of 12 nonsmoking volunteers rose from 1.6 to 2.6 percent in 78 minutes.[9] Additional data of this kind are needed to assist in the evaluation of the extent to which exposure to cigarette smoke constitutes a risk to nonsmokers. Elsewhere, it was noted that public places can accumulate significant concentrations of the carcinogen benzopyrene, which is also a constituent of cigarette smoke.

It was thought until recently that the only danger from carbon monoxide at sufficiently high exposure was asphyxiation. More recently, a number of other results have been derived both from human and animal studies, indicating that more subtle effects are possible at much lower exposures. There have been reports that a 3 to 5 percent concentration of carboxyhemoglobin can effect the central nervous system, as a result of which an individual begins to show tendencies towards inattentiveness. This has not been fully evaluated, but the effect, if confirmed, could be important in relation to the need for vigilance during the operation of a motor vehicle. More significantly, it has also been shown that the onset of angina pectoris, the pain resulting from the restricted blood flow through the heart arteries, is hastened by exposure to carbon monoxide at the concentrations found in heavily trafficked areas.[10,11]

Taking all things into consideration, a 1974 study of the National Academy of Sciences concluded that the literature supports the federal ambient air-quality standard of nine parts per million for eight hours, and a maximum of 35 parts per million for one hour.[12] This is equivalent to a carboxyhemoglobin saturation of about 1.5 percent in a normal resting person, compared to 5 to 6 percent in a cigarette smoker. Carboxyhemoglobin is present in only minute amounts (about .4 percent) if the individual has not been exposed, due to production of carbon monoxide within the body. Carbon monoxide is also produced by a number of natural sources which contribute more to the atmosphere then does the combustion of fossil-fuel.[13] However, the carbon monoxide of natural origin is produced so diffusely that it is of far less significance to health than the more concentrated anthropogenic sources.

Because the carbon-monoxide concentration of street-level air tends to be associated with traffic density, the levels to which people are actually exposed are likely to be quite variable, depending on the movements of the person for the preceding two or three hours. A person living in a suburban area may have only minimal exposure, whereas an individual working or shopping in a heavily trafficked area may be maximally exposed. Moreover, it is known that carbon monoxide emitted from automobiles is quickly diluted with altitude, so that a person would be more heavily exposed at street level than on the upper floors of an apartment house or office building. The best of air-monitoring networks include only a few scattered instruments to measure carbon monoxide. These are not adequate for estimating the exposure of persons who are moving from place to place in the course of a day.

To overcome this problem, studies have been conducted in which people have been used as "air samplers." Since the carboxyhemoglobin quickly reaches equilibrium, blood sampling and analysis should provide an indication of the levels to which a person has been exposed during the hours prior to blood sampling. In one study of 17 urban areas throughout the country and several small towns in New Hampshire and Vermont, place of residence made little difference in the mean concentrations of carboxyhemoglobin in the blood of 31,000 nonsmokers studied.[14] Thus, New York City residents averaged 1.42 percent compared to 1.36 percent for residents of small towns in Vermont and New Hampshire. For some unexplained reason, the highest concentrations were found among residents of Denver, Colorado.

A second such study was undertaken of 16,000 blood samples from smokers and nonsmokers in St. Louis.[15] Among nonsmokers, the mean carboxyhemoglobin level averaged .8 percent compared to 1.4 percent for those classified as "industrial workers." This showed that industrial sources of exposure tend to elevate the average carboxy hemoglobin concentrations of industrial workers, generally. These values are lower then those found in smokers, as noted earlier.

In the first study, which included a number of cities having vastly different demographic and meteorological characteristics, no substantial difference was found in the mean concentration of carboxyhemoglobin in the blood samples taken from individuals living in the various cities studied. The second study did find a gradient between rural and urban residents, but it was not significant, and the investigators concluded that the differences did not support the conclusion that uptake from carbon monoxide from urban air is important from a physiologic point of view.

From a study of the levels of carboxyhemoglobin in the blood of Paris policemen, it was found that the blood levels of traffic officers frequently decreased after they reported to their stations. This was because they smoked before they went on duty, but were prohibited from doing so thereafter. This is not a surprising observation, in view of the well-known fact that the carboxyhemoglobin level in smokers is higher than would be expected among persons living or working in the most heavily trafficked city areas.[16]

There is no doubt that carbon monoxide can be harmful to health at concentrations much lower than were thought harmful as recently as ten years ago. However, the actual significance of these findings on public health is still not clear. One obvious conclusion is that more facts are needed. Considering the widespread concern over carbon-monoxide pollution, it is difficult to understand why there have not been more human measurements like those summarized. In addition, the possibility that low levels of exposure can affect the central nervous system needs further study and, if confirmed, their implications for public health require evaluation. Of greater importance, the relationship between carbon monoxide exposure and coronary disease must be further explored.

THE NITROGEN OXIDES

Modern processes for burning fossil fuel operate at so high a temperature that atmospheric nitrogen can be oxidized to any one of several compounds. Although as many as six such oxides can be identified in the atmosphere, nitric oxide (NO) and nitrogen dioxide

(NO_2) are of principal importance. Nitric oxide is discharged in the emissions from automobile tailpipes and, to a lesser extent, from power plant stacks. However, the NO soon oxidizes to NO_2, which is the more toxic of the compounds and is also significant because it reacts with atmospheric hydrocarbons under the catalytic action of sunlight to produce a complicated group of compounds generally classed as "oxidants."

Nitrogen dioxide is the anhydrous form of nitric acid and is markedly toxic in sufficiently high concentrations. This gas is much to be feared and has been known to cause deaths among employees accidentally exposed to its fumes. Fatal exposure can result from accidental spills of concentrated nitric acid, which liberates nitrogen dioxide as a brown fume. The lung is irritated to such an extent that fluids infiltrate the alveolar spaces causing pulmonary edema, a condition in which one can literally drown in one's own body fluids. This can occur if an individual is exposed to concentrations of about 500 parts per million or higher. Severe, and sometimes fatal, effects have also been reported at 50 to 300 parts per million, and exposure to 25 parts per million has been known to result in bronchitis and bronchopneumonia, with full recovery in a matter of weeks. These are the effects of severe short-term exposure associated with industrial accidents.

Until comparatively recently, there was no suggestion that nitrogen dioxide was toxic at concentrations below one part per million, such as exist in urban atmospheres. However, studies were undertaken in 1970 of the effects of nitrogen dioxide on the health of school children in Chattanooga, Tennessee living in the vicinity of an industrial source of this pollutant. The investigators reported a higher incidence of respiratory infections among those exposed to more than .05 parts per million.[17,18] This observation has been questioned on the grounds that it was not possible to rule out other factors that could have been responsible for the observed effect.[19] Nevertheless, on the basis of these observations, the federal government established .05 parts per million as the ambient air-quality standard for nitrogen dioxide. This standard has been the subject of much controversy, and nitrogen dioxide was identified in 1973 as being "unquestionably the most controversial of all air pollutants for which standards have been set. . . ."[20]

Apart from the fact that the epidemiological basis for this standard has been questioned, there is also uncertainty about the method by which nitrogen dioxide is measured in air. New air-sampling methods, like other laboratory procedures, have traditionally been required to stand the test of peer review before being adopted for general use. A scientist who develops a new method of analysis normally submits a description of the procedure to a scientific journal for publication, following which the manuscript is scrutinized by reviewers selected by the journal editor before actually being accepted for publication. After publication, analytical chemists then subject the new method to laboratory evaluation. However, in the frenetic pace at which environmental matters have been decided in recent years, there has been little time for these time-consuming but necessary evaluations. In April 1971 the United States Environmental Protection Agency (EPA) specified the official method for analyzing nitrogen oxides. The method was not subject to the normal scientific review, but was "promulgated" by publication in the *Federal Register* for April 30, 1971. This document, published each day by the federal government, is a dreary compilation of the various laws, rulings, promulgations, and other legal forms of governmental edict. The *Federal Register* is well known to lawyers, and recently has become a sort of quasiscientific journal in which highly technical matters are published by regulatory agencies of the federal government.

Analytical chemists immediately found fault with the EPA procedure, but there was little that could be done because it was "official" and could not be replaced by an alternative method unless it could be demonstrated to the administrator's satisfaction to have a consistent relationship to the specified method.[21] The main problem with the method was that it seriously overestimated low concentrations of the nitrogen oxides. Air sampling by the reference method indicated that many places in the United States had concentrations higher than the ambient air quality standard of .05 parts per million, but upon reexamination it was found that the data were in error because of the faulty analytic procedure. By June 1973, the method was rescinded and the federal government relaxed its requirements for nitrogen dioxide control.

HYDROCARBONS AND OXIDANTS

During the early 1940s, residents of southern California began to be aware that an air-pollution problem was developing that was characterized by reduced visibility, damage to vegetable crops, and eye irritation.[22] The pollution came to be called "smog," the century-old term coined by the British to describe the characteristic mixtures of smoke and fog that were so prevalent in many of their cities. However, the southern California smog was different in both appearance and chemical characteristics. It was more of a haze than a fog, with a slight touch of brown color. Unlike the reducing characteristics of sulfurous smogs, the California haze was found to have marked oxidizing qualities which were ultimately found to be due to the presence of ozone (O_3) and a complex mixture of other oxidizing substances. Ozone is a form of oxygen that has powerful oxidizing characteristics. Its odor can be detected at concentrations as low as .02 parts per million.[24] Nose and throat irritation occur at about .3 parts per million, and the eyes of many people begin to smart at about .1 parts per million. The oxidants in California smog contain about 90 percent ozone. The fact that ozone was present with the California smog in relatively high concentrations (several tenths of a part per million) was particularly perplexing because no obvious source of ozone existed. The mystery was ultimately solved in 1953, when it was demonstrated that ozone and other oxidants are produced by the action of sunlight on an atmospheric mixture of hydrocarbons and nitrogen oxides.[23] The smog problem increased in severity with the passage of years, and by the late 1950s, "photochemical oxidants" were words well entrenched in the vocabulary of the air-pollution specialist. The problem was particularly acute in the Los Angeles basin as a result of the dense vehicular traffic, and a combination of meteorological factors, including the tendency for atmospheric inversions to develop frequently below the levels of the surrounding mountain ridges.

In a review of the effects of oxidants on human health, Carroll concluded in 1973 that "despite widespread public annoyance with oxidant air pollution and its obvious irritant effects on the eyes, clearly-demonstrated effects of other types on human health have been surprisingly hard to document."[25] This conclusion also emerges from a reading of a more comprehensive (more than 700 pages) report on oxidants from the National Academy of Sciences.[26] Complaints of eye irritation (but not other respiratory symptoms) can be correlated to increasing oxidant concentrations over the range of .1 to .55 parts per million. Studies of daily mortality among elderly residents in Los Angeles demonstrated a striking correlation with temperature,[27] but not to the levels of oxidants. The mortality records in Los Angeles have also been studied to determine if an effect on the long-term death rate can be demonstrated. Comparisons of the death rates in Los Angeles County with metropolitan areas known to have much lower levels of oxidant pollution failed to demonstrate any difference.

Los Angeles hospital records have also been studied to determine if admissions for respiratory or other relevant diseases might be correlated with the oxidant levels. No such relationship was demonstrable. Studies have also been undertaken to ascertain if absenteeism due to respiratory illness in elementary schools can be related to oxidant levels. Here again, no significant association was found. One study of the effects of the smog on high-school long-distance runners showed a striking inverse correlation between performance and oxidant concentrations over a three-year period of study.

The national primary standard for photochemical oxidants requires that the maximum one-hour average concentration not exceed .08 parts per million on more than one day per year. This standard is exceeded on 200 to 300 days per year in Los Angeles, sometimes by a factor of 7.[12] The Los Angeles experience is far more severe than that of New York City, where the hourly maximum was exceeded on 41 days during a one-year period, with the maximum hourly reading being about three times the standard. Whereas in Los Angeles and certain other western cities the photochemical oxidant problem is both visibly obvious and annoying, it is not apparent that the problem exists to a major extent in the large eastern cities. In 1973 a committee of health specialists who considered the problem in New York was of the opinion that a photochemical oxidant problem did not exist in that state.[28] Moreover, it is not apparent that the oxidants in New York State originated from the automobile.

Concentrations in excess of the national air-quality standard are commonly found in rural areas.[29] It is known that hydrocarbon vapors emanate from many natural sources,[30] and the high rural oxidant levels may be due to photochemical reactions involving naturally occurring hydrocarbons. An alternative explanation is that the smog ingredients are introduced into the atmosphere in urban areas, but require so much time to react that the oxidants are not formed until the air mass has been carried some distance away.

PROGRAMS FOR THE CONTROL OF AUTOMOBILE EMISSIONS

While quantitative data may not have been available to the extent one would wish, it was clear by the midsixties that something had to be done about automobile emissions in California. Accordingly, the state officials established ambient air-quality standards at a one-hour maximum of .10 parts per million for photochemical oxidants, ten parts per million as a 12-hour average for carbon monoxide, and .25 parts per million as the one-hour maximum for nitrogen dioxide.

These ambient air-quality goals were then translated into "emission standards" which are useful and necessary to regulatory agencies. However, the relationship between ambient air quality and emissions is not a simple one. The tailpipe emissions from automobiles are diluted by atmospheric turbulence, and the individual chemical constituents react with each other and with other chemicals present in the atmosphere. The pollutants can be washed out of the atmosphere by rain, or they can attach themselves to atmospheric dust, which then settles out. The concentration at any given distance from a point source of emissions is thus related to the distance from the source, as well as to the rate of emission, its height above ground, factors involving atmospheric chemistry, and the meteorological conditions prevailing at the time. These relationships, although complicated, are reasonably well understood and can be expressed in mathematical equations that are useful to air-pollution meteorologists.

The control plan, developed originally in California, involved increasingly stringent reductions in emissions from new vehicles, as well as a program for emission controls on used vehicles. The essential elements of control for the new vehicles required the elimination of the crankcase vent through which gasoline vapors that slip past the piston rings are discharged into the atmosphere; the elimination of evaporation of gasoline from the carburetor and gas tank; and the improvement of the oxidation of hydrocarbons and carbon monoxide, with the goal of discharging carbon dioxide and water while at the same time reducing nitrogen dioxide and carbon monoxide.

In 1970, the federal government passed a Clean Air Act which specified emission controls on a national basis that would be made increasingly stringent to the extent that tailpipe emissions would be reduced 90 percent by 1975. These federal requirements were far more stringent in some respects than those adopted in California, and they immediately precipitated a debate between industry and the environmentalists. Industry originally contended that the goal could not be achieved, but the environmentalists, supported by the Congress, held fast. During more than a decade of intense debate, the advocates for more stringent automobile-emission control have come mainly from the ranks of the environmental activists. They were motivated largely by a desire to clean the air because it was perceived by them to be the correct thing to strive for on social grounds. It was their deep-rooted belief that somehow the cost of clean air could be more than justified by health and economic factors.

The automobile manufacturers fought the federal requirements for clean air on the grounds that the strict standards were not necessary, technologically unfeasible, and overly expensive. At the request of the federal government, the question of technical feasibility was examined by committees of the National Academy of Sciences. They concluded that, with the exception of the nitrogen oxides, the goals established by the Clean Air Act of 1970 were in fact feasible, although there were many unanswered questions, including those concerned with the durability of the required mechanical changes, the nature and magnitude of the hazards to health posed by the pollutants released in automobile emissions, and the high cost of achieving the goals. In the NAS report, it was noted that there was great uncertainty about the costs of the emissions-control program, and that the annual expenditure could run as high as $23.5 billion.[12] Less than two years later, the cost estimate was reduced to an upper limit of $11 billion per year.[31] This still is a very high price to pay in view of the uncertainties of the benefits to be achieved. The benefits, in the form of reduced discomfort, improvement in visibility, better health, and less damage to vegetation were estimated to be in the range of $2.5 to $7 billion annually, with the most likely value to be $5 billion. If expenditures of this magnitude are to be undertaken to benefit the public health, the advantages should be more obvious. There are certainly many other public health problems of far greater urgency that could be solved at less cost. The idea that human discomfort or health impairment can be equated to dollars is repugnant to many. Nevertheless, expenditures to improve the public health must be allocated according to some rational scheme of priorities. Among the factors that must be considered are the severity of the health effects, the number of people affected, and the cost of eliminating or minimizing the effect. Some public health problems are more serious than others, and can be controlled for less money. Automobile emissions are expensive to control, but their effects on health are minor compared to others that could be controlled for far less money, such as the poisoning of infants by peeling lead paint.

During the early 1970s the arguments of automobile manufacturers were reinforced by a number of developments. It was found that the measures required to reduce carbon monoxide and hydrocarbons actually increased the emissions of nitrogen oxides. In addition, the interim control measures required by the Clean Air Act increased the consumption of gasoline by 10 to 15 percent. Since 1975 the automobile industry has been relying primarily on a device, known as a "catalytic converter," for improving the oxidation of hydrocarbons and carbon monoxide. The catalyst is a blend of two noble metals, platinum and palladium. The Environmental Protection Agency stipulated that the catalytically equipped cars must meet the interim standards after 50,000 miles of use, allowing for one catalyst change. An interesting side effect of the decision to use the noble metal catalysts was that the metal lead is known to poison such systems and therefore had to be eliminated as a gasoline additive. Thus, environmentalists achieved for unrelated technological reasons a goal that they had sought to justify on grounds of public health protection.

Another side effect, probably of little significance, is that the relatively modest sulfur content of gasoline which, up to the present time has not been an important source of atmospheric sulfuric acid, is found to be oxidized by the catalytic converter to sulfur trioxide. This, in turn, reacts with atmospheric water to form sulfuric acid. The rate of emission of sulfuric acid in exhausts from cars equipped with catalytic converters is nine to 50 times higher than from other cars.[32]

The Clean Air Act of 1970 required that by 1975 automobile emissions be reduced by 90 percent of their 1970 values. Introduction of the catalytic converters in 1975 models reduced the emissions by 83 percent. One body of opinion has held that, all aspects considered, further regulation should be held in abeyance to avoid the inefficiencies of rushing headlong into a less-than-optimal program for eliminating the remaining 7 percent. A National Science Foundation study conducted by three universities recommended that the ultimate standards be delayed to avoid locking automobile manufacturers into the catalytic system to such an extent that they could not afford to convert to more efficient and more effective methods.[33]

The Clean Air Act of 1970 was very specific in the goals to be achieved for control of automobile emissions: a 90 percent reduction in carbon monoxide and hydrocarbon emissions of the 1970 models, to be achieved by 1975, and a similar reduction in the nitrogen oxide emissions of the 1971 models, to be established by 1976. However, the EPA administrator was authorized to extend each deadline by one year, and the 1975 standard was so extended in 1973.[34] Additional extensions and the authorization for setting more moderate interim standards were permitted by a series of amendments to the 1970 Clean Air Act passed in 1974 as a result of the fuel emergency that arose from the war in the Middle East. Partly as a result of increased use of gasoline by the requirements for emission control, but also because of the finding that the sulfuric acid mist was being produced by engines equipped by catalytic converters, the EPA administrator suspended the 1975 emission standards for hydrocarbons and carbon monoxide.[34] It is likely that additional postponements will be authorized by either the EPA or the Congress, but substantial progress towards reducing tailpipe emissions has already been made and emissions from six prototype 1976 cars tested by the EPA were found to be within the federal limits.[35] However, the extent to which the protective devices will deteriorate with road use is not known.

There are many defects in the logic of the emissions-control program. There is no question but that the automobile is a major environmental hazard. It causes accidents that kill nearly 50,000 people per year and injures millions more. Automobiles also clutter our streets and expose us to noise.[36] The traffic congestion in our large cities is a cause of enormous economic waste that increases the cost of doing business, in addition to adding to the frustrations and irritations of urban living. It is the automobile itself that should be controlled on environmental grounds, not the tailpipe emissions.

A logical course of action would require that our automobile economy gradually use smaller cars. The average weight of American-made cars had increased by 300 to 1,000 pounds during the period between 1962 and 1974. A major reduction in the horsepower and weight of cars would make sense from many points of view: The tailpipe emissions would be reduced greatly; the cars would use less gasoline; and, in the course of their manufacture, the cars would require less steel, rubber, copper, and other raw materials. Gradually, the weight of cars is being reduced, but Americans continue to prefer large cars.

Another alternative for achieving the air-quality goals of the federal government, as pointed out by the 1974 NAS report, would be a "two-car strategy," by which cars would be manufactured to one standard of performance for those parts of the country that have air-quality problems, and to a less-stringent standard for other, less affected parts of the country.[12] This would present obvious problems, but would avoid penalizing most of the United States population who purchase cars for use in places that do not have photochemical smog or carbon monoxide problems.

Alternatives such as these notwithstanding, the country has embarked on a hasty program of required emission controls that will be expensive to accomplish, will have only marginal health benefits, and will not affect the most important of the environmental effects of automobiles—congestion, noise, and death and injury by accident.

NOTES

1. National Academy of Sciences. "The Relationship of Emissions to Ambient Air Quality," "Air Quality and Automobile Emission Control: A Report by the Coordinating Committee on Air Quality Studies," *Serial No. 93–24* (Sept., 1974), Vol. III.
2. Kleinman, M. T. "The Apportionment of Sources of Airborne Particulate Matter," Doctoral Dissertation, New York Univ., New York (1977).
3. Cleary, G. and C. Blackburn. "Air Pollution in Native Huts in the Highlands of New Guinea," *Arch. Environ. Health* 17:785 (1968).
4. Westerlund, K. and H. von Ubisch. "Carbon Monoxide from Small Camping Appliances and from Stoves without Chimney Connection," *Nord. Hyg. Tidskr.* 53:26–33 (1972).

5. *Environmental Health and Safety News.* "Burning Charcoal Brazier Causes Carbon Monoxide Fatality," Vol. 19, Nos. 10, 11, and 12, Univ. of Washington, Seattle, Washington (October, November, December, 1971), 19:10, 11, 12.

6. Eisenbud, M. and L. Ehrlich. "Carbon Monoxide Concentration Trends in Urban Atmospheres," *Science* 176:193 (1972).

7. U.S. Public Health Service. "Smoking and Health: Report of the Advisory Committee to the Surgeon General of the Public Health Service," Public Health Service *Publication No. 1103,* U.S. Government Printing Office, Washington, D.C. (1964).

8. Sebben, J., P. Pimm and R. J. Shephard. "Cigarette Smoke in Enclosed Public Facilities," *Arch. Env. Health* 32:53–58 (1977).

9. Russell, M. A. H. "Absorption by Non-Smokers of Carbon Monoxide from Room Air Polluted by Tobacco Smoke," *The Lancet* (March 17, 1973).

10. Goldsmith, J. R. and L. T. Friberg. "Effects of Air Pollution on Human Health" in: "Air Pollution," 3rd ed. (A. Stern, ed.), Academic Press, New York (1977), Vol. II.

11. Aronow, W. S., C. N. Harris, M. W. Isbell and S. N. Rokaw. "Effect of Freeway Travel on Angina Pectoris," *Ann. of Internal Med.* 77:669–676 (1972).

12. National Academy of Sciences. "Air Quality and Automobile Emission Control: A Report by the Coordinating Committee on Air Quality Studies," *Serial No. 93–24.* I. Summary Report. II. Health Effects of Air Pollutants. III. The Relationship of Emissions to Ambient Air Quality. IV. The Costs and Benefits of Automobile Emission Control.

13. Babich, H. and G. Stotsky. "Air Pollution and Microbial Ecology," *Critical Reviews in Environmental Control* 4:353–421 (1974).

14. Stewart, R. D., E. D. Baretta, L. R. Platte, E. B. Stewart, J. H. Kalbfleisch, B. Van Yserloo and A. A. Rimm. "Carboxyhemoglobin Concentrations in Blood from Donors in Chicago, Milwaukee, New York, and Los Angeles," *Science* 182:1362–1364 (1973).

15. Wallace, N. D., G. L. Davis, R. B. Rutledge and A. Kahn. "Smoking and Carboxyhemoglobin in the St. Louis Metropolitan Population," *Arch. Environ. Health* 29:136–142 (1974).

16. Chovin, P. "Carbon Monoxide: Analysis of Exhaust Gas Investigations in Paris," *Env. Res.* 1:198–216 (1967).

17. Shy, C. M., J. P. Creason, M. E. Pearlman, K. E. McClain, F. B. Benson and M. M. Young. "The Chattanooga School Study." Effects of Community Exposure to Nitrogen Dixoide. II. Incidence of Acute Respiratory Illness, *Journal of the Air Pollution Control Association* 20:582–588 (1970).

18. Shy, C. M., J. P. Creason, M. E. Pearlman, K. E. McClain, F. B. Benson and M. M. Young. "The Chattanooga School Children Study." I. Methods, Description of Pollutant Exposure, and Results of Ventilatory Function Testing, *Journal of the Air Pollution Control Association* 20:539–545 (1970).

19. Warner, O. and L. Stevens. "Revaluation of the 'Chattanooga School Children Study' in the Light of Other Contemporary Governmental Studies: The Possible Impact of These Findings on the Present NO_2 Air Quality Standard," *Journal of the Air Pollution Control Association.* 23:769–772 (1973).

20. National Academy of Sciences. *Proceedings of the Conference on Health Effects of Air Pollutants*, National Academy of Sciences, *Serial No. 93–15*, U.S. Government Printing Office, Washington, D.C. (November, 1973).

21. Saltzman, B. E. "Analytical Methodologies for Nitrogen Oxides in Perspective," *Proceedings of the Conference on Health Effects of Air Pollutants*," National Academy of Sciences, *Serial No: 93–15*, U.S. Government Printing Office, Washington, D.C. (November, 1973), p. 317.

22. Stephens, E. R. "Photochemical Formation of Oxidants," *Proceedings of the Conference on Health Effects of Air Pollutants*, National Academy of Sciences, *Serial No 93 15*, U.S. Government Printing Office, Washington, D.C. (November, 1973), p. 465.

23. Haagen-Smit, A. J., C. E. Bradley and M. M. Fox. "Ozone Formation in Photochemical Oxidation of Organic Substances," *Ind. Eng. Chem.* 45:2086–2089 (1953).

24. Balchum, O. J. "Toxicological Effects of Ozone, Oxidant, and Hydrocarbons," *Proceedings of the Conference on Health Effects of Air Pollutants*, National Academy of Sciences, *Serial No. 93–15*, U.S. Government Printing Office, Washington, D.C. (November, 1973), p. 489.

25. Carroll, R. E. "Epidemiologic Studies of Oxidant and Hydrocarbon Air Pollution," *Proceedings of the Conference on Health Effects of Air Pollutants*, National Academy of Sciences, *Serial No. 93–15*, U.S. Government Printing Office, Washington, D.C. (November, 1973), p. 541.

26. National Academy of Sciences. "Ozone and Other Photochemical Oxidants," Committee on Medical and Biological Effects of Environmental Pollutants, Washington, D.C. (1977).

27. Oechsli, F. W. and R. W. Buechley. Excess Mortality Associated with Three Los Angeles September Hot Spells, *Environ. Res.* 3:277–284 (1970).

28. N.Y. State Health Planning Commission-New York State Health Planning Advisory Council. "Recommendations and Report on the Need for a Reevaluation of Federal Environmental Strategies," submitted to Gov. Nelson A. Rockefeller, Joint Planning Committee on Health Aspects of the Environment, Albany, N.Y. (April, 1973).

29. Coffey, P. E., and W. N. Stasuik. "Evidence of Atmospheric Transport of Ozone into Urban Areas," *Env. Science & Tech.* 9(1):59 (1975).

30. Stutzky, G. and S. Schenck. "Volatile Organic Compounds and Microorganisms," *CRC Critical Reviews in Microbiology* (May, 1976), pp. 333–382.

31. National Academy of Sciences. "The Costs and Benefits of Automobile Emission Control," in: "Air Quality and Automobile Emission Control:

A Report by the Coordinating Committee on Air Quality Studies," *Serial No. 93–24* (September, 1974), V. IV.

32. PAT Report (Practical, Available Technology). "Key to Reliable Exhaust Emissions Data," *Environ. Sci. & Tech.* 8(9):790 (1974).

33. Holden, C. "Auto Emissions: EPA Decision Due on Another Clean-Up Delay," *Science* 187:181 (1975).

34. Council on Environmental Quality. "Sixth Annual Report," U.S. Government Printing Office, Washington, D.C. (1975).

35. Council on Environmental Quality. "Seventh Annual Report," U.S. Government Printing Office, Washington, D.C. (1976).

36. Grad, F. P., A. J. Rosenthal, L. R. Rockett, J. A. Fay, J. Heywood, J. F. Kain, G. K. Ingram, D. Harrison and T. Tietenberg. "The Automobile and the Regulation of Its Impact on the Environment," Univ. of Oklahoma Press, Norman, Ok. (1975).

CHAPTER 13

Nuclear Power

It was concluded earlier, that of the various energy options available to modern civilization for the foreseeable future, only coal and nuclear fuels can provide adequately for the energy needs of the most countries, including the United States. However, the development of nuclear energy has been impeded by persistent controversies that have arisen out of a lack of confidence in the safety of nuclear-power generation.

The subject of nuclear-reactor safety cannot be approached without first recognizing that many people are understandably influenced by the psychological association between nuclear power and nuclear weapons. After all, the older generation first learned about nuclear energy when the two atomic bombs were dropped on Japan during World War II. These were dramatic events that could not help but leave memories of death and devastation in many minds. Many people, old and young alike, cannot be expected to dissociate production of energy by nuclear fuels from their uses in bombs for mass destruction.

There are also other reasons why so many people fear nuclear power. Some are concerned about the gaseous and liquid radioactive wastes discharged during normal operation of the nuclear plants while others stress the potential dangers from the radioactive clouds that could escape in the event of a catastrophic nuclear accident. Additional reasons for concern are the difficulty of disposing of the huge quantities of radioactive wastes, the extraordinary toxicity of plutonium, and the possibility that plutonium might be diverted clandestinely and fabricated into nuclear weapons.

NUCLEAR REACTORS

The first nuclear reactor was operated briefly in Chicago in December, 1942, about three years after the discovery of fission. Less than one year later, a research reactor was put into operation at Oak Ridge, Tennessee at an initial power level of 700 kilowatts.[1] Even more remarkable, the first of several reactors designed for production of plutonium, a transuranic element capable of fission, was put into operation at Hanford, Washington in 1944 at an initial power level of 250 megawatts. At present, more than 700 land-based reactors have been operated or are under construction in various parts of the world. In addition, about 330 ships of the American and Russian navies are now powered by nuclear reactors.[2]

The present generation of reactors is fueled with natural uranium that has been slightly enriched with the isotope U-235 and fabricated into uranium oxide pellets. The most abundant form of uranium in nature is U-238. It is present in natural uranium in a concentration of 97.3 percent but is not fissionable in light-water reactors. U-238 is partially converted to an artificial fissionable nuclide, plutonium-239, in the course of reactor operation. The plutonium can be recycled through light-water reactors or it can be used with U-238 in the "breeder," a reactor that actually produces more fissionable material than it uses. Both plutonium recycling and the breeder are necessary to extend the life of our uranium reserves.

Nuclear-power plants, like those that use coal, oil, or gas, produce electrical energy by making steam that drives a turbine connected to an electric generator. The present generation of nuclear-power reactors in the United States are of the so-called light-water types (LWRs), which can be subdivided into the pressurized water reactor (PWR), and the boiling-water reactor (BWR). The basic operating features of the two systems are shown in Figures 13–1A and 13–1B.

In the boiling-water reactor, the water is converted directly to steam, which is passed through a turbine that drives the electric generator. The steam discharged from the turbine is then condensed and the water is returned by pumps to the reactor.

In the pressurized-water reactor, the water that passes through the reactor is maintained under high pressure so that steam is not immediately produced. The heat is then transferred to another water loop within the boiler in which the steam is produced that drives the turbogenerator.

The amount of energy that can be derived from uranium is enormous. One ton of fuel releases the energy equivalent of more than 5 million tons of coal, or more than 20 million barrels of fuel oil. A reactor capable of producing 1,000 megawatts of electricity (MWe) has a core that contains about 100 tons of uranium, sufficient for a year or two of power generation. The uranium fuel is in the form of

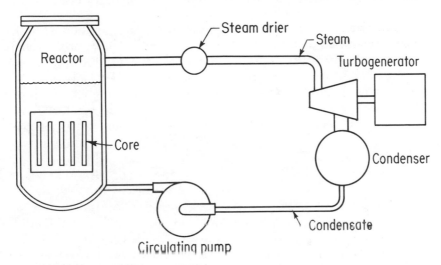

Figure 13-1A. Schematic diagram of a boiling-water reactor (BWR). Water is pumped through the reactor core where it is heated to form steam that drives the turbogenerator.

uranium oxide pellets that are stacked within zirconium or stainless steel tubes arranged within a stainless steel vessel (Figure 13–2).

At the beginning of the reaction, the uranium fuel elements are only slightly radioactive and can be handled safely with bare hands. However, as the fuel ages, enormous quantities of radioactive elements (fission products) are produced by the splitting (fissioning) of uranium nuclei.

The uranium used in light-water reactors has been slightly enriched with the isotope U-235, which is readily capable of fission. Although the other form, U-238, is not capable of fission, it does capture neutrons and transmutes to a fissionable isotope of plutonium, much of which is actually consumed by the fission within the reactor, thereby contributing to the energy produced. However, some of the plutonium is present in the fuel when the U-235 is so depleted that the core must be replaced.

Because U-238 is not fissionable, light-water reactors are inherently inefficient and convert only 1 to 2 percent of the potentially available energy of uranium into heat. Enormous quantities of U-238 are being accumulated both as a by-product of U-235 enrichment, and in the chemical processing of spent reactor fuel. The United States is currently storing 200,000 tons of depleted uranium, a resource that can be utilized for fuel in the "breeder reactor" which has the ability

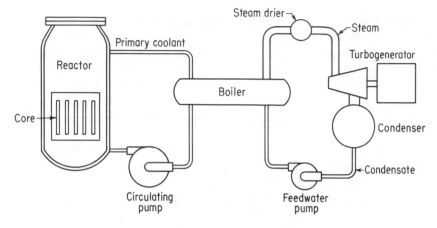

Figure 13-1B. Schematic diagram of a pressurized water reactor (PWR). As in the BWR, water is pumped through the core but is maintained under pressure, so that it does not form steam. The heat is transferred in the heat exchanger to the secondary loop, where steam is formed that drives the turbine generator.

to convert U-238 to fissionable form, and increase the efficiency of uranium use from less than 2 percent to as much as 75 percent. The breeder actually produces more fissionable material than it uses, and thus offers a means to substantially extend the lifetime of existing uranium resources. The fissionable material produced by the breeder is plutonium, a substance that has stirred wide controversy.

Most of the radioactive elements produced by the fission of uranium have such short radioactive half-lives that they disappear in a matter of hours. Others, such as strontium-90 and cesium-137, have half-lives of about 30 years. In addition, a number of "transuranic" elements are produced by neutron irradiation of the uranium. These are heavier than uranium and do not exist in nature in significant amounts. They include such elements as plutonium, neptunium, americium, and curium. Some of these have isotopes with very long half-lives. The most important of these is plutonium-239, which has a half-life of about 25,000 years.

In addition to the fission products and transuranic elements, a third group of radioactive substances originates from substances used in the construction and operation of the reactor. Some of these absorb neutrons and are transmuted to radioactive forms that are called "activation products."

When the fuel has been in the reactor for a year or two, the accumulation of fission products and the depletion of U-235 require that the fuel be replaced. The fuel elements have not changed in external appearance, but they have become intensely radioactive and must now be handled with great care by remote control machinery. After removal from the reactor vessel, the fuel elements are stored for a minimum of a few months in water-filled tanks to permit the shorter-lived radionuclides to decay, and to provide for dissipation of the heat produced during the early decay process. The water also serves to shield nearby workers from the intense radioactivity. After several months, the radioactivity of the fuel elements has been reduced so that they can be loaded into casks and shipped to fuel-reprocessing centers. These have in the past been located on government reservations in the states of Washington, Idaho, and South Carolina, and at a privately owned plant in upstate New York. The latter facility has been shut down, but a second private plant is ready to begin operation in South Carolina. The purpose of fuel reprocessing is to separate the uranium and the newly produced plutonium from the highly radioactive fission products which constitute the waste products of the nuclear-power process.

The plutonium can be returned to the fuel-fabrication plants and mixed with enriched uranium to produce new fuel for light-water reactors; or, it can be fabricated directly into fuel for the breeder reactor, as was discussed earlier. The controversial subjects of the breeder and fuel reprocessing will be discussed again later in this chapter.

BIOLOGICAL EFFECTS OF EXPOSURE TO RADIATION

The discovery of X-rays in November, 1895, and of radioactivity one year later, were milestones in scientific history. Reports of radiation injury began to appear in the literature with tragic rapidity, and the very first volume of the *American X-ray Journal,* in 1897, included a compilation of 69 X-ray injuries that had already been reported from laboratories and clinics in many countries. The very earliest injuries were relatively minor skin lesions, and it was not known until several years later that many of these lesions would develop into cancers that would cause death or disfigurement.

Experience with radioactivity in the form of radium was similarly tragic, but the hazardous properties of this substance did not become apparent until about 1925, when bone cancers were reported among former workers in shops in which luminous watch dials were manufactured, mostly during World War I. The luminous paint used on the dials contained radium, which was applied to numerals and letters with camel-hair brushes. The workers (mostly young girls) were paid by the piece, and they found that their productivity improved if they pointed the brushes with their lips. Unfortunately, this practice caused them to ingest traces of radium, which has chemical characteristics like those of calcium and tends to deposit in the skeleton when absorbed into the body. In addition, radium was administered to patients during a period of radiation faddism in the 1920s. Nearly a hundred cases of bone cancer developed in these two groups.

The injuries resulting from the use of X-rays and radium compounds serve to illustrate the two basic ways in which people can be exposed. In the case of X-rays, the source is external to the body. In the case of the radium dial painters, the source of radiation was deposited within the body.

The misuses of radium were studied thoroughly, and by 1940 it was learned that the dial painters developed bone cancer only if the radium deposited in their bodies exceeded a certain amount. This knowledge made it possible to establish work procedures to limit the amounts of radiation received by the skeleton. The standards thus established have been used to protect those exposed, not only to radium, but to other radioactive substances as well, many of which have been artificially produced in the atomic energy programs of the world. It is now more than 35 years since the initiation of the atomic-energy program during World War II, and at any one time in the United States as many as 200,000 employees have been involved in the production of the radioactive equivalent of many billions of pounds of radium. Whereas only 2 pounds of radium extracted from the earth's crust during the first 40 years of this century caused the deaths of about 100 people, there have been no known deaths due to the artificial radioactive materials in the numerous industrial, research, and medical plants and laboratories in which the radioactive substances have been used.

Regrettably, among about 2,000 uranium miners in the western United States, there have been more than 200 deaths from cancer due to excessive exposure to radon, a radioactive gas produced by the radioactive decay of radium.[3-6] The lung cancers among these miners are all the more tragic because an earlier experience in European mines was not heeded. Mines in Central Europe had long been worked for precious metals, and it had been known for centuries that the men who worked in these mines developed a fatal lung disease known locally as "Bergkrankheit" (mountain sickness). In the early part of this century it was realized that the disease was lung cancer, and it was later concluded that the cancers were caused by a radioactive gas, radon, that had been emanating from radioactive ores present in the mines. The same mines, which for centuries had been mined for gold, silver, and platinum, were later to be exploited for their uranium.

A standard to protect workers against the effect of radon was recommended just prior to World War II, but not adhered to in the uranium mines in the United States. Predictably, a high incidence of lung cancer began to develop in the early 1950s. (Improved ventilation in these mines has now probably eliminated the danger.) The fact that adequate precautions were not taken initially was the result of a tragic bureaucratic blunder. The Atomic Energy Act of 1946 assigned the U.S. Atomic Energy Commission (AEC) responsibility for protecting the health and safety of the workers and the public from the dangers of radiation produced by the atomic-energy program. This preemption was unusual, because industrial safety had until then been the responsibility of the individual states. The law required that the AEC assume this responsibility because of the complexity of the industry, its unusual risks, and because the required highly specialized knowledge then resided only within the federal government. However, the 1946 law specified that the AEC's responsibility began *after* the raw material had been removed from the ground. Because of this stipulation, the mining activities remained the responsibility of the individual states, which apparently lacked either the will or the means to deal with the hazards involved. Had the responsibility been given to the federal government originally, the epidemic of radiation-induced lung cancer that developed among the uranium miners would have been prevented.

Apart from the cases of lung cancer among uranium miners, there have been seven deaths in the atomic-energy program due to accidental massive exposure to external bursts of radiation, mostly in experimental laboratories. These were due to accidental nuclear reactions, and must be distinguished from the more-common type of exposure that results from day-to-day handling and use of radioactive substances and absorption of radioactive substances into the body. At this writing, the last of these fatalities is recorded as having occurred in 1962.[7] In the absolute sense, these seven deaths were tragic accidents that could have been prevented—but only in a less-imperfect world. It helps to place the radiation-induced deaths in perspective in order to compare them to the total number of work-related deaths that have occurred among atomic energy employees due to causes other than radiation. These have totalled more than 300 in the government atomic-energy program alone, which employs about 100,000 workers. This is a large number, yet the occupational-fatality rate in the atomic-energy program of the United States has been excellent (about one-half of the national industrial average).

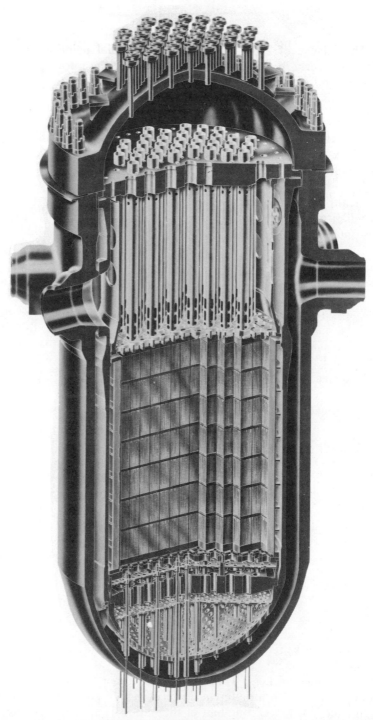

Figure 13-2. General configuration of a nuclear reactor. (*Courtesy of Babcock & Wilcox.*)

The misuse of 2 pounds of radium for dial painting and medical purposes early in this century thus provided the information that has made it possible to develop safe procedures to protect hundreds of thousands of people who work with radioactive substances.

THE USE OF RADIATION PROTECTION STANDARDS

The term "ionizing radiation" includes X-rays as well as all types of radioactivity emitted by radioactive processes. (The "nonionizing" radiations include all others in the electromagnetic spectrum such as visible light, infrared, ultraviolet, radio waves and the radar frequencies.) The unit of ionizing radiation dose is the rem, or the millirem (one thousandth of a rem). The rate at which radioactive substances emit ionizing radiation is measured in curies.

A dose of 150 rem administered quickly would result in the death of about 50 percent of the people irradiated, and 100 percent mortality would be expected for doses above 1,000 rem. Except for the use of atomic bombs in World War II, and the few accidents mentioned earlier, sudden massive exposure to such high doses of radiation has fortunately been rare.

In addition to the effects that become apparent within a short period after heavy irradiation, some other effects may not appear for many years. Leukemia and other forms of cancer are of the greatest concern in this regard. Radiation exposure can also cause genetic effects in future generations, but this danger has proved to be very much less than was thought to be the case in the decade following World War II. For this reason, discussions of the dangers of radiation exposure are currently limited mainly to the risk of cancer.

Most of the information on which our knowledge of radiation effects is based involves relatively high doses (greater than 10 rem) administered over a short period of time. The information comes from human experience (the medical uses, industrial experience, and the Japanese bombings) as well as a vast number of animal experiments. For the reasons given in Chapter 8, it is not a simple matter to estimate the effects of low doses of an injurious agent, received over a long period of time, using data obtained at high doses administered in a few minutes or hours. A quart of alcohol consumed within an hour would be disastrous, probably lethal, to most people. This is not so if it is taken at a rate of one ounce per day over a period of 32 days.

There is some evidence that the risk of cancer following radiation exposure is proportionate to dose, is independent of the rate at which the dose is received, and that there is no threshold. However, there is also evidence that the opposite is true, that is, that the dose-response curve is S-shaped, indicating that the risk diminishes when the dose is protracted, and that there is a threshold.[8] The first set of assumptions have been adopted in the interest of safety, but most reports on the subject note that the estimates of the risk that are based on these assumptions are conservative in that they define the *upper* limit of risk. The actual risk could be lower if the dose-response relationship is not linear, or if the response is reduced when the dose is protracted.[9,10]

Relying upon all available human data, most of which have been obtained at relatively high-dose rates, and accepting the assumptions with respect to linearity of response and independence of dose rate, we can estimate that a dose of 1 rem delivered to 1 million persons will cause about 100 cancers over a 20-year to 30-year period.

Standards for protection against the effects of ionizing radiation[11] were among the first occupational health standards to be adopted. The International Commission on Radiation Protection was formed in 1928 and has operated ever since in close cooperation with national standards-setting organizations and the World Health Organization. In the United States, the National Council on Radiation Protection and Measurements (NCRP), which was formed in 1929, is the national counterpart of the ICRP. It was located in the Bureau of Standards until 1964, but is now an independent organization operating under a Congressional charter. The recommendations of ICRP and NCRP were originally intended for the protection of workers exposed in industry, laboratories, and the medical professions.

The ICRP and NCRP have for many years taken the position that, in the interest of safety, it should be assumed that there is no radiation threshold and that the response to radiation is linear in relation to dose. They therefore specify that the radiation dosage should be kept as far below the recommended limit as is practicable. The U.S. Nuclear Regulatory Commission has found that, at least in the operation of light-water reactors, it is practicable to maintain the dose to the general population at a maximum of 5 to 10 millirem, and light-water reactor operations are required to limit their radioactive releases accordingly.[12]

NATURAL RADIOACTIVITY

It helps put the subject of radiation exposure in perspective if we can understand the levels of natural radioactivity to which humans have always been exposed.[4,13] Radioactive substances of natural origin are in the air we breathe and the food we eat. These radioactive elements become incorporated into our tissues to such an extent that, on average, the atoms of which our bodies consist are disintegrating at a rate of about 500,000 per minute, due to the presence of naturally radioactive species of carbon, potassium, and other elements. In addition, we are exposed externally to cosmic rays from outer space, to gamma radiation being emitted by the earth's crust, and to the radioactivity normally present in the atmosphere and which is inhaled into our lungs.

The total body irradiation received by humans in most parts of the world results in a dose of about .1 rem per year, a figure that varies somewhat from place to place, depending on the altitude above sea level and the composition of the rocks and soils. The dose increases as one moves to higher altitudes because the atmosphere becomes thinner and serves as a less effective shield against cosmic rays. The dose from the rocks and soils varies from place to place, depending on the amounts of radioactive minerals present. The greatest exposure from terrestrial sources occurs normally from granitic rocks. Major deviations from the norm occur in places where the thorium or uranium content of the rocks and soils is greatly elevated, as in parts of Brazil and India.[4,14] Plants that grow on a hill in Minas Gerais, Brazil absorb so much radioactivity from the soil that it is possible for them to be autoradiographed by simply placing their leaves in contact with X-ray film (Figure 13-3).

The lungs normally receive a higher radiation dose than the rest of the body due to the natural presence of atmospheric radon, the concentration of which varies from place to place. Radon exposure in many localities will deliver a dose of about .2 rem per year to the linings of the bronchial passageways, which are of particular interest because this is where most lung cancers originate. Dose rates of nearly ten times this value are encountered indoors, particularly when the building is made of materials with a high content of natural radium.

Radon, which has a half-life of 3.8 days, decays progressively through several shorter-lived radioactive species to Pb-210, which has a half-life of 22 years, and this radioactive substance ultimately deposits on the earth's surface. It, in turn, decays to polonium-210, which is a highly radioactive substance. Only in the last few years have we begun to appreciate that humans have always been subject to this form of natural fallout and that broad-leafed plants contain relatively high concentrations of this isotope because of deposition of Pb-210 on their foliage. It is well known that polonium-210 is present on the leaves of tobacco plants and becomes incorporated into cigarettes. Measurements of the polonium-210 activity in human lung tissue

Figure 13.3. Autoradiograph of the fern, adiantium, from the Morro do Ferro in the state of Minas Gerais, Brazil. The presence of radioactivity is due to absorption from radioactive minerals in the soil. (*Courtesy of Dr. Eduardo Penna Franca.*)

have shown that the dose received by smokers from this source is five times that received by nonsmokers.

RADIOACTIVE RELEASES FROM NUCLEAR REACTORS DURING NORMAL OPERATIONS

The fuel used in nuclear-power reactors accumulates enormous amounts of radioactivity. The radioactive substances are mainly fission products, but "activation products" produced by neutron bombardment of corrosion products in the circulating water are also present, as are plutonium and other transuranic elements formed by neutron bombardment of the uranium-238. Under normal operating conditions, most of the accumulated fission products are immobilized within the ceramic uranium oxide fuel in which they were formed. The fuel pellets are encased in a zirconium alloy that is impervious to most substances, but radioactive forms of the noble gases krypton and xenon, as well as radioactive iodine and cesium, have the ability to diffuse from the fuel and pass to the circulating water through minor imperfections in the zirconium cladding. As a result, the circulating water accumulates radioactive substances in small but readily measurable quantities.

Commercial power reactors are equipped to keep the level of radioactivity in the circulating water below predetermined levels. The coolant is filtered to remove radioactive particulates and the radioactivity in solution is removed by ion-exchange resins. The dissolved gases are separated and passed to storage tanks where the shorter-lived constituents are permitted to decay. The remaining gaseous radioactivity is due mainly to krypton-85, which has a half-life of about ten years and which can be released to the atmosphere under controlled conditions.[15] Sometimes other shorter-lived nuclides are released, including iodine-131 and various species of xenon and krypton. Small quantities of radioactivity in liquid form are also released under controlled conditions.

The quantities released in gaseous and liquid form are monitored, for the releases must be maintained below limits imposed by the regulatory authorities. These regulations take into consideration that certain of the nuclides can be absorbed into food chains that ultimately reach humans, and that biomagnification may occur. Thus, a relatively low concentration of radioiodine in air can deposit on grass and then ultimately appear in concentrated form in the milk of grazing cows. The permissible amount of liquid that can be released must also take into account that shellfish feed by filtering huge quantities of water, and thus have the ability to concentrate trace substances.

The operators of light-water reactors in the United States are required to limit their discharges so that no individual will be exposed to more than .01 rem per year. In the case of airborne releases, the maximally exposed individual is usually a hypothetical person who is assumed to sit at the site boundary 24 hours per day throughout the year. If there is a dairy nearby, the maximally exposed individual may be a child who obtains all its milk from grazing cows from the one location. In other cases, the maximally exposed individual might be a person who consumes fish or shellfish from the location where an exposure to the liquid releases is maximum. Hence, if such an individual does not exceed the permissible dose, others will receive much less.

It may help to place the subject in better perspective by relating the maximum permissible radiation dose for the general population (.010 rem per year) to the dose ordinarily received from natural sources. In New York City, there is a difference of .015 rem per year between the dose received by most residents of Brooklyn compared to those living in Manhattan: Most of Brooklyn is built on sand, which has a lower level of natural radioactivity than the rocky terrain of Manhattan.

The power plants in the United States have been operated, without significant exception, well within the limits imposed by the regulations. It has been estimated that by the year 2000, when several hundred reactors will exist in the United States, the average dose received by the population will be .0002 rem per year, or about .2 percent of the dose received from natural radioactivity.[16] This estimate includes exposure from the entire industry, including fuel reprocessing, waste disposal, and transportation, in addition to reactor operation. A radiation dose such as this is well within the range of variability in the dose received from nature, depending on the altitude at which one lives, the materials of which one's home is constructed, or whether one lives on sand, loamy soil, or near granite outcrops. The radiation exposure due to nuclear power production is insignificant insofar as the public health is concerned. The maximum dose to any exposed individual should rarely exceed 5 to 10 millirem per year.

NUCLEAR REACTOR ACCIDENTS

Since 1942, about 1,000 reactors of all kinds have been built in the world. This total includes not only power plants, but research reactors and more than 300 reactors operated aboard ships of the United States and the Soviet navies.[1,17]

There have been no accidents involving nuclear power plants in which significant amounts of radioactivity have been released to the environment. Only one accident, in 1957, has involved a reactor of any kind (a plutonium-production reactor in Windscale, England). Significant quantities of radioactive materials were released, but no injuries resulted from that accident.[4] There have also been a few accidents involving research reactors (one in the United States) in which employees were killed or injured, but environmental contamination did not occur. However, the possibility of a catastrophic accident involving a commercial nuclear-power reactor does exist, and has received continuing attention throughout the years. Although the probability of such an accident is slight, its consequences, should it occur, could be extremely severe.

The danger that preoccupies reactor designers and operators, the regulatory authorities, and nuclear opponents is the "loss-of-coolant" accident.[4] The circulating water (coolant) passes in and out of the reactor pressure vessel through large stainless steel pipes. If one of these pipes should suddenly rupture, the coolant would be lost and the reactor core would begin to overheat from the energy released by radioactive decay of the enormous amount of fission products. The fuel could overheat to such an extent that the cladding and fuel would melt, releasing great quantities of radioactivity in a volatile form. If provision is not made to contain the radioactivity in the event of such an accident, there could be massive casualties from the radioactive cloud at great distances from the plant, and large areas of land would become contaminated to such an extent that there could be severe economic consequences. It is thus essential that the reactor systems be designed in such a way that the probability of a loss-of-coolant accident is vanishingly small. However, after having designed the reactor in such a way that there is an exceedingly small probability of occurrence, it is then assumed that core destruction does occur and that exposure to nearby populations must be controlled by equipment designed specifically for that purpose.

The probability that a pipe would fail has been estimated to be less than one per million reactor-years of operation.[18] The chemical industry has had long experience with many miles of pressurized stainless steel pipe, and that experience is applicable to nuclear reactors. To guard against the consequences of a pipe failure, an emergency core-cooling system is provided to inject cooling water into the reactor core and prevent overheating. However, assuming that the system will not operate properly and that the core will partly melt, an airtight containment building is provided to contain the released vapors. In addition, nozzles are installed within the containment building to spray water that would condense the steam and wash the radioactive materials from the atmosphere into a sump within the sealed building. Chemicals are added to the sprays to wash radioactive iodine vapors from the air which, as an added precaution, can be recirculated through filters to remove any radioactive particulates and vapors that remain airborne.

A proposal to build a nuclear-power reactor must be accompanied by a safety-analysis report of many volumes, in which are included complete descriptions and safety assessments of the site, including the local hydrology, geology, meteorology, and ecology. The design of the reactor must be described in detail. The safety-analysis report is then reviewed by the Nuclear Regulatory Commission (NRC) staff and by the Advisory Committee on Reactor Safeguards (ACRS). This committee, composed of a group of specialists, was established by the Congress, and because it reports directly to the commissioners, the ACRS analysis is carried on independently of the Nuclear Regulatory Commission staff. If the reviews of the NRC and the ACRS are favorable, the Commission will usually hold a public hearing before ruling on the license application. Massive technical reports are generated in the course of the licensing procedures, and these are available for public review. The accumulation of hearing transcripts and technical reports during one licensing procedure is shown in Figure 13–4.

While there has been general agreement that the probability of a major nuclear accident is very low, there has been disagreement about the effectiveness of the methods used by reactor designers to mitigate the consequences of such an accident. In 1972, the Nuclear Regulatory Commission contracted with the Massachusetts Institute of Technology to undertake a detailed assessment of accident risks in commercial nuclear-power plants in the United States. This study took three years to complete, cost $4 million, and resulted in a technical report of about 3,000 pages that was published in October, 1975. The main conclusions of that report are summarized in Figures 13–5 and 13–6.[19] The estimated frequencies of occurrence of nuclear accidents and the actual frequencies of nonnuclear accidents are plotted, in the figures, in relation to the number of fatalities that would be caused. The data were derived after assuming that 100 nuclear-power plants would be operating. The probability that a nuclear accident would kill more than 100 people is estimated to be about one per 100,000 per year. Stated another way, 100 reactors would, on the average, be expected to operate for 100,000 years before an accident occurred which would kill more than 100 people. Figure 13–5 compares the probabilities of nuclear accidents of various severities with manmade catastrophes, including dam failures, air crashes involving people on the ground, releases of chlorine, and fires. The probability

Figure 13-4. Mass of technical reports and hearing transcripts accumulated in the course of application for a power reactor operating license. Shown are the Preliminary Safety-analysis Report, Final Safety-analysis Report, Environmental Reports, and transcripts of the construction license hearings. (*Courtesy of Consolidated Edison Company.*)

that a nonnuclear accident would kill substantial numbers of people is seen to be many orders of magnitude higher than that of nuclear power. Figure 13–6 presents the comparative information for nuclear versus natural catastrophes. The probability that a member of the public will be killed by a nuclear-power plant accident is roughly the same as for being killed by a meteorite, one of the rarest of all natural causes of death. If the calculations are in error by even a factor of 1,000, the risk of death by nuclear accident would still be less than that of being killed in a tornado or a hurricane!

MANAGEMENT OF NUCLEAR WASTES

The problems associated with the transportation, processing, and storage of intensely radioactive waste products has also been a major

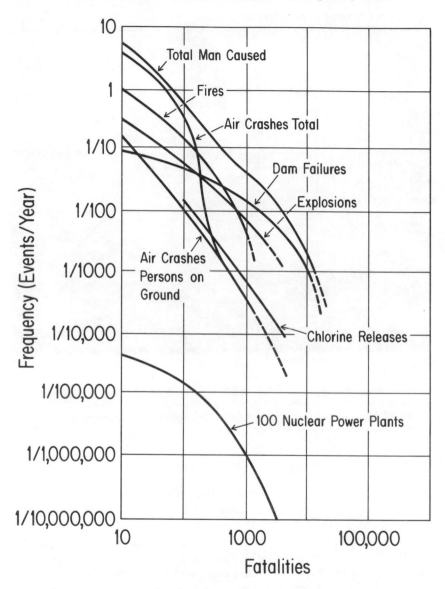

Figure 13-5. Frequency of accidents of varying severity due to man-caused events.[33]

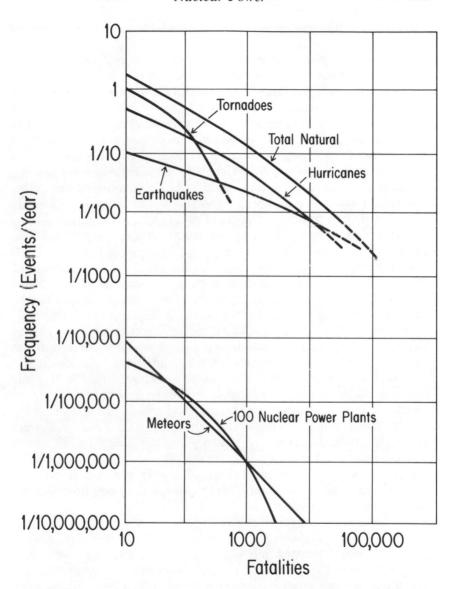

Figure 13-6. Frequency of natural events that cause varying numbers of fatalities.[33]

issue in the dialogue between the opponents and proponents of nuclear power.

It is necessary to differentiate between low-level and high-level wastes. The former are generated in the course of normal reactor operation and take the form of contaminated rags, mechanical equipment, and ion-exchange resins containing traces of radionuclides separated from the coolant system.[4,20] Low-level wastes do not present a great problem, and in general have not been a matter of contention. It is common practice to mix them with concrete in 55-gallon steel drums within which they are immobilized when the concrete hardens. The concrete shields the radiation, and the drums can then be shipped to storage centers where they are ordinarily buried in shallow trenches. The exact procedures are regulated by the government and the low-level wastes are much less of a problem than many of the ordinary chemical wastes encountered in industry.

Of far greater concern, however, are the problems of managing the highly radioactive wastes generated by processing spent fuel. Some opponents of nuclear power argue that the industry should not be allowed to develop further until the waste-management problems have been solved. The highly radioactive waste products are largely long-lived fission products in which are included small but significant amounts of transuranic elements such as plutonium, americium, and curium. During World War II these wastes were generated at only one location, near Richland, Washington, and huge quantities were stored as liquids in underground pits. Later, steel and concrete tanks were constructed for their containment. These systems of storage were relatively crude, but were satisfactory for the quantities of wastes being generated for the military program. The civilian power program will produce far greater quantities of highly radioactive nuclear wastes, and it will be necessary to make more-elaborate provision for their permanent storage.[20]

An important requirement of the Nuclear Regulatory Commission is that the highly radioactive waste products of the fuel-reprocessing plants must be converted into solid form within five years after they are produced.[21] The advantage of conversion to solid form is that the volume is reduced enormously and the wastes can be managed more readily. Several processes exist for this conversion, and their feasibility has been demonstrated on a pilot-plant scale. However, commercial full-scale operation has not as yet been undertaken. It has been shown to be feasible to convert the wastes into a glasslike substance, with a solubility not much different from that of quartz, the principal mineral in beach sand. The wastes, once solidified, must then be packaged in containers preparatory to "permanent" storage.

The total volume of the solidified wastes will be very small. It is estimated that by the year 2000 the volume of total national inventory will be about 600,000 cubic feet which, if stored in one place, would constitute an 85-foot cube.

There is widespread concern about the potential for hazards to future generations because some of the waste products have very long half-lives. The transuranic elements, which have half-lives longer than 10,000 years, must be stored securely for hundreds of thousands of years before their radioactivity will decay to harmless levels. Considering that written history is only a few thousand years old, we are presented with a problem for which there is no precedent: Social systems come and go on time scales measured in centuries, but methods must be devised by which the hazards from radioactive wastes can be controlled in perpetuity. The radioactive wastes must be managed in a way that will not create a hazard, even assuming, for example, that our land surfaces are someday covered by glacial sheets which destroy all traces of our present society, but which will eventually melt and allow repopulation in 50,000 or 100,000 years.

There are many opportunities for permanent storage of nuclear wastes in geological formations that are so deep underground that they could not be accessible, even to extreme glacial scouring.[23] For example, the wastes could be placed in cannisters and buried in natural salt deposits. Salt beds have a number of advantages, including geological stability. The deep-sea bed has also been considered, as have proposals to store the wastes in deep rock caverns that are known to be stable on time scales of hundreds of millions of years. Many such geological formations are known to exist. A sensible approach would seem to involve storage in any convenient and safe manner during the next several decades, pending a decision as to the best means of ultimate disposal. The question at this time is not whether radioactive wastes can be managed safely, but which of the several options is best from the point of view of economics and operating convenience. The volumes of wastes will be relatively small and could easily be accommodated on an interim basis at any of a number of government reservations.

PLUTONIUM TOXICITY

Although minute traces of plutonium exist in nature, it is for all practical purposes a manmade element produced by neutron capture in uranium-238. Of the several isotopes of plutonium produced in reactors, plutonium-239 occurs in the greatest abundance. This is a fissionable material that was separated from spent nuclear-reactor fuel during World War II. It was used in the bomb dropped on Nagasaki.

Animal studies have shown that plutonium-239 is a highly toxic substance.[22,24] When it is inhaled or ingested in the soluble form, it tends to deposit in bone, where it can produce cancer. If inhaled as an insoluble particle, it can produce cancer by irradiating the lung.

The concern over the toxicity of plutonium can be understood by comparing it to radium, which is a highly toxic, naturally occurring substance. Expressed per unit of radioactivity, the maximum permissible concentration of soluble plutonium in air is one-fiftieth of that permitted for radium-226. It is estimated that since World War II, the United States has produced, separated, and fabricated about 200,000 kilograms of plutonium.[22] On a mass basis, this is about 200,000 times as much radium as has been produced. It is reassuring that although many thousands of persons have worked with plutonium, the precautions have been such that no cases of cancer due to plutonium exposure are known to have occurred in the more than three decades since the first plutonium was produced.

Plutonium is now a manmade contaminant of the general environment, having been dissipated in the upper atmosphere in the course of nuclear weapons tests. A total of about 500,000 curies of plutonium-239 (7,500 kilograms) are now widely distributed, and traces of plutonium can be found in human tissues. The plutonium standards recommended by the International Commission on Radiological Protection and the National Council on Radiation Protection and Measurements have not been shown to be unsafe on the basis of many years of experience, but scientists with strong antinuclear persuasions have suggested that the recommended values are too high by the enormous factor of 115,000 and should be reduced accordingly.[25] Based on this widely disputed line of reasoning, it has been predicted that approximately 1 million cases of lung cancer will be caused in the Northern Hemisphere in the next 30 years as a result of exposure to the plutonium dust dispersed in weapons tests.[26] These arguments have been thoroughly examined and rejected by the NCRP, the British Medical Research Council, and the National Academy of Sciences. Expert committees of all these organizations have issued reports that strongly disagree with the suggestion that the plutonium standard should be reduced.[24,27,28]

The disagreement stems from the fact that inhaled insoluble plutonium, is particulate in nature. These particles emit alpha radiations that have a very short range in tissue. Moreover, a few particles in the lung do not irradiate the lung uniformly, but rather in "hot spots." Some have argued that alpha radiation from plutonium is more carcinogenic than heretofore assumed because it is mainly the cells near the surface of the bronchial tree that are susceptible to development of lung cancer, and alpha-emitting dust particles that deposit along the bronchial passageways therefore irradiate the most sensitive part of the lung. However, one can argue on radiobiological grounds that the risk is not increased but, rather, is substantially decreased by the particulate nature of the source. This is the position taken by the three scientific organizations that considered the matter.

SABOTAGE AND THEFT OF FISSIONABLE MATERIALS

The possibility that fissionable materials could be stolen for illicit uses has been recognized ever since the earliest days of the atomic energy program, but the concern then was that the material would fall into the hands of unfriendly governments who would in this way obtain the means to fabricate nuclear weapons. This was more than a quarter of a century ago, when extreme forms of violence were exclusively within the domain of governments and were not being practiced by groups of terrorists to the extent that has since become relatively commonplace. During the past ten years, restive students in the United States and other parts of the world have from time to time bombed university buildings and other structures, and the hijacking of aircraft by political terrorists is not uncommon. Terrorists have invaded embassies, and have bombed banks and airport terminals. In this climate, it is only proper that there should be an increasing concern about the possibility of nuclear sabotage.

Nuclear sabotage can take the form of an assault on a power plant for the purpose of destroying it (and releasing fission products to the environment), theft of plutonium or other fissionable material for the purpose of fabricating a bomb, or theft of plutonium or other radioactive materials for the purpose of extortion.

In recent years, the security forces available at reactor sites have been greatly strengthened in response to the rise of terrorism. However, it is always possible to hypothecate that a terrorist attack will be mounted in sufficient strength, and with so high a level of technical sophistication, as to overwhelm the defenders and defeat the technical safeguards that have been provided. One can always conceive of newer and better methods of defense, but stronger means can always be devised to overwhelm the strongest defenses. The problem is not confined to nuclear reactors: If terrorists can destroy power plants, they can also destroy dams or poison the water supplies of major cities.

Most of the concern over nuclear sabotage has to do with the possible theft of plutonium for the purpose of fabricating a nuclear weapon. Modern nuclear weapons are sophisticated devices that are difficult to build. They are relatively small, powerful, and efficient. But in the hands of a potential saboteur, even a crude device transportable in an automobile or truck could yield explosive energy equivalent to several tons of TNT, thereby wreaking great mischief. Such an explosion, if it took place in the garage below a skyscraper, could cause its total collapse, with the deaths of thousands of people. Or, such an explosion could cause valleys to flood if dams were destroyed; and the largest of suspension bridges could be destroyed.[29]

Is it possible for a few people to fabricate such a weapon? The experts disagree, and when there is disagreement on a matter of this kind it is necessary that we assume it can be done. At one time, the information needed to construct a nuclear bomb was highly secret. Even to this day, much of the information needed to produce modern nuclear weapons is classified, but the basic information has gradually been released in a variety of forms. As noted by others, much of the essential information is contained in an article as readily available as one in, say, *Encyclopedia Americana,* prepared by one of the nation's foremost nuclear weapons experts.[30] In that article, the basic physical reactions are explained, the required masses of plutonium-239 are given, and the basic method of initiating the explosion is described. It has been concluded, in one of the most thoughtful examinations of the risks of nuclear theft, that the design and fabrication of a simple, transportable fission explosive is not a difficult task.[29] If a few kilograms of plutonium were available, a crude bomb could be fabricated by a knowledgeable person with the aid of readily available machine tools.

Such a bomb could be made from plutonium-239 or from highly enriched uranium. (Uranium used in light-water reactors is not sufficiently enriched for this purpose.) The only place in the nuclear industry from which plutonium might be stolen would be at the fuel-reprocessing plant. If plutonium is to be recycled through light-water reactors, it could be stolen while it was being transferred between the fuel-reprocessing plant and the fuel-fabrication plant where the plutonium is combined with uranium oxide. It is not reasonable to assume that plutonium could be extracted clandestinely from reactor fuel because this would require facilities far beyond the capability of even the most sophisticated terrorists. Thus, so far as light-water reactors are concerned, the possibility of diversion would be greatly minimized if the plutonium that had been separated at the fuel-reprocessing center could be immediately combined with the slightly enriched uranium and converted to the form in which it would be used in the light-water reactors. This would be a far safer procedure than past practice, in which the plutonium left the fuel-reprocessing plant in the form of a plutonium nitrate solution which could be readily converted into plutonium oxide or plutonium metal. Such shipments have formerly been sent over the open road in unescorted trucks driven by a single, unarmed driver.[29] The trucks sometimes carried more than 50 kilograms of plutonium, enough to make perhaps ten bombs. As a result of prodding from outside the government, the Nuclear Regulatory Commission has begun to take far more stringent security precautions. It would certainly seem desirable in the future that the opportunities for theft of plutonium be minimized by reducing the need to transport it, and by reducing the time during which it exists in a form that could be fabricated into a bomb.

The fear of plutonium diversion and concern about the hazards of inhalation have resulted in opposition to two major developments concerning nuclear power. A demonstration breeder reactor scheduled for completion in Tennessee in the early 1980s has been strongly opposed by public interest groups who are perhaps less concerned over the safety of the demonstration breeder itself than the possible hazards of an economy based on breeder power in the next century.[31,32] The position has been taken that the demonstration unit should not be built until the generic hazards of the breeder have been fully analyzed. This presents a difficult problem for the government because much of the information one needs in order to quantify the risks of a breeder-based energy industry can only be obtained by operating a demonstration unit. It is not logical to expect that an agency can realistically describe the risks of the second or third generation of units to be constructed in the next century when these have not yet been designed, and can not be, until the required engineering experience has been obtained by operating the demonstration unit. Additional opposition came from President Carter early in 1977 because of his concerns about the role of plutonium in the breeder. The President also opposed the plan to recycle the plutonium produced in light-water reactors.[33] The President believes the risk of nuclear-weapons proliferation will be reduced by our unilateral decision concerning the breeder and fuel reprocessing. The only modern fuel-reprocessing plant in the United States has been completed in South Carolina, but it has not been authorized to begin operation due to our national posture. However, this has not affected the policies of other nations. Russia and France, among others, are now actively operating breeder reactors. Spent Japanese reactor fuel that was to be reprocessed in the United States will be accommodated in the United

Kingdom, and a nuclear complex in Brazil that was to have been built by industries of the United States (and which would have included a reprocessing plant) will now be supplied by West Germany.

Since plutonium is produced in light-water reactors and is separated in the course of fuel reprocessing, the life of existing uranium reserves would be greatly extended if the plutonium could be recycled in the proposed manner. Estimates of uranium reserves are an important ingredient of the decision-making process. An underlying assumption is that solar energy, and possibly fusion, are likely to become available sometime in the next century, and that the need for nuclear energy will thereafter become less important. If the reserves of uranium prove to be sufficient until new energy options become available, it would be possible that we would not have to rely on the breeder and the necessity of plutonium recycling. Spent fuel could be stored permanently, not reprocessed. Uranium reactors would be fueled from reserves that would last until fusion systems and solar systems are developed.

The problem is that many experts do not agree that the uranium reserves are adequate to supply the nation's energy needs beyond the end of the century. If they are correct, and if solar and fusion sources are not available, an unacceptable energy shortage could develop. The other school of thought takes a more conservative approach, on the assumption that it is better to have the breeder and not need it than to need it and not have it.

SOME COMMENTS ON THE SAFETY OF THE NUCLEAR INDUSTRY

Despite its seemingly formidable hazards, the nuclear industry is one of the safest of all industries from the point of view of employee protection. To date, no member of the public has been injured by the handling of the industry's copious quantities of radioactive materials. In the highly polarized debates, the antinuclear factions have failed to

Figure 13-7. The safety of a spent fuel shipping cask was put to a severe test when a rocket-propelled locomotive was crashed into it at more than 80 miles per hour. The 28-ton cask received minor surface dents during the test, but did not crack open or leak any contents. The cask at impact was knocked into the air, and bounced twice on the ground before coming to rest between the rails of the track. The front half of the locomotive was totally crushed by the impact, and the trailer on which the cask was mounted was bent around the locomotive in a U-shape. These results confirmed predictions made earlier by project engineers on the basis of scale model tests. (*Courtesy of U.S. Department of Energy.*)

acknowledge that the industry is now 35 years old, that its safety record is unsurpassed, that the federal government has instituted elaborate regulatory procedures, and that the nuclear industry was 25 years ahead of others in requiring detailed safety analyses, careful environmental monitoring, and strict adherence to standards set, in many cases, by international agreement.

One example of the extremes to which the federal government goes to assure the safety of its operations is shown in Figure 13–7, which is a sequence of photos showing a locomotive crashing, at 81.5 miles per hour, into a container used for shipping spent fuel. The container survived the tests with only minor dents.

Nuclear power looks particularly attractive when a comparison is made of the relative risks of generating electricity by nuclear or fossil means. A number of writers have made such comparisons, in which they have taken into account the risks to both employees and the general public due to each step, from mining to waste management.[34–39] In all studies, the effects of nuclear power on health are lower, by factors ranging from 10 to 100,000, than those from use of fossil fuels. There are great uncertainties in these estimates because of the difficulties involved in comparing the effects of low-level radiation exposure to those from sulfur dioxide, nitrogen oxides, and particulates. Uncertainties in making such estimates have been emphasized many times in the course of this book.

Until recently, there was general acceptance of the original decision by Congress, in 1946, that health and safety in the atomic energy industry should be preempted by the federal government. However, as time passed, the states have developed varying degrees of competence and in recent years have wanted to play a role in decisions relating to nuclear safety. This, despite the fact that uranium mining, which was controlled by the states, has resulted in the deaths of more than 200 miners because the need for proper mine ventilation was not enforced in any of the states within which uranium mining took place. Moreover, it has come to be recognized that the states have failed generally to provide adequate industrial safety, and it is for this reason that the federal government has created such organizations as the Occupational Safety and Health Administration (OSHA) and the Mining Enforcement and Safety Administration (MESA). It is generally accepted that the public receives unnecessary exposure to radiation from the improper use of medical X-ray equipment, which many states fail to regulate properly. Common sense would seem to dictate that the nuclear industry should be regulated by a single set of standards, and that if the record of federal regulation has been excellent for over 30 years, the existing system should be continued.

However, there are powerful pressures in the opposite direction, and it is difficult to forecast whether the present system of nuclear-energy regulation will be continued.

NOTES

1. Hewlett, R. C. and O. E. Anderson. "The New World" (Volume I of a History of the Atomic Energy Commission), Pennsylvania State Univ. Press, University Park, Pa. (1962).
2. Joint Committee on Atomic Energy, Congress of the United States. Hearing before the Subcommittee on Legislation, JCAE, 94th Congress, "Naval Nuclear Propulsion Program-1976," Testimony of Admiral H.G. Rickover. Washington, D.C. (March 18, 1976), p. 3.
3. Archer, V. E., J. D. Gillam and J. K. Wagoner. "Respiratory Disease Mortality among Uranium Miners," *Annals of the New York Academy of Sciences* 271:280–293 (1976).
4. Eisenbud, M. "Environmental Radioactivity," 2nd ed., Academic Press, New York (1973).
5. Lorenz, E. "Radioactivity and Lung Cancer: a Critical Review in Miners of Schneeberg and Joachimstahl," *J. Nat. Cancer Inst.,* 5:1 (1944).
6. Lundin, F. E., J. K. Wagoner and V. E. Archer. "Radon Daughter Exposure and Respiratory Cancer Quantitative and Temporal Aspects," National Institute for Occupational Safety and Health–National Institute of Environmental Health Sciences Joint Monograph No. 1, U.S. Public Health Service, U.S. Department of Health, Education and Welfare, National Technical Information Service, Springfield, Va. (1971).
7. U.S. Atomic Energy Commission. "Operational Accidents and Radiation Exposure Experience within the United States Atomic Energy Commission 1943–75," *WASH-1192,* Division of Operational Safety, National Technical Information Service, Springfield, Va. (1975).
8. National Council on Radiation Protection and Measurements. "Review of the Current State of Radiation Protection Philosophy," *NCRP Report No. 43,* Washington, D.C. (1975).
9. National Academy of Sciences-National Research Council. "The Effects on Populations of Exposure to Low Levels of Ionizing Radiation," National Academy of Sciences-National Research Council, *BEIR Comm. Report* (1972).
10. United Nations Scientific Committee on the Effects of Atomic Radiation. 27th Session, Supplement No. 25 (A/8725), United Nations, New York (1972).
11. Taylor, L. S. "Radiation Protection Standards," Chem. Rubber Publ. Co., Cleveland, Ohio (1971).
12. U.S. Atomic Energy Commission. "Code of Federal Regulations," Title 10, Part 50, Appendix I. U.S. AEC, Washington, D.C. (1969).
13. National Council on Radiation Protection and Measurements. "Natural

Background Radiation in the United States," *NCRP Report No. 45*, Washington, D.C. (1975).

14. *Proceedings of the International Symposium on Areas of High Natural Radioactivity* (Pocos de Caldas, Brazil, June, 1975), Academia Brasileira de Ciencias, Rio de Janeiro, Brazil, 1977.

15. National Council on Radiation Protection and Measurements. "Krypton-85 in the Atmosphere," *NCRP Report #44*, Washington, D.C. (1975).

16. Environmental Protection Agency. "Estimates of Ionizing Radiation Doses in the United States 1960–2000," Report of Special Studies Group, Division of Criteria and Standards, Office of Radiation Programs (1972).

17. International Atomic Energy Agency. "Power Reactors in Member States, *STI/PUB/423*, IAEA, Vienna (1975).

18. Bush, S. H. "Reliability of Piping in Light Water Reactors," *Nuclear Safety* 17:568–579 (1976).

19. U. S. Nuclear Regulatory Commission. "Reactor Safety Study: an Assessment of Accident Risks in U.S. Commercial Nuclear Power Plants," Main Report, *WASH-1400*, National Technical Information Service, Springfield, Va. (1975).

20. U.S. Energy Research and Development Administration, *Proceedings of the International Symposium on the Management of Wastes from the LWR Fuel Cycle, CONF-76-0701*, National Technical Information Service, Springfield, Va. (1976).

21. *Federal Register* 10 CFR 50, Appendix F, "Policy Relating to Siting of Commercial Fuel Reprocessing Plants and Related Waste Management Facilities" (November 14, 1970).

22. Electric Power Research Institute. "Plutonium: Facts and Inferences," *EPRI EA-43-SR*, Special Report. Palo Alto, Cal. (1976), p. 6–3.

23. de Marsily, G., E. Ledoux, A. Barbreau and J. Margat. "Nuclear Waste Disposal: Can the Geologist Guarantee Isolation?" *Science* 197:519–527 (1977).

24. Medical Research Council (Great Britain). "The Toxicity of Plutonium," Her Majesty's Stationery Office, London (1975).

25. Tamplin, A. R. and T. B. Cochran. "Radiation Standards for Hot Particles. A Report on the Inadequacy of Existing Radiation Protection Standards Related to Internal Exposure of Man to Insoluble Particles of Plutonium and Other Alpha-Emitting Hot Particles," National Resources Defense Council, Washington, D.C. (1974).

26. Gofman, J. W. "The Plutonium Controversy," *Journal of the American Medical Association* 236:284 (1976).

27. National Council on Radiation Protection and Measurements. "Alpha-Emitting Particles in Lungs," *NCRP Report No. 46*, Washington, D.C. (1975).

28. National Academy of Sciences-National Research Council. "Health Effects of Alpha-Emitting Particles in the Respiratory Tract," National Academy of Sciences-National Research Council, Washington, D.C. (1976).

29. Willrich, M. and T. B. Taylor. "Nuclear Theft: Risks and Safeguards," A Report to the Energy Policy Project of The Ford Foundation, Ballinger Publishing Co., Cambridge, Mass. (1974).

30. *Encyclopedia Americana*, Grolier, Inc., New York (1973), Vol. 20, pp. 520–522.

31. U.S. ERDA. "Liquid Metal Fast Breeder Reactor Program," Public Hearing Record (held May 27–28, 1975) on the Proposed Final Environmental Statement *WASH*-1535, Washington, D.C. (June, 1975), Vol. 1.

32. U.S. ERDA. "Light Water Breeder Reactor Program: Summary and Background," *ERDA 1541*, Washington, D.C. (June, 1976).

33. U.S. Nuclear Regulatory Commission. "Reactor Safety Study: an Assessment of Accident Risks in U.S. Commercial Nuclear Power Plants," Main Report, *WASH-1400*, National Technical Information Service, Springfield, Va. (1975).

34. Lave, L. B. "Health Effects of Electricity Generation from Coal, Oil and Nuclear Fuel," in: "Energy and the Environment" (H. Ashley et al., eds.), Pergamon Press, New York (1976), p. 63.

35. Sagan, Leonard. "Public Health Aspects of Energy Systems," in: "Energy and the Environment" (H. Ashley et al., eds.), Pergamon Press, New York (1976), p. 87.

36. Sagan, L. A. "Human Costs of Nuclear Power," *Science* 177:487–493 (1972).

37. Martin, J. E., E. D. Harward and D. T. Oakley. "Comparison of Radioactivity from Fossil Fuel and Nuclear Power Plants," 91st U.S. Congress, 1st Session, Joint Committee on Atomic Energy, Hearings on Effects of Producing Electric Power, U.S. Government Printing Office, Washington, D.C. (1969), Vol. 1, pp. 773–809.

38. Hull, A. P. "Radiation in Perspective: Some Comparisons of the Environmental Risks from Nuclear-and Fossil-fuel Power Plants," *Nuclear Safety* 12(3):185–196 (May–June, 1971).

39. Ellett, W. H. M. and A. C. B. Richardson. "Estimates of the Cancer Risk Due to Nuclear-Electric Power Generation," *Tech. Note ORP/CSD-76-2*, Environmental Protection Agency, Office of Radiation Programs, Washington, D.C. (1976).

CHAPTER 14

Some Long-range Atmospheric Effects of Pollution

There are pollutants present in the atmosphere which are of themselves harmless to human health, and which are perceptible neither by odor nor sight. Such pollutants may nevertheless affect human well-being by changing the pattern of global climate, or by altering the spectral qualities of solar radiation reaching the earth's surface.

POSSIBLE EFFECTS ON GLOBAL CLIMATE

The possibility that human beings may be capable of causing extreme changes in global climate is perhaps the most ominous of the conceivable environmental consequences of human activity. Other catastrophes such as nuclear war could kill a larger number of human beings than did the plagues of the Middle Ages. However, the effects of an all-out nuclear war would be reversible over a period of time, depending on the level of hostilities, probably not in excess of a century or two. It is also possible that a new chemical introduced to the environment could have unforeseen consequences on human health or on lower forms of life, but it is inconceivable that such effects, however severe, would not be reversible on a time scale measured in decades. However, should human activity somehow trigger a major climatic change, the consequences could literally be cataclysmic on a global scale, and might not be reversible for many tens of thousands of years.

A remarkable characteristic of the world's climate is that despite such wide variability from place to place and from time to time, the global means of the important parameters are relatively constant from year to year.[1,2] However, the mean values do vary over longer

336

periods of time, and relatively modest variations in the annual mean values can have profound effects on human well-being. For example, serious drought conditions can be caused by comparatively small reductions in the seasonal precipitation. The U.S. Weather Bureau defines a drought as a condition in which the rainfall is 30 percent of normal for a period of 21 days, but on a longer time scale, annual precipitation that is less than 75 percent of normal can have serious consequences over a period of several years.

An annual difference of as little as 1 to 2°C on a global scale can have major effects on agricultural productivity. The global mean in recent decades has been warmer than during the past century, but there is evidence that a cooling trend began about 30 years ago, which could be an ominous development in view of the precarious balance between supply and demand of food. The world's reserves of food are already so low that they could be wiped out by only one year of poor weather on a global scale.[3]

The sun is the primary source of the energy that drives the global climatic system. The intensity of solar radiation at the upper bounds of the atmosphere is relatively constant, and is not believed to vary by more than 2 percent. However, observations of this type have been made for only a comparatively short period of time, and it is conceivable that larger variations have occurred in the past and may occur in the future.

Some of the energy received from the sun is reflected back into space. This reflected energy is known as the earth's "albedo," and it can vary from as high as 90 percent for areas covered by snow to 10 percent for black soil. If polar ice should melt as a result of a rise in temperature, it would expose previously ice-covered land or water. The earth's albedo would be decreased, and this, in turn, would increase the amount of energy absorbed by the earth, thereby further increasing its mean temperature. This increase could cause additional snow cover to melt, which would cause a further decrease in albedo and a further increase in temperature. This could proceed until all of the earth's snow cover melted, and, if the process continued to the extent that the Greenland icecap also melted, the level of the world's oceans could rise by about 160 feet. This would have disastrous consequences for many coastal areas. Conversely, it has been calculated that a 1.6 percent decrease in incoming radiation (corresponding to a 5 to 10 percent increase in the earth's albedo) could result in another period of glaciation.

Systematic meteorological observations have been possible only since the development of instrumentation in the early part of the seventeenth century. Thus, only about 350 years of actual measurements are available anywhere in the world, and for much of that period the small number of meteorological stations were concentrated in Western Europe. Written historical records of various kinds go back about 2,000 years and have yielded a variety of information that has been useful in reconstructing climate history in many parts of the world. However, even 2,000 years is but a mere "instant" of time since life first developed more than 3 billion years ago. To reconstruct climatic history over longer periods of time, scientists have had to resort to geological studies and the reconstruction of botanical history by studies of the fossils in peat, ocean sediments, and other accumulations of the debris of the past.

One of the major changes that has taken place over a period of tens of millions of years is that the axis of the earth has shifted, due to the gradual drifting of land masses at the rate of a few centimeters per year. This change in the orientation of the earth's axis with respect to the sun, although gradual, has been so pronounced that about 60 million years ago, the North Pole was located north of eastern Siberia—22° south of its present position. This, with possibly other factors, has in turn resulted in subtropical climates as far north as the 60th parallel, which is close to Anchorage, Alaska, the southern tip of Greenland, and Oslo, Norway. The gradual shift in the orientation of the earth's axis brought the North Pole to its present position about 10 million years ago, with no perceptible change since.

The last million years has been characterized by four or five periods of extensive glaciation in the Northern Hemisphere. Each glaciation lasted for about 100,000 years, during which time the ice accumulated for about 90,000 years, followed by comparatively rapid melting over a 10,000-year period. During the ice ages, the average thickness of the ice sheet was about 1,200 m, with a maximum of 4,000 m. In the United States, the glaciers penetrated as far south as Kentucky bulldozing enormous volumes of soil, gravel, and rock. This debris came to rest in the terminal moraines which today mark the farthest penetration of the glaciers before they began to recede. Many rolling hills in New York, New Jersey, and other Middle Atlantic states, are the monuments to the farthest penetration of the glaciers. So much of the earth's water was then stored in the huge ice domes that existed during the last ice age (about 20,000 years ago) that the ocean level was lowered by an estimated 100 m.

These periods of glaciation suggest the existence of very long-term oscillations in the earth's climate. Based on studies of climate since the last glaciation, it appears that shorter-term oscillations exist on a time scale in the range of 100 to 1,000 years. The 400-year period, from 1300 to 1700 A.D., was characterized by a cooling trend of such severity in northern Europe as to result in it being called the "little ice age." That period was preceded by 400 years of warm weather, from 800 to 1200 A.D., which permitted the establishment of Viking settlements in Iceland and Greenland. These cycles of climatic change occur on such a long time scale as to escape observation in the period of a single lifetime.

For the past 100 years, it has been possible to compile average temperatures on a global basis, and these data indicate a slow rise at a rate of about .008°C per year until 1945. Since then the global temperature has been dropping at a somewhat more rapid rate.[4] The reasons for these periodic fluctuations in the climate of the past are not known. However, there are two ways by which human beings might affect climate in the future: by producing carbon dioxide to such an extent as to increase the mean global temperature; and by discharging dust into the global upper atmosphere and thereby reducing the mean temperature.

The earth is constantly receiving energy from the sun, but thermal equilibrium is maintained by energy returned to outer space in the form of infrared radiation. One of the properties of carbon dioxide is that, although it readily passes visible light, it absorbs infrared light. The process of reradiating the energy received from the sun can thus be gradually impeded by increasing amounts of atmospheric carbon dioxide. This would result in a gradual temperature rise. In this respect the carbon dioxide acts like the glass in a greenhouse, and it is for this reason that the effect has been called the "greenhouse effect." We saw earlier that this phenomenon has been used to capture solar energy for heating buildings.

The higher temperature would increase the amount of moisture in the atmosphere and, since water vapor also absorbs infrared radiation, further heating would occur. However, the increased moisture might result in greater cloud cover and, because clouds play a major role in reflecting solar radiation, increased cloud cover could increase the earth's albedo and lower the surface temperature. This example serves to illustrate the complexities of the various actions, reactions, positive feedback, and negative feedback that make climatology a complex science.

The concentration of carbon dioxide in the atmosphere was a little less than 300 parts per million in 1860, but it has been increasing ever since. It has been estimated that the average surface temperature of the earth would increase by 2°C if the concentration of carbon dioxide were to double. World-wide production of carbon dioxide in 1860 was about 100 million metric tons, but this amount has been increasing exponentially, due primarily to the combustion of fossil fuels.

Carbon-dioxide emissions are currently estimated to be about 5 billion metric tons per year and, based on an annual increase of 4.3 percent in fossil-fuel consumption, would reach about 14 billion metric tons in 2000 and 42 billion metric tons in 2025.[5] Since 1860, the concentration of atmospheric carbon dioxide has increased about 10 percent to 330 parts per million, and it has been predicted that if we continue to increase the use of fossil fuel, the concentration of carbon dioxide could be increased seven to eight times the amount that existed in 1860. This might take place between 2075 and 2275.

Whereas the carbon dioxide produced by human activity tends to warm the atmosphere, dust has a cooling effect.[3] The concentration of atmospheric dust is far more variable than that of carbon dioxide. Carbon dioxide is relatively inert chemically and tends to distribute itself throughout the atmosphere of both hemispheres. There is ample time for it to reach an equilibrium. However, the mean residence time of a particle of dust in the lower atmosphere is only a week or two, and since the time required for lateral atmospheric mixing is much longer than this, the dust content of the atmosphere tends to be higher in latitudes in which the dust is produced. However, when dusts or other forms of aerosols are introduced into the stratosphere, they take much longer to be removed. Based on studies of radioactive debris introduced into the stratosphere by the testing of nuclear weapons, it has been determined that the residence time of dust can be as long as several years, depending on the height to which it is projected.

There are many natural sources of atmospheric dust, including salt from sea spray, windblown soil (particularly in arid regions), smoke from forest fires, and volcanic debris. It is estimated that in an average year, as many as 2 billion tons of dust can be produced by these mechanisms. Manmade emissions are estimated to range up to .4 billion tons per year.[4] Volcanic eruptions of unusual violence can inject so much dust into the stratosphere that solar radiation will be attenuated before it reaches the lower atmosphere. The 1912 volcanic eruption of Mt. Katmai caused a 20 percent reduction in the intensity of solar radiation throughout Europe. During the mid-1600s and again in the early part of the nineteenth century, there were extended periods of volcanic activity which seem to have been associated with cooling trends.

It is estimated that by the end of this century human activity may increase the particulate content of the stratosphere by 50 percent during periods of low volcanic activity.[4] It is not thought that this would be important so far as the radiation balance of the atmosphere is concerned and, in any case, the contribution of human origin would be overwhelmed from time to time by intrusions of volcanic dust.

Studies of solar radiation made at the Mauna Loa Observatory in Hawaii indicate that no discernible change due to human activities occurred from 1958 to 1970. However, the 1963 eruption of Mt. Agung in Bali resulted in a sudden drop in solar radiation, which did not return to normal levels for six years. The relatively long recovery time indicates that the dust from this eruption was projected high into the stratosphere.[6]

It is not possible at the present time to state unequivocally that human activity can provoke changes in the global climate. However, there is reason to be concerned with the fact that we do not as yet have a quantitative understanding of how the major geophysical parameters interact with each other to cause climatic changes. We do not know why the periods of glaciation occurred, or why they ended. However, we do know that human activity has the capacity to affect the global heat balance. Since there is evidence that climate is unstable and has, in fact, been subject to enormous fluctuations in the past, the prospect that human activity might trigger a major climatic change must be considered very seriously. The present situation was referred to succinctly in the *Study of Man's Impact on Climate*, a report sponsored by the Massachusetts Institute of Technology:

> We have a conviction that mankind *can* influence the climate, especially if he proceeds at the present accelerating pace. We hope that the rate of progress of our understanding can match the growing urgency of taking action before some devastating forces are set in motion—forces that we may be powerless to reverse.[4]

If a future ice age should follow a pattern similar to those of the past, it would develop very slowly, over a period of several tens of thousands of years, so that the changes would be imperceptible within any one generation. When the last glaciation occurred, the world's population was probably no more than a few tens of millions of people. This was during the hunting and gathering phase of human development, and when the weather began to deteriorate, the populations were able to drift gradually toward warmer climates without really being aware of what was happening. However, the world's population is now so large and so heavily concentrated in the temperate zones that mass migration is no longer possible. The climate, far in advance of the return of glaciers, would be so cool as to make agriculture impossible. In the event of an actual glacial age, many of the major cities of the world could be scoured from the face of the earth and transported as rubble for hundreds of miles, in some cases to end up as a contribution to a new line of terminal moraines.

If it is possible that human activity can inadvertently trigger an ice age, it is also possible that in time we will understand the mechanisms by which climatic alteration takes place. Humans might then have within their control the means to abort the onset of future glaciations.

On a much smaller scale, scientists have already spent two decades studying the techniques by which local weather can be modified. The economic incentives for this are substantial: Between 1965 and 1967, hurricane damage alone averaged about $500,000,000 per year in the United States. Agricultural losses due to hail averaged about $300,000,000 per year, and the annual loss to airlines due to fog has been estimated at $75,000,000 per year.[7] All of this is in addition to the even greater losses from crop failures which were brought about because the temperature was too high or too low, or the rainfall too much or too little.

Techniques already exist by which it is possible to increase precipitation and temper the winds of hurricanes. However, even on this relatively small scale, there are many unknowns, and attempts to modify weather have resulted not only in uncertainty as to whether or not the desired effect was produced, but also as to whether the resultant side effects were desirable or undesirable. There have already been a number of lawsuits involving weather modification. In none of these cases were financial damages assessed against those involved in weather modification, but in some cases restraining orders were issued against further attempts at modification.[7]

Another major long-range concern of geophysicists is whether the heat resulting from the production and use of energy is capable of global climatic modification. At the present time, the solar energy absorbed at the surface of the earth is about 10,000 times greater than the amount of heat released by the operation of power plants, factories, automobiles, home heating units, and all other heat-producing human activities. However, there are economists who believe that if sufficient energy could be made available, the earth could support a population of 20 billion or more people, each person consuming four or more times the amount of energy now being consumed in the United States of 10 kilowatts.[8] A global society that uses energy to such an extent could alter the thermodynamics of earth's atmosphere and possibly have profound and unpredictable effects on the world's climate, a matter that will deserve careful study in the decades ahead. We can only hope that the required studies are undertaken in time for the information to become available as needed. This may or may not be a real problem: The necessary research must be undertaken.

Small-scale weather effects are known to take place in urban, and near-urban, and industrialized areas.[1] Retention of heat due to the relatively low albedo and the large capacity of built-up surfaces to absorb heat causes cities to be warmer than the countryside. The particulates introduced to the atmosphere by combustion processes, together with the considerable amount of moisture also produced, tend to increase cloudiness and possibly cause greater-than-normal precipitation. (The haze due to pollution also reduces visibility to a marked degree.)

For the time being, these local effects do not affect regional weather patterns. However, with the increase of urban areas, such as the gradual coalescence of built-up areas between Boston and Washington in the United States, the climatic effects of urbanization may produce effects that cannot now be foreseen.

Although the whole subject of climate modification seems speculative at this time, certain conclusions seem to be in order. First, it appears reasonable to assume that unrestricted discharges of carbon dioxide from the combustion of fossil fuels *could*, in time, alter the heat balance to such an extent that gradual climatic changes would be brought about on a cataclysmic scale. While it is not possible to say unequivocally that this can happen, the consequences of climatic modification are very severe. Therefore, we must improve our understanding of atmospheric geophysics to the extent required to understand the relationships between pollution and climate. We are dealing here with a very long-range problem and a major objective of international scientific collaboration should be the development of the science of climatology to the extent that a quantitative understanding of the consequences of our actions can be achieved.

Second, while it may be speculative whether man can affect climate, it is clear that climate can affect man, and in disastrous ways. The prospect that an extended period of severe and widespread drought might occur, or that the temperature might decrease or increase by a few degrees, is sufficiently probable that the human race should undertake climatological studies at a greater level of effort. As a result, with international collaboration, it may become possible eventually to stabilize climatic changes with full confidence that undesirable side effects will not occur. Compared to research budgets in other fields, measured in billions of dollars per year, the research expenditure for weather control is as yet miniscule.

STRATOSPHERIC OZONE DEPLETION

Although solar radiation makes life on earth possible, there is a region of the ultraviolet spectrum that is known to cause human skin cancer and to be harmful to vegetation. The potentially harmful radiation impinges on the top of the earth's atmosphere, but is absorbed by the gas ozone, which is normally present in the stratosphere.

The subtle relationships between chemical emissions to the atmosphere and geophysical processes are illustrated by the possibility that using such diverse inventions as the supersonic transport (SST), spray cans in the home, and nitrate fertilizers on the farm may eventually decrease the ozone content of the stratosphere. Because of the role of ozone in absorbing ultraviolet radiation, its decrease could allow an increase of the intensity of ultraviolet light at the earth's surface. This in turn, can possibly cause a higher incidence of skin cancer in humans.

Ozone may also play another important role. It is believed that the temperature of the stratosphere is controlled by the extent to which ozone absorbs solar radiation. A change in the stratospheric content of ozone could possibly affect the stratospheric temperature and, in ways that are not fully understood, could result in small but significant changes in the temperature of the earth's surface. There might also be small but significant effects in rainfall.[9]

The possibility that the ozone content of the stratosphere can be reduced by human activity was first raised in connection with the proposal that the United States government support the development of supersonic aircraft. Subsonic commercial and military aircraft have been flying in the lower stratosphere (up to about 40,000 feet) and supersonic aircraft operated by European countries are cruising at up to about 54,000 feet. It is possible that in the future, routine commercial transport could extend up to about 100,000 feet.[9] If this were to be the case, the stratospheric ozone content would be affected by chemical reactions with the nitrogen oxides in the jet-engine exhaust gases. It is estimated that ozone reduction could amount to 3 to 6 percent per hundred aircraft of the projected supersonic types. On the assumption that 300 or 400 SSTs would operate throughout the world, a 10 percent decrease in stratospheric ozone would result from this fleet alone, causing an estimated 20 percent increase in skin cancer.

The fluorocarbons, which have been identified as another possible agent for ozone depletion, are known widely under the trade name "Freons." These compounds contain carbon, fluorine, and chlorine in many combinations, and possess properties of chemical inertness and low toxicity that have encouraged their use for many purposes.

For several decades the Freons have been the principal gas used in home refrigerators and air-conditioning systems. More recently, they have been widely used as the propellant gas in aerosol cans that dispense insecticides, hair sprays, antiperspirants, and deodorants.

These uses have been increasing rapidly, and it is estimated that 3 billion aerosol cans containing 1.7 billion pounds of Freon were sold in 1973.[10]

Because of the chemical and biological inertness of the fluorocarbons, and due to their low solubility in seawater, they literally have no place to go, except to remain in the atmosphere. In the lower atmosphere (the troposphere) they are capable of doing no harm, but these compounds gradually diffuse to the stratosphere, where conditions are favorable for their dissociation by photochemical processes. It is postulated that the chlorine thus released reacts with, and reduces, the concentration of stratospheric ozone. Ozone, as was mentioned above, plays an important role in absorbing the harmful ultraviolet component of sunlight. As a consequence, the intensity of the ultraviolet radiation would increase at the earth's surface.

The possibility that stratospheric ozone is reduced from the use of nitrate fertilizers was first proposed in 1971.[11] [10] It has been postulated that when these fertilizers are applied, some of the nitrogen is released from the soil as nitrous oxide. When this gas diffuses to the stratosphere, it becomes involved in chemical reactions that reduce the concentration of ozone.

The argument that an increase in ultraviolet intensity would increase the incidence of human skin cancer is based on strong epidemiological evidence.[14] Skin cancer is relatively prevalent among people who work out-of-doors, such as sailors and farmers. Also, skin cancers usually develop on the exposed portions of the body. Finally, the incidence of skin cancer is higher among those who live at lower latitudes, where the intensity of sunlight is highest. Skin cancers are relatively rare in individuals with heavily pigmented skin, but is common among fair-skinned people.

It is estimated that there are about 300,000 cases of skin cancers per year in the United States, with a fatality rate of about 1 percent. The common types of skin cancers (basal cell carcinomas) produced by sunlight are far less life-threatening than melanomas, a form of skin cancer that, fortunately, occurs far less frequently, but which is associated with high mortality.

It has been estimated, based on purely theoretical considerations, that the fluorocarbon that has been released to the atmosphere in the past may have already reduced the stratospheric ozone content by between .5 and 1 percent, and that even if no more Freon is released, the ozone in the stratosphere would continue to diminish for about a decade, by which time as much as 3 percent of it will have been removed. Recovery to normal values would be a slow process that could take a century or more. If fluorocarbons are released into the atmosphere at the same rate they were in 1973, it is estimated that the reduction of ozone could be about 7 percent.[2,10]

Studies of the incidence of skin cancer in relation to variations in latitude suggest that about 6,000 new cases of skin cancer, other than melanoma, would occur among fair-skinned individuals in the United States for each 1 percent reduction in the ozone concentration. A 7 percent reduction in ozone could thus result in an additional 42,000 cases of skin cancer in contrast to the 300,000 cases that have developed every year in the past.

It was not until 1974 that the consequences of the reduction of ozone from the use of fluorocarbons were first discussed in scientific literature.[15] Many thoughtful scientists have considered this problem, and although there is no unanimity of opinion, the U.S. Environmental Protection Agency in 1977 decided, quite properly, to ban the use of fluorocarbons for many purposes. However, the matter is by no means closed, since it requires action on an international scale, and there is no mechanism as yet for obtaining the required cooperation of other countries. The example of fluorocarbon serves to illustrate the subtlety with which chemical innovations can have an impact on human well-being.

Concern about human interference with atmospheric processes has developed very recently. Although Tyndall and Arrhenius independently speculated about the possible effects that a blanket of carbon dioxide would cause in the last century, the matter did not receive serious attention until after World War II.[16] The potential dangers of stratospheric ozone depletion were not recognized until about ten years ago, and the possible effects of the Freons were not discussed until 1974. In addition, a more recent publication suggests that krypton-85, a radioactive noble gas released by the nuclear fuel cycle, may influence atmospheric electrical processes which, in turn, could cause weather changes.[17] It seems clear that the impact of atmospheric pollution on atmospheric phenomena will receive increasing and imaginative attention.

NOTES

1. Landsberg, Helmut E. "Man-Made Climatic Changes," *Science* 170:1265–1274 (1970).
2. Schneider, S. H. "The Genesis Strategy: Climate and Global Survival," Plenum Press, New York/London (1976).
3. National Academy of Sciences. "Climate and Food," a Report of the

Committee on Climate and Weather Fluctuations and Agricultural Production, Washington, D.C. (1976).

4. Study of Man's Impact on Climate. "Inadvertent Climate Modification," MIT Press, Cambridge, Mass. (1971).
5. Baes, C. F., H. E. Goeller, J. S. Olson and R. M. Rotty. "The Global Carbon Dioxide Problem," *ORNL-5194*, National Technical Information Service, Springfield, Va. (August, 1976).
6. Ellis, H. T. and R. F. Pueschel. "Solar Radiation: Absence of Air Pollution Trends at Mauna Loa," *Science* 172:845–846 (1971).
7. Fleagle, R. G., J. A. Crutchfield, R. W. Johnson and M. F. Abdo. "Weather Modification in the Public Interest," Univ. of Washington Press, Seattle, Wash. (1974).
8. Kellogg, W. W. and S. H. Schneider. "Climate Stabilization: For Better or for Worse?" *Science* 186:1163–1172 (1974).
9. National Academy of Sciences. "Environmental Impact of Stratospheric Flight: Biological and Climatic Effects of Aircraft Emissions in the Stratosphere," Climatic Impact Committee, NRC-NAS-NAE, Washington, D.C. (1975).
10. Council on Environmental Quality. "Fluorocarbons and the Environment: Report of Federal Task Force on Inadvertent Modification of the Stratosphere (IMOS)," Federal Council for Science and Technology, U.S. Government Printing Office, Washington, D.C. (June, 1975).
11. Crutzen, P. J. *J. Geophys. Res.* 96:7311 (1971).
12. McElroy, M. B. and J. C. McConnell. *J. Atmos. Sci.* 28:1095 (1971).
13. Nicolet, M. and E. Vergison. *Aeronom. Acta* 90:1 (1971).
14. U.S. Department of Health, Education and Welfare. "Measurements of Ultraviolet Radiation in the United States and Comparisons with Skin Cancer Data," *DHEW No. (NIH)76-1029*, U.S. Government Printing Office, Washington, D.C. (November, 1975).
15. Crutzen, Paul. "A Review of Upper Atmospheric Photochemistry," *Can. J. Chem.* 52:1569–1581 (1974).
16. Plass, G. N. "The Carbon Dioxide Theory of Climatic Change," *Tellus* VIII (2):140–154 (1956).
17. Boeck, W. L. "Meteorological Consequences of Atmospheric Krypton-85," *Science* 193:195–198 (1976).

PART IV

Where Are We, and Where Are We Going?

CHAPTER 15

The Environmental Movement and Human Health: An Overall Appraisal

The modern environmental movement has achieved much success but has also suffered many failures. On the positive side is the fact that unprecedented support for environmental protection has developed: the public now understands that its resources in land, fuels, raw materials, air, and water are finite and must be protected and conserved. A series of tough laws has defined the federal government's environmental objectives and reorganized the government agencies to facilitate achievement of ambitious goals. In this process, the Congress created the Environmental Protection Agency to consolidate various functions formerly scattered throughout many government agencies. A new post of Assistant Secretary of Labor for Occupational Health and Safety was created, and the National Institute of Occupational Health and Safety was established within HEW to serve as the research arm of the Department of Labor. A major new institute, the National Institute for Environmental Health Sciences, has also been established. Considering also that the private and public sectors have demonstrated their willingness to spend huge sums of money to achieve the objectives of the new laws, there was every reason to believe, by the late 1960s, that environmental protection was assured.

Progress has indeed been made, but there have also been failures. Priorities have become disordered, and many illogical decisions have been made, all of which has resulted in a considerable amount of disheartenment among many who at first welcomed the environmental movement. This has been particularly true among scientists, physicians, and engineers who were already working in the field of environmental health. Implementation of the complex environmental

351

protection laws that were passed in the late 1960s and early 1970s should have utilized the scientific and technical talent that already existed in the federal and state governments. Instead, much of what existed was destroyed.

DISORGANIZATION THROUGH REORGANIZATION

When public interest in the environment developed in the mid-1960s, there were complaints that the Public Health Service (PHS), the Federal agency that was the most responsible for air and water pollution control, had not done its job properly. Progress towards clean air and clean water was certainly slow, but the fault was not with the personnel of the PHS, nor with their counterparts in the state and local governments. The staffs of the existing agencies knew what needed to be done, and they knew how to do the job, but they were not given the means. The legal authority to initiate new environmental-health programs resided with the states, and the PHS could take little action unless it was requested to do so. Moreover, the funds made available to PHS were grossly insufficient.

Because of his dissatisfaction with the performance of the PHS, Senator Edwin Muskie of Maine introduced a bill in 1963 that, among other provisions, was to transfer federal authority for water-pollution control to a new organization, the Federal Water Pollution Control Administration (FWPCA), to be established within the Department of Health, Education and Welfare (HEW). The PHS was then more than a century old, and its staff enjoyed a fine reputation, mainly because of its accomplishments in the field of communicable disease control. However, the basic talent needed to meet the new challenges existed within PHS despite the fact that it was badly financed and lacked the legal authority to take the required initiatives. When the decision was made to transfer these career officers to another organization within HEW, many sought other assignments in the PHS, and others took early retirement. As a result, no less than 50 percent of the seasoned staff of water specialists were lost with the signing of the Muskie legislation by the President in October, 1965. On that occasion, President Johnson stated: "Today we proclaim our refusal to be strangled by the wastes of civilization. Today we begin to be master of our environment."[1] It was ironic that the first major step in the process of environmental rehabilitation, the transfer of the water-pollution control function from the PHS, resulted in loss of so many of the badly needed staff.

During this period, the Department of Interior was headed by a secretary who had identified himself with the environmental movement and who campaigned actively for air- and water-pollution control to be transferred to his department. As a result, the newly created FWPCA was transferred from HEW to the Department of Interior only a few months after its removal from PHS. This action once again resulted in a loss of key staff.

The reorganizations also affected the Public Health Service's National Center for Air Pollution Control, which was transferred into a new Consumer and Environmental Health Service within HEW, a step that once again resulted in loss of many career Public Health Service officers.

A few years later, on January 1, 1971, these two important organizations, responsible for air and water pollution, were once again transferred, this time to the newly created Environmental Protection Agency. The physicians, scientists, and engineers of the federal government, who had fought for years for clean water and clean air, but who for decades had been neglected by the Congress and the agency chiefs in the executive branch, were thrown repeatedly into confusion by these successive reorganizations.

By the late 1960s the governmental responsibilities for air- and water-pollution control were so greatly expanded that large numbers of additional personnel were required. The relatively few well-trained highly disciplined professional environmentalists who were then in government were soon outnumbered by scientists, lawyers, and public-spirited citizens who had little training or experience in the environmental field, and who were impatient with what they perceived to be the stodgy and conservative attitude of the incumbents. There was insufficient consideration given to the fact that many of the environmental problems involved a combination of technical, administrative, and legal complexities that argued against hasty decisions. Control over the environmental agencies passed rapidly from the hands of many officials with long experience in the field to those who often had neither the experience nor the training.

The exodus from government was precipitated not only by confusion within the agencies, but also by new opportunities that were developing outside government. Most of the environmental scientists who left government were quickly relocated in rewarding positions in academe or industry. This, of course, could have helped the national environmental effort, except for the fact that the governmental agencies had become depleted of seasoned professionals far beyond the point where the agencies could continue to be effective.

THE CONFUSION BETWEEN HUMAN HEALTH AND ECOLOGY

One of the effects of the reorganizations was to remove responsibility for environmental health from health agencies. At the federal level, responsibility passed from PHS and other agencies within HEW to the EPA, and in New York State the units responsible for air- and water-pollution control were moved from the Health Department to the Department of Environmental Conservation. Similar reorganizations were made in other states. It was illogical that responsibility for environmental health should have been allowed to transfer from the health agencies to governmental units that were in many cases oriented towards protection of wildlife.

It has not always been a simple matter to distinguish between environmental issues that involve public health and those that do not. There are some clear-cut exceptions, as the case of the snail darter, a three-inch perch that achieved national prominence in 1977 when a U.S. Court of Appeals ordered a halt to the construction of a dam being built by the Tennessee Valley Authority.[2] This action was taken because it was discovered that the dam would destroy the only habitat of this species of fish, of which it was estimated there were no more than 15,000 in existence. The dam was more than 90 percent complete when construction was halted, and if the decision is not eventually reversed, it will cost about $120 million to protect that small fish population.

Purists will argue that in nature, everything is connected to everything else, and that the well-being of human beings will suffer in some subtle way whenever natural systems are affected adversely. Nevertheless, even most ecologists would be willing to argue that the snail darter decision carried things too far.

Until the mid-1960s, there was little connection between the "environmentalists" concerned with protecting wildlife, and those concerned with the public health. Relatively few biologists called themselves ecologists, and those who did were concerned with basic biological processes such as the flow of energy through ecosystems, physiological adaptation to the environment, and population dynamics. The Ecological Society of America had long published an excellent journal in which the papers were concerned with ecology as a basic science, but environmental toxicology and problems of environmental protection were rarely if ever included. Biologists employed by the wildlife services of the federal or state governments concerned themselves with management of fish and game and functioned as applied ecologists, although they may not have identified themselves as such. In addition, large organizations of concerned citizens, such as the Sierra Club, the Audubon Society, and the National Wildlife Federation, carried on valuable educational activities that were limited largely to land management and the preservation of wildlife.

In contrast, the subject of environmental health was traditionally the professional domain of physicians, engineers, and scientists employed mainly by the U.S. Public Health Service, state or local health departments, and universities. The health professionals had little contact with the wildlife specialists in the past. This was unfortunate, because the interests of the two groups often do overlap. For example, if a chemical proves to be injurious to wildlife, it may also be damaging to human health.

In recent years, classical ecologists and wildlife protectionists have often introduced questions of public health into environmental debates. The DDT hearings were such an example. Although control of the insecticide could probably have been justified by its effects on wildlife alone, the question of the carcinogenicity of DDT was introduced at the hearing, was greatly emphasized, and was ultimately cited as a major reason for the ban that was ordered by the EPA. (See Chapter 9, page 233.) Although there were exceptions, the main thrust of the argument that DDT was a public health hazard did not come from the health specialists. As noted by others, the pollution movement was driven by a loose amalgam of pro-environmentalists, conservationists, and consumers who had little alliance with the health professions.[3] This was unfortunate, because most of the more drastic and expensive of the pollution laws were designed to protect the public health. In many of these debates, the professional ecologists have received strong support from popular organizations originally concerned with wildlife protection, the burgeoning membership of which have become concerned with the health aspects of food additives, agricultural chemicals, nuclear energy, and air and water pollution. The publications of these organizations are widely distributed and receive far more publicity than the journals published by the health scientists, who have failed to muster the kind of public support received by the conservationists.

THE PROBLEM OF MISINFORMATION

The early years of the new environmental movement have been characterized by a widespread popular apprehension. The words "environmental crisis" have permeated the popular vocabulary and have been perceived by many to mean literally that human existence is being jeopardized. That this sense of crisis should have developed to such a marked degree has reflected the mood of the popular literature. Books, such as *Vanishing Air, Silent Spring, Famine, 1975!, The Population Bomb, Water Wasteland,* and others with equally sombre titles, found nothing reassuring about the state of the twentieth-century environment. Much of the world has been inundated by wave after wave of acute concern about pollution problems, including lead, mercury, DDT, PCBs, carbon monoxide, photochemical smog, nuclear wastes, and sulfur dioxide. Unfortunately, the information about these problems has not been presented to the public in a balanced manner.

We have seen from earlier chapters that most of the issues of environmental health are technically complex, and frequently involve difficult political and legal ramifications. The evaluation of the risk from a pollutant may depend on an interpretation of ambiguous laboratory or epidemiological data. More often than not, the evaluation of risk requires an understanding of the tradeoffs that must be made. For example, a food additive may have been shown to be capable of producing cancers in laboratory animals when administered in large amounts. However, it may not produce cancer in humans who ingest small amounts of it. From the point of view of the public interest, it may also be wise to accept a small risk that a few people may develop cancer in exchange for substantial benefits to the population as a whole.

To evaluate problems such as these requires a willingness to examine a great volume of scientific literature, the discipline necessary to assure that the literature is being reviewed in an objective manner, and experience in the rigorous analysis of a complex set of information. Unfortunately, only a small fraction of the population has either the time or the training required to arrive at a competent and independent assessment of such issues. For this reason most people obtain their information about these subjects from newspapers, television, magazines, and popular books. This, of course, is true not only of information about the environment, but of other matters of public policy as well.

Regrettably, the media has not always exercised a degree of self-discipline commensurate with its enormous responsibility. In general, much of the media reporting is of a high quality, but this has not been true of its coverage of environmental issues. A serious impediment to objective reporting is that reporters prefer to work with scientists who speak in readily comprehended, albeit unqualified language. It is confusing to be presented with a series of equivocations. A scientist who speaks to reporters in somewhat catchy, unequivocal phrases is much more likely to attract them than one who speaks in the more disciplined language of the laboratory.

The public does not understand the need for scientists to equivocate. The image of the scientist is that of a person who has the ability to sort truths from untruths unerringly. To the contrary, there is a large area of middle ground where the scientist must employ judgment and intuition in coming to a conclusion. Two scientists will often come to different conclusions when presented with the same set of data. An objective scientist will often refuse to take a clear-cut position in response to a question.

Scientists who are trained to think in a highly disciplined manner within their areas of specialization may not apply the same standards of intellectual rigor when they become involved with matters outside their specialties. This is also not understood by the public. When scientists leave the laboratory bench, they are subject to the same emotions, prejudices, and frailties that characterize human beings generally. The public does not understand that a scientist who speaks on a subject outside his area of expertise is speaking as a layman.

The newspapers, radio, and television stations comprise one of the largest industries in the United States, and they should develop the standards of self-discipline that they expect of other industries. The work of individual reporters and TV program directors should be reviewed periodically either by their peers or panels of outside experts. The work should be graded for accuracy of the facts presented, and the findings should be published prominently. This would give the media free rein to interpret the facts, but not to misstate them.

POLITICIZATION AND POLARIZATION

The high degree of polarization of views about environmental matters and the extremes of advocacy that exist in the environmental field are everywhere in evidence. Polarization has developed hand in hand with politicization, and together they have often prevented programs of environmental protection from evolving along rational lines.

Polarization exists because of many factors. The environmental movement has produced a major revolution in the way our economy operates. In any other period, the changes would have been brought about gradually by a program of well-ordered priorities based on careful analysis of the environmental needs. Unfortunately, the tendency of the press and the more strident environmentalists to present the sensational sides of the environmental issues resulted in an exaggerated sense of imminent catastrophe, and public-opinion polls taken during the late 1960s began to report that people were placing "environment" on the top of their list of public issues. This was hard to understand when one considers that the polls were taken during the Vietnam war, at a time when racial violence was revealing deep-rooted problems that had to be faced, and when physical deterioration of many cities was already well underway. The public suddenly began to believe that pollutants such as sulfur dioxide, carbon monoxide, mercury, lead, and asbestos were new problems. It was not explained that many of the problems had existed for a long time and were being brought under control. Most assuredly, new pollution problems were appearing on the scene, and far greater efforts were needed to deal with them, but a calm approach along traditional lines was indicated. In the agitated atmosphere that existed, anyone who spoke reassuringly, to even the slightest degree, was regarded as an opponent of the environmental crusade. There was a clamor for immediate action and important, expensive decisions were made in an atmosphere of urgency that in many cases was not justified by the facts.

It was only natural that scientists, who were known to disagree with a particular position taken by a regulatory agency, should in some cases be retained as advisors to industry. Industry has always needed the same kind of help that government has traditionally obtained from university scientists or other knowledgeable persons. Polarization began to develop because scientists who disagreed with the environmentalists on an issue often found themselves in disfavor with both the environmentalists and the government agencies. Organizations such as the National Academy of Sciences suddenly found that the objectivity of their advisory committees were being questioned because some of the members had previously served as consultants to industry. Once begun, this process of polarization was not easily stopped.

It was also natural that many politicians (some elected and some appointed to public office) should have sided with the environmental activists. Some did so because they were genuinely concerned about the environment, and believed what the activists were telling them. Others joined with them simply out of political opportunism. For whatever reason, the popular environmental movement spread to the political ranks.

It has been stated frequently in the preceding pages that the major laws passed by the Congress during the late 1960s and early 1970s were important forward steps that, if anything were long overdue. However, as a result of popular pressure on the Congress, many laws were passed that required too much too soon. Any sincere suggestion that a particular deadline should be deferred, or that a proposed amendment should be eliminated because the costs did not seem justified by the public-health benefits, was interpreted as being "anti-environmental" by badly informed citizen groups. Environmental extremists have been unrelenting in the pressure they have placed on industry for neglect of their environmental responsibilities, and on legislative bodies and officials of executive agencies because they are too protective of industry.

The extremists have adopted some of the methods of muckrakers. This is unfortunate, because the extremists have made it possible to achieve what could not have been accomplished by those of more moderate temperament. But the time has come for a more deliberate approach to environmental problems. One cannot help but remember Theodore Roosevelt's aphorism: "The men with muck rakes are often indispensable to the well being of society; but only if they know when to stop raking the muck."

THE NEED FOR REALIGNMENT OF ENVIRONMENTAL HEALTH PRIORITIES

The costs for air-pollution control in the decade between 1975 and 1984 are estimated at about $137 billion. The cost of water-pollution control is expected to be about $110 billion, and the cost of all-pollution control required by federal environmental legislation beyond those that would have been made in the absence of such legislation is expected to be nearly $258 billion.[4] The bulk of these expenditures will be made to protect the public health.

The large sums required for air-pollution control are largely for desulfurization of fuels, installation of power-plant flue-gas scrubbers, and control of automobile emissions. The review of these subjects in Chapters 11 and 12 gives little reason to believe that the health benefits from either program will result in health benefits commensurate with the costs. When expenditures are made on so huge a scale, the health effects should be readily identifiable.

One example of the extent to which money has been wasted in the field of water-pollution control is illustrated by a sewage-treatment plant being built in New York City on the west side of Manhattan. A well-operated modern secondary sewage treatment plant can remove more than 90 percent of the biochemical oxygen demand (BOD).* The

* Modern sewage-treatment plants provide primary and secondary treatment. The primary treatment removes about 50 percent of the organic solids by screening, flotation, or sedimentation. The remaining organics, when discharged as finely divided solids or in solution, impose a BOD on the

required degree of BOD removal depends on the characteristics of the receiving water, which at the proposed plant site (the west side of Manhattan) were such that the elimination of 67 percent of the BOD would be sufficient. The plant was designed accordingly, and construction was scheduled to begin in the mid-1960s at an estimated cost of $250,000,000.

The environmental movement was getting underway at this point, a development that coincided with the Congressional election of 1966. When it came to the attention of certain candidates that the plant was being built to eliminate 67 percent of the BOD only, politicians seeking election began to press for higher efficiency on the grounds that New York City should have the best plant that technology can offer. As a result, the plant was redesigned to eliminate 90 percent of the demand, despite the fact that competent technical advisors concluded that a capacity to remove 67 percent of the demand would be quite sufficient and that to go from 67 percent to 90 percent would be very costly because of the exponential relationships between the capacity and plant costs. Nevertheless, the plant was redesigned, and construction began in 1970 at an estimated cost of about $1 billion! The additional several hundred million dollars that will be thus spent on this plant will have no perceptible benefit, either on the local ecology or on the public health.

The kind of money being invested to achieve some of the contemporary environmental objectives could be better spent elsewhere. Thus, expenditure of public funds to improve substandard housing has virtually ceased. Funds have been curtailed for neighborhood health centers that could assist disadvantaged people to deal with their special environmental problems. The lead-poisoning problem in dilapidated tenements could be eliminated for all time everywhere in the United States for a fraction of the money being put into over-designed facilities for water-pollution control. Above all, we could spend money on health education that would achieve dramatic results in the next generation.

THE NEGLECTED SOCIO-BEHAVIORAL ENVIRONMENT

The human environment includes social and behavioral factors that have an important influence on health. Yet these factors have not

---

receiving body of water. The demand can cause the water to become oxygen deficient, which would have a damaging effect on aquatic life. In the secondary treatment system, the effluent from the primary process is aerated for a sufficient length of time to oxidize much of the remaining organic material before it is discharged.

been sufficiently emphasized in the contemporary environmental movement.

Cigarette smoke is the most noxious of all the pollutants to which humans are exposed in the United States. The government has taken some steps to reduce the prevalence of cigarette smoking, such as by banning advertising in the electronic media and by use of warning labels on each pack of cigarettes. In addition, local governments are, to an increasing degree, placing restrictions on smoking in public places. However, government also encourages the use of tobacco: price supports per pound of tobacco increased 35 percent from 1964 to 1974, and the federal tax on tobacco has not changed since 1952, despite a substantial increase in its market value. It is estimated that in 1973 the national bill for health care costs resulting from cigarette smoking was 11.5 billion dollars.[5] The enormous power of the environmental movement should be directed towards creation of massive educational programs beginning with preschool children, so that the harmful effects of cigarette smoking can be understood by children during the formative years. It would seem logical that we should spend at least as much money advertising the dangers of smoking as is spent by the tobacco companies in the printed media. Perhaps advertising tobacco products in the printed media should be banned. The main point here is that a major environmental problem has not attracted sufficient attention from the environmentalists.

Another major environmental problem that has been neglected is the condition of the cities. They are functionally inefficient, are physically deteriorating, and are suffering the malignant effects of the conditions under which millions of disadvantaged people must live. The efforts of environmentalists to improve the urban environment have not considered all relevant factors. For example, traffic congestion is not being attacked because it results in multilating accidents, is economically wasteful, and subjects city dwellers to high noise levels. Instead, people have been made to fear automobile traffic because of the air pollution it causes. Solution of the traffic problem can be far better justified on other grounds.

The fact that the environmental problems of the impoverished have not attracted sufficient attention from organized environmentalists is shortsighted, for reasons of self-interest. Poverty impacts in many ways on the fortunate as well as on the disadvantaged. Poverty in modern times, as all through history, breeds deviant behavior, and the crime associated with drug addiction is particularly costly in body, limb and purse. Taxpayers complain about the high cost of the welfare system, but do not realize the extent to which they are paying also for less obvious costs such as those due to criminal activity, the increased cost of fire, police, and sanitation services in the ghettos, and the high cost of medical services to people sickened by malnutrition, alcohol, drugs, and ignorance of the basic principles of hygiene. The modern environmental movement should be more concerned with the environmental problems of the poor—not only in the cities, but in the rural areas as well. In many cities of the eastern United States, a vicious cycle of cause and effect is now in progress. The deviant behavior of the chronically underprivileged raises the crime rate to such an extent that life becomes a fearsome experience for many, and those who can leave the city for the suburbs continue to do so. The homes and apartments they vacate then tend to be occupied by the less fortunate, and the neighborhood deteriorates because of a further increase in crime, the inability of landlords to collect adequate rents, and their resulting failure to maintain the buildings properly. Often the buildings are abandoned because the landlord cannot derive sufficient income to pay taxes and to maintain the buildings according to the local building codes. The buildings are then occupied by squatters, who are unable either to maintain the building properly or maintain discipline among themselves. Plumbing and electrical fixtures will sooner or later be stolen, and eventually the buildings will be set afire either accidentally or deliberately. This process has been going on for so long and at so rapid a rate in many parts of the eastern cities that some neighborhoods are reminiscent of what was seen in wartime Europe and Japan.

Enormous efforts are sometimes directed at stopping construction of a power plant because it will spoil a scenic shoreline. High priority goes to eliminating chemical pollution that reduces the reproductive potential of sports fish, or gaseous emissions that produce a haze over our cities. The modern environmentalists have spawned thousands of organizations, have attracted millions of dollars of support, and have used their resources to obtain legislation requiring the expenditure of tens of billions of dollars annually. All of this was accomplished in a decade of inflationary pressures, at a time when the federal and state governments found it necessary to curtail expenditures for rat control, child care, family planning, and other less-glamorous health programs. Not many modern environmentalists think of rats as an important environmental problem: rats do not seem as relevant as PCBs or ozone.

The environmental movement has developed at a time in the history of public health when very little progress is being made in extending the length of useful living. The average child born in 1900 in the United States had a life expectancy of 47.3 years, compared with 70.9 in 1970, an increase of nearly 25 years. This improvement was due almost entirely to the reduction in childhood mortality. A white male who is now 40 years of age enjoys a life expectancy that is only four years longer than it would have been at the beginning the century. The various vaccines, antibiotics, the higher level of nutrition, and the greater availability of medical care, have succeeded in eliminating most of the infectious diseases, but the degenerative diseases of adulthood have not been controlled and now dominate the statistics of life and death despite the best efforts of the medical scientists. Pneumonia, influenza, and tuberculosis were the leading causes of death in 1900. By 1970, they were replaced by heart disease, cancer, and stroke, which now account for about two-thirds of all deaths.

In the years immediately ahead, any extension in the length of useful life is not likely to result from treatment of disease, but from changes in the way individuals behave, that is, relate to their physical and social environment. There are widespread complaints about the deteriorating environment (everything in the world but ourselves), but we as individuals are our own worst enemy. Of the 2 million deaths that occur in the United States each year, about 25 percent occur prematurely from causes related in one way or another to the manner in which we behave. Cigarette smoking causes about 80,000 cases of lung cancer per year, and probably in excess of 100,000 deaths from cardiovascular disease. Automobiles kill nearly 50,000 persons per year, and disable about 2 million. There are annually an additional 60,000 deaths and 8.5 million disabling injuries due to nonvehicular accidents including about 24,000 in the home. More than 30,000 deaths per year result from cirrhosis of the liver caused primarily by excessive consumption of alcohol. Each year there are about 25,000 suicides and 20,000 homicides.

Coronary heart disease continues to be virtually nonexistent in most of the world, but since the beginning of the twentieth century has reached epidemic proportions in Western countries. There are about 750,000 deaths from heart disease annually in the United States, of which no less than 100,000 are due to cigarette smoking. Of the remainder, an unknown fraction is believed to be associated with dietary habits, particularly excessive consumption of animal fats. There is no agreement as to the number of deaths that are due to dietary factors, but mortality from coronary disease has increased to such an extent, presumably from some environmental cause, that an estimate of 100,000 deaths per year due to dietary factors is certainly not unreasonable.

The effects of relatively simple daily habits (eating at regular hours, moderate exercise, adequate sleep, no smoking, no excessive alcohol consumption, and weight control) on life expectancy were investigated among Californians with astonishing results. It was found that male life expectancy at age 45 varied from 22 to 33 years depending on adherence to these habits.[6] In other words, all the gains we have won to improve public health since the beginning of this century have succeeded in extending life expectancy of adult men by only about four years, whereas factors more within the control of the individual than the physician can extend life by 11 years!

The influence of these factors on longevity is illustrated by studies of the relative health status of residents of Utah and Nevada. The population of Utah still consists largely of Mormons who abstain from alcohol and tobacco, and cultivate habits of great regularity and moderation. The inhabitants of Utah are among the healthiest in the United States,[7] whereas in the adjacent state of Nevada, known for its more hedonistic lifestyle, the health of the residents is among the poorest in the nation. Cirrhosis of the liver occurs in Utah with less than half the frequency it does in Nevada and California.

Why is it that there is so little concern with the influences from our social environment that we know to be the causes of death and disease? Isn't violence on television as much a controllable part of our environment as water pollution? Isn't a health education program aimed at teaching young mothers to eliminate accident hazards in the home at least as important a public health matter as low level chemical pollution of water, a subject that receives continuing prime time attention by the media? Why is it that so much attention has been given to the need to eliminate lead from gasoline to reduce exposure to airborne lead, which has not been identified as a health hazard, whereas peeling lead paint in ghetto buildings has received only minor attention, despite the fact that we know infants are being poisoned? Rickety home ladders, improperly maintained automobiles, and slippery bathtubs are examples of environmental problems that may be less exciting scientifically than the presence of DDT in mothers' milk or benzopyrene in city air. However, the former are hurting and killing people in great numbers, whereas the hazards of the latter are far less, if they exist at all. These are not only environmental problems that are hurting and killing people; they are problems that in many cases can be corrected. Not until 1978 did the New York City building code require that outside protective bars be placed on apartment house windows. Hundreds of children have fallen to their deaths from windows over the years. The new regulation will eliminate this source of danger if enforced.

The environmental movement has given its attention primarily to the out-of-doors, whereas people spend most of their time indoors.

From late fall through winter, for nearly five months, most people leave the shelter of their buildings or vehicles for only brief periods. Even farmers in the United States spend more than half their time within buildings or vehicles.

The physical environment of the poor has a major effect on health, and it is not to the credit of our nation's health planners that an adequate analysis of the causes of sickness and death among the disadvantaged does not exist. However, we can conclude from scattered bits of information that the poor person, whether in rural areas or in the inner cities, has a higher death rate and is subject to more frequent periods of incapacity. We have noted frequently throughout this book that there is an inverse correlation between socioeconomic status and health. Substandard housing, poor nutrition, the mental stress associated with poverty, and perhaps the heavy physical demands of their jobs in some way affect the health of the poor, whose death rates are relatively high from birth to late adult life. Former Vice President Nelson A. Rockefeller is one of the statesmen of the modern environmental movement. His views on environmental matters were contained in a book in which he discussed the familiar problem areas—clean air, pure water, and open spaces.[8] The book also included a chapter entitled "The Arts" and the "Quality of Life." The book, which was written when he was Governor of the State of New York, deals nowhere with what was unquestionably the state's number one environmental problem: the dreadful conditions of the urban poor. The President's Council on Environmental Quality, in its 1975 and 1976 reports, which are extensive assessments of the state of the nation's environment, give no attention to the environmental problems of the millions who inhabit the nation's urban ghettos, the camps of the migrant farm workers, or the chronically dependent in rural areas.

THREE BASIC MISTAKES

The failures of the modern environmental movement during its first decade can be attributed to a lack of historical perspective, improper assignment of priorities, and attempts to do too much too soon.

The lack of historical perspective that exists within the environmental movement has been emphasized repeatedly in this book.

Improper assignment of priorities (and the failure to differentiate health from ecological issues) is in part due to the subjective nature of the subject. Insofar as human health is concerned, it has been a mistake to allow the development of policies that have required vast expenditures of funds (sulfur oxide and automobile-emissions control, for example) while neglecting others (such as safety in the home). The former policies are not likely to result in a measurable benefit to health. A home-safety program would save lives and reduce disabilities on a readily measurable scale, and such a program could be effective for a modest expenditure.

The attempts to do too much too soon result from the lack of historical perspective, the disordered priorities, and failures of judgment on the part of those charged with implementing the environmental policies. The policies to control sulfur dioxide and automobile emission are likely to become classic examples in which not only were the wrong goals selected, but the timetables for achieving them were extravagantly unrealistic.

Many groups must share responsibility for the disorder that has beset the environmental field. Industry, organized science, the Congress, agencies of the executive and judicial branches of government, the media, and the general public have all contributed to the confusion.

Industry must bear a major share of the blame for several reasons. The industrial organizations have consistently blocked progress. Beginning with industry's opposition to workmen's compensation insurance near the turn of the century, there has been a steady pattern of opposition to needed environmental legislation of all kinds. What industry would not do voluntarily they are now required to do by law. As the price of decades of intransigence, they are now required to do too much. When the pendulum swung, it swung too far.

Organized science has done relatively little to assist in the resolution of scientific disputes being waged in the public forum. Scientists with extreme views have tended to dominate the scene, and the scientific societies have until now failed to develop any effective means to set forth the facts in an objective form for consideration by the public.

Congress has been too restrictive of its appropriations for basic scientific research. When Congress finally decided to move ahead with long overdue environmental legislation, it expected too much in too short a time, and, instead of approaching crucial environmental matters in a bipartisan objective manner, it became increasingly responsive to environmental extremists and the subject became highly politicized. Because Congress did not provide adequately for the research needs, many decisions are being made on the basis of insufficient information.

The agencies of the executive branch of government have also been too responsive to the demands of extremists. For lack of leadership, the staffs have been demoralized and deprofessionalized. This trend was accelerated by the tendency of the Congress to require frequent reorganizations in the environmental agencies.

The judicial branch of government has also been at fault. The U.S. Supreme Court in a unanimous opinion in April, 1978 admonished the lower federal courts for their tendency to "second guess" decisions of the Congress and executive agencies in matters concerning construction of nuclear power plants. Appeals court judges were accused in the ruling of "engrafting their own notions of proper procedures upon agencies entrusted with substantive functions by Congress."[9] The courts are not designed to resolve complicated technical disputes. There has been recent discussion of the need for establishment of a form of "scientific court" in an atmosphere in which the scientists, the judges, and lawyers would guide the procedures but not control them in the traditional manner.

Finally, as noted earlier, the media must take its share of the blame because it allowed its reporting of environmental issues to distort the facts and misinform the public.

During the first decade of the environmental movement there has been too much emphasis on the "evil by-products" of technology, with no counterbalancing acknowledgement of the benefits received. This has contributed to the distorted picture that has emerged. Technological innovation has significantly lengthened life expectancy and has eliminated many forms of disability due to disease. Technology has increased human productivity to such an extent that the workday has become shorter and in most developed countries an additional day of leisure has been added to the traditional one day of rest. The increase in productivity made possible by technology has made it possible for an average person to enjoy a standard of living that was impossible, even for the wealthy, a century or two ago. World travel, private means of transportation, television, and a profusion of books and music to suit the tastes are being enjoyed on an unprecedented scale.

For the benefits of technology we must pay a price. For every technological innovation there are risks as well as benefits. Society must decide in each case if the risks are justified by the benefits to be achieved. It is usually possible to reduce the risks by one means or another, but for each unit of risk reduction, additional costs must be incurred. Because the law of diminishing returns applies, a decision on the acceptability of a risk must be based on the benefits to be achieved in relation to the cost of risk reduction.

Decisions as to the acceptability of risk were in the past based on the intuition of decision makers in government or industry. The consumer was not represented, and the decisions were made in the absence of any body of quantitative information by which relationships between risks and benefits could be evaluated. These are now deficiencies of the past. Environmental-impact analyis is now required as a test of the acceptability of new technological developments. New chemicals will no longer be introduced in the workplace or discharged to the environment without prior toxicological studies. The new federal laws are designed to protect the public and, if administered wisely and efficiently, they will safeguard society from a repetition of mistakes that were made in the past. Properly implemented, laws such as the National Environmental Policy Act, the Toxic Substances Control Act, and others mentioned in earlier chapters, should make it possible for society to enjoy both the benefits of new technological development with confidence that the pleasures and safety of protected environment can be enjoyed.

NOTES

1. Davies, J. C. and B. S. Davies. "The Politics of Pollution," 2nd ed., Pegasus (Bobbs-Merrill Co., Inc.), Indianapolis, Ind. (1975), p. 34.
2. Sixth Circuit, Hill vs. TVA, 76–2116 (Decided Jan. 31, 1977).
3. Burger, E. J. "Protecting the Nation's Health," D.C. Heath and Co., Lexington, Mass. (1976).
4. Council on Environmental Quality. "Seventh Annual Report," U.S. Government Printing Office, Washington, D.C. (1976).
5. Walker, W. J. "Government-Subsidized Death and Disability," *Journal of the American Medical Association* 230:1529–1530 (1974).
6. Fogarty, John E. International Center for Advanced Study in the Health Sciences, and American College of Preventive Medicine. "Preventive Medicine., USA," Prodist, New York (1976).
7. Fuchs, V. R. "Who Shall Live?" Basic Books, New York (1974).
8. Rockefeller, N. A. "Our Environment Can Be Saved," Doubleday and Co. New York (1970).
9. *New York Times* (April 4, 1978), p. 1; (April 16, 1978), Section 8, p. 1.

Subject Index

369

Author Index

381